The American Century

Consensus and Coercion in the Projection of American Power

Edited by
David Slater and Peter J. Taylor

BLACKWELL
Publishers

Copyright © Blackwell Publishers Ltd 1999
Editorial apparatus and arrangement copyright © David Slater and Peter J. Taylor 1999

First published 1999

2 4 6 8 10 9 7 5 3 1

Blackwell Publishers Ltd
108 Cowley Road
Oxford OX4 1JF
UK

Blackwell Publishers Inc.
350 Main Street
Malden, Massachusetts 02148
USA

British Library Cataloguing in Publication Data

A CIP catalogue record for this book is available from the British Library.

Library of Congress Cataloging-in-Publication Data

The American century : consensus and coercion in the projection of American power / edited
   by David Slater and Peter J. Taylor.
     p.   cm.
   Includes bibliographical references and index.
   ISBN 0-631-21221-3 (alk. paper). – ISBN 0-631-21222-1 (alk. paper)
    1. National characteristics, American. 2. Civilization, Modern – American influences.
  3. Americanization – History – 20th century. 4. United States – Foreign economic relations – 20th century. 5. United States – Foreign relations – 20th century. 6. Imperialism – United States – History – 20th century. I. Slater, David, 1946– . II. Taylor, Peter J. (Peter James), 1944– .
  E169.1.A471966 1999
  973.92 – dc21
                                                   99-14781
                                                    CIP

Typeset in 10¹⁄₂ on 12 pt Sabon by Best-set Typesetter Ltd., Hong Kong
Printed in Great Britain by MPG Books Ltd, Bodmin, Cornwall

This book is printed on acid-free paper.

'We will not abandon our opportunity in the Orient. We will not renounce our part in the mission of our race, trustee under God, of the civilization of the world.'
Senator Albert Beveridge, 1900, quoted in D. L. Larson, (1966), 'Objectivity, Propaganda and the Puritan Ethic,' in Larson (ed.), *The Puritan Ethic in United States Foreign Policy*, Princeton, NJ: Van Nostrand Co., p. 12

'almost simultaneously with the appearance of the United States as a conquering and colonising power, Europe has been appalled by the sight of America bursting her bonds and stepping armed cap-à-pie into the arena as an industrial giant of almost irresistable power, with the openly proclaimed determination to conquer the world's markets and gain universal commercial superiority.'
Sir Christopher Furness (1902), *The American Invasion*, London: Simpkin, Marshall; Hamilton, Kent & Co., p. 1

'. . . we can make a truly *American* internationalism something as natural to us in our time as the airplane or the radio . . . the belief – shared let us remember by most men living – that the 20th century must be to a significant degree an American Century.'
Henry Luce (1941), *The American Century*, New York: Farrar and Rinehart, pp. 26, 31 (emphasis in the original)

'The American idea is founded upon an image of man and of his place in the universe, of his reason and his will, his knowledge of good and evil, his hope of a higher and a natural law which is above all governments, and indeed of all particular laws: this tradition descends to Americans as to all Westerners, from the Mediterranean world of the ancient Greeks, Hebrews and Romans.'
Walter Lippmann (1943), *U.S. Foreign Policy and U.S. War Aims*, New York: Overseas Editions Inc, p. 252

'There is, in short, a huge tacit conspiracy between the US government, its agencies and its multinational corporations, on the one hand, and local business and military cliques in the Third World, on the other, to assume complete control of these countries and "develop" them on a joint venture basis. The military leaders of the Third World have been carefully nurtured by the US security establishment to serve as the "enforcers" of this joint venture partnership, and they have been duly supplied with machine guns and the latest data on methods of interrogation of subversives.'
Edward S. Herman (1982), *The Real Terror Network*, Boston: South End Press, p. 3

'America is powerful and original; America is violent and abominable. We should not seek to deny either of these aspects, nor reconcile them.'
Jean Baudrillard (1988), *America*, New York: Verso, p. 88

'. . . much of the conversation between the United States and Europe before and after 1945, has been characterized more by an exchange of metaphors than by a sharing of information. For many Europeans, "America" was and is a symbol; a receptacle for fears and fantasies; a state of mind rather than a real country.'
Richard Pells (1997), *Not Like Us*, New York: Basic Books, p. 2

'America today is effectively the centre and driving force of one of the greatest social, economic, political and cultural revolutions the modern world has seen. It is the first revolution in history to have reached the entire planet, casting doubt on established certainties and undermining ancient institutions, even the sovereignty of nation states; including the United States of America.'
A. G. A. Valladão (1998), *The Twenty-First Century will be American*, London: Verso, p. xv

Contents

List of Contributors

Roberto S. Bartholo Jr, COPPE, UFRJ, Brazil.

Berta K. Becker, Instituto de Geociencias, UFRJ, Brazil.

Morag Bell, Department of Geography, University of Loughborough.

Alan Bryman, Department of Social Sciences, University of Loughborough.

David Campbell, Department of Politics, University of Newcastle.

Tim Cresswell, Department of Geography, University of Wales, Lampeter.

Peter Daniels, School of Geography and Environmental Sciences, University of Birmingham.

Peter Dicken, Department of Geography, University of Manchester.

Klaus Dodds, Department of Geography, Royal Holloway, University of London.

Marcus Doel, Department of Geography, University of Loughborough.

Pablo González Casanova, Director of the Centre of Interdisciplinary Research for Science and Humanities, National Autonomous University of Mexico, Mexico City.

Michael Heffernan, Department of Geography, University of Loughborough.

Mark Holmes, Department of Geography, University of Loughborough.

Brian Hoskin, Department of Geography, University of Wales, Lampeter.

Ron Johnston, Department of Geography, University of Bristol.

Claudia de Lima Costa, Universidade Federal de Santa Catarina, Florianópolis, Brazil.

Kate Manzo, Department of Geography, University of Newcastle.

Eric Pentecost, Department of Economics, University of Loughborough.

James Sidaway, School of Geography, University of Birmingham.

David Slater, Department of Geography, University of Loughborough.

Michael Smith, Department of European Studies, University of Loughborough.

Michael Taylor, Department of Geography, University of Portsmouth.

Peter Taylor, Department of Geography, University of Loughborough.

Gillian Youngs, Centre for Mass Communication Research, University of Leicester.

Preface

In the post-Cold war era, now more than ever, the twentieth century appears to have been the 'American Century'. This concept, first formally enunciated in 1941 by Henry Luce, begs the question as to the nature of American power. Written in wartime, Luce was well aware of the coercive potential of the American state, but he was more concerned with economic and cultural power and their underlying consensual basis. But, above all, he was interested in the scope of American power – whether military, political, economic or cultural. He saw the USA as an exceptional state with a capacity to project its power in the global domain. Clearly, the continuing popularity of the concept of American Century shows Luce's vision to have been most prescient.

No doubt many publications on the American Century will appear at about the same time as this one, for there are many different ways in which US power can be analysed. What makes our book distinctive is that we focus upon the projection of that power from the perspective of its 'receivers' both as beneficiaries and as victims. This is not to neglect the constitutive dynamic of the society projecting its power, for clearly we cannot possibly understand the American Century without a broad understanding of the social forces that have fuelled the drive to expand. Rather, our intention is to focus on the effects of the projection of American society on the rest of the world. In this book we analyse the varied impact of a society that constructs its destiny through projection of its own capacities and desires onto the territories of other cultures and polities. We will chart the shifting modes of penetration and appropriation, and of exclusion and inclusion that help define the encounter with the USA. In this transaction the receivers are far from being passive or docile; local and national social practices interact with the intruding discourses of a globalizing power to create worlds of 'hybrid Americas' – multiple ensembles of meanings and practices which intertwine the inside and the outside.

This 'geographical dialogue' has taken two general forms. In the metropolitan countries of the world-system the projection of American power has been predominantly conceived as the diffusion and adoption of the 'American way of life', of mass consumption and material prosperity. The seductiveness of the

American Dream and the perpetual drive to affluence give body to the notion of the 'Americanization' of these societies. In contrast, beyond these favoured metropolitan countries, American power has been deployed in much more coercive forms, whereby the desire to discipline, to restore order and to counter subversion have been intrinsically bound up with the will and capacity to wage war. This is often referred to as 'American imperialism'. The two forms are by no means mutually exclusive as the seductiveness of images of affluence, freedom and progress also permeates the societies of the South, as partly reflected in the continuing movement of migrants to the heartlands of the North, especially the USA.

The literatures on these two projections of power are usually separate and do not reference each other even though it is the same nation which is projecting its power. This book brings together original studies of Americanization and American imperialism as two interwoven strands of a single pursuit of power. In the contemporary world the capacity to be effective in war is increasingly connected to the power over information technology, and the effectiveness of military action is indisputably linked to the capacity to influence and mould the media presentation of war as event. Communications, technology, war and information coalesce to constitute one crucial arena in which American projections of its power and destiny take shape, whilst the passion to possess, the desire to consume and be consumed by the benign power of the market capture that other 'American way of life'.

By drawing on authors from outside the USA, this book follows a precedent set by Franz M. Joseph's 1959 collection of essays *As Others See Us*. He brought together a group of elites from different countries to evaluate ways in which the USA influenced their respective countries and how their fellow citizens were responding to this. In fact, this was a very US-inspired exercise, a sign of American concern to be admired by the outside world. Our book is completely different, and not just because our authors would not appreciate the description 'elites'. We have a narrower range of authors geographically but we have organized their contributions by subject to create a wide spectrum of views covering many different topics. We think this bringing together of quite disparate subjects works well because of the mutual concern for the projection of American power. And this common thread brings out comparisons which would never be thought of in any other publishing context. We draw these together in our concluding chapter; here we will briefly introduce the contributions.

We have organized the book quite conventionally into three substantive parts covering economic, political and cultural dimensions of US power. Contributors have been selected who can critically engage in the vast literatures on these topics to present short essays which describe, analyse and enlighten. Before these essays there is an introductory section in which both editors 'locate the American Century' using their respective approaches to social science. As well as being fun for readers to spot differences between us, these two complementary essays have the purpose of providing some basic foundations for what follows.

In Part II our contributors explore the material bases of American influence on the twentieth century. This begins with an essay by Peter Dicken in which he draws out the American dimension from his well-known work on 'global reach'. This study of US multinational corporations is neatly complemented by Michael Taylor, who traces the managerial practices which contributed to US firms getting their global economic edge in the first place. Peter Daniels then concentrates on service multinationals, the sector most closely associated with the Americanization of our world in leisure, food and entertainment. But what of the macroeconomic context within which these American economic agents have been able to operate? Mark Holmes and Eric Pentecost provide a succinct economic history and analysis of the 'almighty dollar'. Finally, Kate Manzo explores one of the key concepts of the American Century, the idea of development for poor countries. Her focus on the 'Washington consensus' reminds us that however hard people have tried to create a development economics, its subject can never be divorced from issues of political power. This chapter, therefore, acts as a transition argument to the next section.

In the third part of our volume, Mike Heffernan's historical essay on key features of the early twentieth century US–European encounter provides an appropriate opening to a wide-ranging series of contributions dealing with the political capacities of American power. The encounter with Europe is also the central focus of Michael Smith's succinct contribution, which illuminates the changing nature of the relationship – reflex, resistance and reconfiguration – between the drive toward European integration and the pervasiveness of US influence. In the next two chapters the focus shifts to the terrain of US foreign policy in the post-Second World War period, where questions of the 'diffusion of democracy' and interventionism in the Third World loom large. Ron Johnston emphasizes the role of the United States in effectively exporting a particular vision and practice of Western democracy, whilst Klaus Dodds draws the reader's attention to the invasiveness of US power in Third World contexts. In the next chapter, Berta Becker and Roberto Bartholo, with Brazil as their focus, scrutinize the modalities of US influence in relation to the making of the Amazon Heartland. Moving back from Brazil to Portugal, the next chapter by James Sidaway traces out certain salient features of the often neglected interaction between US strategy and the specificities of Portuguese colonialism. This section is then brought to a close by two chapters which consider different but connected dynamics of US power in the international arena. Gillian Youngs critically assesses the trajectory of mainstream international relations theory and the role therein of the American model of modernity, and David Campbell, taking Bosnia as a revealing case study, examines certain striking contradictions of the United States as the lone superpower. Both these chapters conjoin the political with the cultural and provide a transition to the next section of the volume, which considers the 'Cultural Capacities' of US power.

In the initial chapter of this section, Marcus Doel develops an historical analysis of the Americanization of world cinema in which he stresses the fact that Hollywood is but one force in the overall historical geography of twentieth-century cinema. For his part Alan Bryman examines Disney theme parks in the context of debates on globalization and hybridization, and he concludes by restating the cultural power of Disney's global presence. Tim Cresswell and Brian Hoskin, in their treatment of American popular music in Britain, argue, like Doel, against over-generalized visions of an all-powerful Americanization process, and underscore the importance of symbolic creativity within any cultural encounter. Equally, Morag Bell, in her investigation of American philanthropy, in which special attention is given to the influences of the Carnegie Corporation in South Africa, also suggests that in general the cultural power of foundations cannot be merely interpreted as part of a broad structure of American global domination. Claudia de Lima Costa, in a contribution which crosses borders, shows the significance of social context when analysing feminist theory and political practice in Brazil. This essay seeks to make enabling connections between North and South, whilst focusing on the problems and opportunities involved in 'travelling feminisms and the Latin American difference'. Finally, in this section Pablo González Casanova provides the transition to our concluding essay by offering the reader a wide-ranging overview of many of the central features of this century's Americanization process. As with the previous chapter, this is achieved with a particularly keen eye on the North–South dimension of the projection of US power.

Clearly the reader will now appreciate that we were not exaggerating when we promised a wide range of topics. It is now up to you to decide whether our claim that the specific focus on the projection of American power does indeed bring these essays together to make a coherent whole. Our argument in favour can be found in a final concluding chapter.

We are both relative newcomers to Loughborough University and its Department of Geography. Obviously, this book would probably never have been conceived were it not for the latter since it brought the two of us together. But it is much more than this, the book would also not have happened without the intellectual stimulus of our fellow geographers at Loughborough. Several are featured in the pages ahead, but all have contributed to making Loughborough such a congenial location to explore ideas amongst friends.

David Slater and Peter Taylor

Part I

Introductions: American Destiny?

Chapter 1

Locating the American Century: A World-systems Analysis

Peter J. Taylor

'... the influence of the United States was more psychological than economic. When Swedes, Germans or Italians spoke of being "Americanized", they were referring to the changes taking place not just in the corporate world but in their own households, in the relationships between husbands and wives, and between parents and children. Here, the American presence was measured less by investments or exports than by the modification of attitudes, expectations, and patterns of conduct. At bottom, imitating the Americans involved more than the purchase of a refrigerator and a new car, or (if one was young) a pair of blue jeans and a black leather jacket. It meant thinking of oneself as a perpetual consumer, soon to arrive in or already a member of the affluent middle class.'

(Pells, 1997, 195–6)

Question: which McDonald's restaurant is the busiest in the world? Answer: the one located on (what was) Red Square in Moscow. Superficially, at least, this simple fact would seem to be a final confirmation that we have indeed been living through a triumphant American Century. But, as always, simple facts can be deceiving. That other quintessential agent of Americanization, Coca-Cola, has recently downgraded its 'domestic' arm to make it just part of a global regional structure, one of six world regions to be precise. Such disregarding of the traditional distinction between 'domestic' and 'foreign' operations points towards a very different conclusion – a condition of globalization in which the USA remains important but no longer dominates (Pells, 1997, 303, 326–7).

It is the purpose of this chapter to introduce world-systems analysis as a fruitful way for locating the American Century geohistorically. This analysis provides a framework for understanding the issues raised by 'Cocacolaization' and 'McWorld', such as those briefly mentioned above, and much else besides concerning the notion of Americanization. The argument proceeds in four stages. First, the relevant premises of world-systems analysis are presented in which the concepts of world hegemony and prime modernity are introduced. In world-systems analysis, the twentieth century corresponds approximately to the American hegemonic cycle (the rise and demise of American world hegemony) as part of which is the spread of American values to other countries

(Americanization) which created a new prime modernity based upon consumption. The specifics of this model are laid out in the second section, which emphasizes the variety of Americanizations. Through time as 'phases' and across countries as 'hybrids', Americanization has meant different things in different periods and places; world-systems analysis offers some order to this diversity. The third section uses a case study, the Americanization of England, to illustrate the framework describing one important hybrid during one key period. Finally, in a short concluding section, I return to the question of how Americanization fits into current ideas concerning globalization.

Some World-systems Premises: World Hegemony and Prime Modernity

In a world-systems analysis the fundamental unit of contemporary social change is not the state and its associated society but rather the wider system of which the state is a part. This is the modern world-system, a hierarchical spatial division of labour which emerged in the 'long sixteenth century' (ca. 1450–1650) and eliminated all other systems to become global in extent by 1900 (Wallerstein, 1979). It follows that in order to understand social change, economic development or political progress we need to study each of these processes at the system scale. In all such descriptions of historical modification of the human condition one feature seems clear: such changes are very uneven over time. In world-systems analysis a large part of this temporal unevenness is interpreted as being cyclical. Several such cycles are identified for the modern world-system, the longest of which are the hegemonic cycles which I focus upon here.

A hegemonic cycle defines the rise, achievement, and demise of world hegemony by a specific state and its associated society (Wallerstein, 1984). There have been three such instances in the history of the modern world-system: the Dutch hegemonic cycle from late sixteenth-century rise to eighteenth-century demise, the British hegemonic cycle from eighteenth-century rise to demise in the decades around 1900, and the American hegemonic cycle from rise in the late nineteenth century to the onset of demise in the late twentieth century. Notice that the cycles overlap during transition periods (in the eighteenth century and around 1900). But the key periods are those of 'high hegemony' at the centre of each cycle when the 'hegemon' is most influential within the world-system. These are usually dated from 1609 to 1672 for the Dutch, from 1815 to 1873 for the British, and from 1945 to 1971 for the Americans. These are periods when the economic prowess of the hegemon is at its maximum: hegemons are the most efficient in agro-industrial production, hence they have the advantage as traders in the world market, which in turn enables their major city to establish themselves as the financial centre of the system (first Amsterdam, then London, now New York). With economic strength comes political power; each hegemon has been the paymaster and arsenal of the winning coalition in the major determining wars of modern world-system: the Thirty Years War

(1618–48), the revolutionary and Napoleonic wars (1792–1815), and the twentieth-century 'German wars' (1914–45, i.e. World Wars I and II interpreted as a single conflict). Notice that for both Britain and the USA, high hegemony is ushered in with victory in what was a 'good war', in economic terms, for the hegemon.

The term hegemony is used in this context to indicate that these three states were much more than simply successful great powers. They had, as Giovanni Arrighi (1990) has termed it, 'something extra' which made them qualitatively different from all other great powers of the time. What is it that the Dutch, the British and the Americans had which the Spanish, the Portuguese, the French, the Prussians, the Austrians, the Russians, the Germans, the Italians, the Japanese, and the Soviets never possessed? The notion of hegemony suggests more than dominance in the strictly political military or even political economic senses. There is a sense of leadership. This implies there will be much consensus mixed in with the usual coercion. As well as being feared for its power, the hegemon is also greatly admired for its achievements. So much so that others come to emulate the hegemon. In this way the hegemon comes to be widely regarded as the model for the future.

Each hegemon has defined the critical dimensions of its age. Emulation of the Dutch ushered in the age of mercantilism, imitation of the British produced the industrial age, and copying the Americans has produced our contemporary age of consumption. Czar Peter the Great, among many others, visited Holland to find the secrets to modernize his country. In the nineteenth century the modern pilgrimage was to northern England as other Europeans and Americans sought to understand the newly invented industrial world. In our century, the American way of life has been relayed to the world through millions of immigrants' letters home but, above all, it has been advertised for all to see in film and on TV. Whether Amsterdam shipyards, Lancashire cotton mills or Los Angeles suburbs, what was being propagated was much more than new technology or larger organization. The changes on offer required profound alterations in all aspects of the social fabric. In short, recognizably new forms of society were created.

One way of interpreting the hegemons is as the most modern society of their times. Modernity is a condition of incessant social change where countries are continually required to keep up to date with the latest way of doing things. Not to do so can be calamitous, as the USSR found to its cost a decade ago (hence the common derogatory description of the country at the time as 'a third world country with rockets'). This obsolescence imperative within modernity is a continuous process but it becomes particularly acute with the rise of a new hegemon. Staying up-to-date and not falling behind now requires massive changes in a world which is 'turned upside-down'. Being merely mercantile in the industrial age, or merely industrial in the consumer age is a recipe for rapid decline. In short, in the new modern world of the hegemon there is little choice but to emulate (Taylor, 1996).

Defining the future for others provides the hegemon with immense cultural power. Thus, as well as military and political-economic power, hegemons possess a critical socio-cultural power which makes them so much more than merely another, albeit powerful, state. Although other great powers have had important cultural influence, for instance the French in many political practices and the Germans in the rise of universities, it is only the hegemons who have created new forms of society with implications for the whole gamut of human activity. These 'new modern worlds' we call 'prime modernities': the Dutch created a mercantile modernity, the British an industrial modernity, and the Americans a consumer modernity. These modernities are considered prime because of their system-wide effects (Taylor, 1998). All countries create their own variant on modernity, but the hegemons at their peak create a new social order to which every other country has to adapt. Americanization is the name given to this process of emulation and adaptation under the condition of consumer modernity.

Americanizations: Phases and Hybrids

The American hegemonic cycle can be sketched very briefly as follows. There were three steps in the rise phase. First, the US overtook the UK as the leading industrial producer in the world-economy in the last quarter of the nineteenth century. Second, in World War I, New York replaced London as the leading world financial centre and subsequent American entry into the war was vital to the victory. However, the resulting potential for world political leadership was not taken up, as isolationism triumphed in domestic politics. Third, World War II destroyed isolationism and the US led another allied victory, but this time they were determined to reap the full benefits of their position. After setting up the UN system of institutions, including the Bretton Woods agreement (World Bank and IMF) which made the dollar the world trading currency, at the end of the war, high hegemony was established through a three-pronged strategy. First, other core countries (western Europe and Japan) had to be rehabilitated economically and politically as strong but subordinate allies – NATO, the Marshall Plan for European economic recovery and the Defence Treaty with Japan were the chief instruments. Second, the USSR and its allies had to be contained in the 'Cold War', leading to supporting anti-communists and sometimes fighting new wars, notably in Korea. Third, the US devised new development plans to dissuade countries of the periphery from joining with the Soviets. These policies succeeded politically and coincided with a massive post-war economic 'boom' which lasted through most of the 1960s. However, three events led to the ending of high hegemony and the US as world political, economic and moral leader. First, in the Vietnam War the US failed to produce a western consensus on the war and finished up losing. Second, resulting from the inflationary pressures of the war, the US dismantled a key part of the Bretton Woods agreement by letting the dollar float in the market in 1971. With the dollar now just another currency, in 1973–4 oil exporting countries used their cartel powers to push through a massive rise in oil prices, precipitating a western economic recession. Although, as an oil producer itself, the US did not suffer as badly as others, it showed a world-economy out of control, beyond direct hegemonic influence. These events set up the demise phase we are in today: the US is still the leading economic and political power but its influence, despite Cold War victory, is far less than the US heyday in the quarter century after World War II.

All the above is the stuff of high politics and macroeconomics which is what constitutes world hegemony. But what of the other side of the process, the prime modernity which impinges directly on the daily lives of ordinary people? This involves not so much the projection of US state power but rather the more subtle projection of American civil society as an ideal to be taken up with enthusiasm by many millions of non-Americans. It is this Americanization which I focus on in the remainder of this section. As the other side of the coin, as it were, from high hegemony, this prime modernity of consumerism was created in three phases broadly matching the hegemonic cycle: incipient Americanization occurs in the rise phase, capacious Americanization is what happens during high hegemony, and the demise phase is marked by a resonant Americanization. But none of these Americanizations, even when capacious, eliminate the cultural distinctiveness of the countries caught up in the process. Rather a hybrid culture is produced in which American influences are clear but where national cultures remain intact. I begin by describing the three phases of Americanization before returning to the question of hybridity.

i) Incipient Americanization

There is a discrepancy between the rise of American state and the influence of its civil society in the first half of the twentieth century. Although both political and economic power was important before 1920, it was not until the 1920s that American cultural influence came to the fore. It was led by American films. According to John Lukacs (1993: 273), 'As early as 1925, millions of people in Europe knew the names and faces of American movie stars while they knew not the name of their own prime minister.' Alongside this 'invasion of Hollywood', America was to make its first major foray into dominating popular music through the international spread of jazz. Important as these cultural products were, there was something even more basic to this incipient Americanization.

It seemed as if, in the 1920s, the USA was creating a new type of society. Lukacs (1993, 145) characterizes the changes in terms of the 'shape' of society from pyramid (i.e. with a base consisting of many poor people) to 'onion-shaped' implying a majority of middle-income people. Inevitably this was associated with fundamental changes in the everyday lives of Americans. Higher wages were reflected in more consumption especially in the home. By the middle of the 1920s over 60 per cent of American houses had electricity and their residents consumed over half the electricity generated throughout the world (Rybczynski 1986: 153). This huge market meant lower prices which stimulated more demand with increasing numbers of domestic appliances. Hence the birth of the new suburban lifestyle centred on the home as the locus of consumer durables. These domestic developments came to the notice of the outside world in two main ways. While other countries were developing their propaganda arm of the state with 'official' cultural institutions, Americanization proceeded through essentially private means; the production and marketing of American corporations and visitor's reports and letters home from immigrants (Duigan and Gann, 1992, 420–1). In fact, the State Department only set up a cultural relations division in 1938 and Voice of America only began broadcasting in 1942 as part of the war effort. Incipient Americanization was clearly not a public project in a period of US political isolationism. But the American message got through nevertheless via the two 'unofficial' media.

First, with increased economic protectionism, the fast-growing American corporations were forced to set up production behind tariff barriers. In Europe, Germany became the main US economic base with ITT, General Motors and IG Chemicals leading the way. Coca-Cola opened its factory in Essen in 1929. The European market for consumer goods still lagged far behind America but the beginnings of the new modernity were being put into place.

Second, visitors to America were struck by its unusual civil society and reported back home accordingly. For instance, Luigi Barzini, an immigrant from Italy in 1925, saw America as new modern world:

> What interested me in the United States . . . was the 'modern' concept, in which all things served to be done in a revolutionary 'modern' way. Everything was the product of fresh thinking, from the foundations up. Everything had been 'improved', and was continually being 'improved' from day to day, almost from hour to hour. The restlessness, mobility, the increasing quest for something better impressed me.
>
> (Barzini, 1959, 73)

As views such as this filtered through to other civil societies, leading intellectuals came to suspect that they were being provided with a glimpse of a new world. Most famously, Antonio Gramsci, writing in 1929, was intrigued by the high wage regime of the USA and asked if this represented 'a new historical epoch' (Hoare and Smith, 1971, 277). He was unsure whether the 'new culture' and 'way of life' represented 'a new beacon of civilization' or merely 'a new coating' on European civilization (317–18). Jean-Paul Sartre had no such doubts: 'Skyscrapers were the architecture of the future, just as the cinema was the art and jazz the music of the future' (Duigan and Gann, 1992, 410). With the benefit of hindsight, we know today that Americanization was only just beginning.

ii) Capacious Americanization

It was in 1941 that Henry Luce, a prominent Republican, proclaimed 'the American Century'. This has been interpreted, quite properly, as a key step in US abandonment of political isolationism. Indeed Luce (1941, 260) called for

'a truly *American* internationalism' in order to influence the 'world environment
. . . for the growth of American life' (23–4). This was part of what Duigan and
Gann (1992, 409) identify as a 'sea change' in American society, with confidence in the 'American Way' replacing an earlier 'mood of cultural deference'
when educated Americans automatically looked to Europe for their cultural
lead. This confidence was quite pervasive. At the end of World War II only the
USA assumed the good times were coming; Ellwood (1992, 21) reports Keynes'
surprise on encountering American optimism at Bretton Woods. As the arsenal
and financier of war victory, America was now in a special position to project
itself across the spectrum of social relations – political, cultural and economic.
This is high hegemony, the period when the rest of the world is offered a comprehensive societal package.

Today, when Americanization is often viewed in narrow cultural terms, it
is important to emphasise that the projection of the American Way of Life at
mid-century was based upon production processes more than anything else.
The technology leadership of US industry – 'American know-how' – meant
that American workers were producing two to five times as much per day as
European workers (Price, 1955, 328). It is not surprising, therefore, that
although the Marshall Plan began by emphasizing the need to raise gross production, it soon changed to focus upon raising productivity. In this way, reconstruction soon came to be viewed as the 'modernization of European industry'
which in turn brought the 'growth idea' to the centre of decision-making. Institutions were set up to take managers and workers on visits to the USA so that
they could learn the new and better ways of production at first hand. It was at
this time (1950–2) that France sent forty such missions to the United States
and, according to Kuisel (1993: 84), discovered 'management' as opposed to
traditional French 'direction'. All this created a new politics based upon economic growth leading to higher levels of consumption and thus voter contentment (Ellwood, 1992, 94). This Americanization was to privilege class
compromise over class conflict in new political reconstruction in all western
European states.

iii)   Resonant Americanization

After 1970 other countries' corporations began seriously to rival the original
US multinationals and the idea of a 'productivity gap' became more likely to
refer to American inferiority than superiority. This coincided with a change in
the way people began to interpret the new modernity. Raymond Aron (1959,
67–8) had earlier wondered whether the new mass culture should be viewed
simply as American culture or more generally as the latest development of material society. Aron proved to be remarkably prescient. In terms of the radical
ideas associated with the 1968 revolution in France, for instance, 'America
became less the perpetuator of some universal crime and more a fellow victim
of a global dynamic' (Kuisel, 1993, 186). Kuisel (1993: 6) identifies 1970 as the
watershed: 'it has become increasingly disconnected from America . . . it would

be better described as the coming of consumer society' (p. 4). But American-ization was too embedded in the everyday lives of Europeans to simply disap-pear with American high hegemony.

The oddity of Americanization is that it is more visible today than during its high point. Whereas a previous generation experienced Americanization at the cinema or on TV, now it can be encountered directly in any high street as numerous burger, chicken or pizza fast-food joints. It is sometimes forgotten how recent this development is: it is only in the 1970s that the now ubiquitous American fast-food franchises begin expanding in Europe. Marling (1993, 15) interprets this contemporary Americanization as 'set firmly in a timewarp': 'it's about nostalgia because it is not America *now* that we're in love with, but America as it was when it first swept us off our feet in the 1950s'. Cowboy films may no longer be popular but things American remain fascinating to new gen-erations of Europeans (Marling, 1993, 7). This is a resonant Americanization reverberating into our present.

Hybrids

We must always be careful not to interpret Americanization as a simple diffu-sion and imposition. Like any package of cultural attributes away from its origins, it interacted with existing patterns of thought and practice to create new mixtures, not simple imitations. In the case of Americanization, although very powerful as the purveyor of a new modernity, it nevertheless resulted in many variants of the new consumerism:

> when Europeans did borrow American ideas or imitate American patterns of behavior, the ideas and behavior were modified to suit the special requirements of France or Britain, Italy or Sweden. America's culture did not cross the Atlantic intact, any more than European culture retained its original shape after it was brought to the New World.
>
> (Pells, 1997, 282)

Hence Pells' (1997, 283) conclusion that '[w]hat emerged was a hybrid culture, part American and part European'.

The best study of the creation of such a hybrid culture is Richard Kuisel's (1993) *Seducing the French*. A generation earlier Raymond Aron (1959, 70) had asked 'Will an industrialized France, concerned with productivity, become a mediocre replica of the United States?' Kuisel's answer is an emphatic 'no'. The fact the French people buy their Big Macs on the Champs-Elysées does not make them any less French. The outcome of the French engagement with American-ization has been neatly summarized as follows:

> Contemporary France is a different society because of changes associated with Americanization . . . [but it] neither obliterated French independence nor smoth-ered French identity. France did change but it remained the same. . . . The history of Americanization confirms the resilience and absorptive capacity of French

civilization. . . . Americanization has transformed France – it has made it more like America – without a proportionate loss of identity. France remains France, and the French remain French.

(Kuisel, 1993, 231, 233, 237)

Such a conclusion does not lessen the systemic importance of Americanization. The key point is that every hybrid has the same common denominator: the projection of American civil society. It was French and American, Swedish and American, Japanese and American, and so on across the world, even including a clandestine Russian and American hybrid society. And in all cases 'the debate over Americanization was an argument about the meaning and consequences of modernity' (Pells, 1997, 202) which resulted in a new prime modernity.

Mention of debate brings us to my final point on the reception of Americanization. It was most certainly not a socially neutral process. There were at least three crucially different patterns of reaction. First, it was ordinary working men and women who warmed first to the promise of an Americanized affluent society – after all these were to be the main beneficiaries. Second, it was the younger generation who were first attracted to the new ways – after all America invented the teenager. Third, it was the intelligentsia who were particularly resistant to what they saw as the shallow culture which was on offer – after all it was their traditional social authority which was being undermined. These differences in reception, although general, were particularly evident in the Americanization of England.

The Americanization of England

The three phases of Americanization can be easily discerned in the English hybrid. During the incipient phase, the English flocked to see American movies (95 per cent of all films shown by the mid-1920s were American (Pells, 1997, 16)) and many American corporations set up factories in England to quickly become 'household names' (Hoover, Gillette, McLeans, Remington, Firestone and, of course, Ford (Marling, 1993, 106)). At the other end of the process, the first McDonald's was opened in 1974 and within a decade high streets throughout the country abounded with American fast-food restaurants. However, this was only a resonant Americanization as part of a wider internationalization of food with traditional English fish-and-chip shops joined first by Chinese and Indian to be followed by a whole gamut of 'regional' food restaurants from all corners of the world by the 1990s. The 'golden arches' might be conspicuous, and conspicuously American, but they are part of a geographically broader process than simply Americanization. Both the incipient and resonant phases are fascinating episodes in the Americanization of England, but I will focus upon the crucial middle phase. It is during capacious Americanization that English society changes fundamentally as it modernizes in response to American high hegemony.

For this case study I use four contemporaneous publications as my sources. These are a review of the Marshall Plan in Britain (Hutton, 1953), a report of mid-century journeys across England to find out the state of the country (Hopkins, 1957), an essay which was commissioned to ascertain British attitudes towards America (Brogan, 1959), and a book which calls itself 'a study of the revolution in English life in the fifties and sixties' (Booker, 1969). These are very different in their attitudes to the USA, largely reflecting their relative timing. The first two are frankly nationalistic and thereby defensive, the third is grudgingly accepting of the influence of America, and the final source openly celebrates that influence.

During the Marshall Plan (1948–52) the Anglo-American Council on Productivity set up 66 groups of managers and trade unionists from different industries to consider how to raise the productivity of British manufacturing. These groups involved meetings and factory visits in both countries and although ostensibly members from both countries were equal, in fact the British visited America to learn, the Americans visited Britain to teach. Each industry group produced its own report and their findings were subsequently brought together for the new British Productivity Council by Graham Hutton (1953) in his aptly named *We Too Can Prosper*. This book is fascinating because it lies on the cusp of profound change, seeing the alternatives but missing the power imperatives.

After confirming the superiority of American manufacturing, Hutton (1953, 18) argues that its efficiency is not 'particularly *American*' (emphasis in original) because the methods leading to high productivity in US factories are familiar to all industrialized countries; it is just that the Americans carry them 'to the highest degree of application' (p. 31). Hence he concludes that:

> Higher productivity can be achieved in any industrial country without having to imitate slavishly every ingredient in 'the American way of life'. . . . [Improved methods of production] do not necessitate the adoption of American social habits, any more than they demand the American legal system or political system for their working out.
>
> (Hutton, 1953, 31)

However, Hutton did see productivity as a social problem as much as an economic one. America enjoyed a 'climate of productivity' (p. 182) in which 'all Americans' were 'more productivity-minded'. He noted that 'Americans can buy more things with their pay; more consumer goods and services are available' and they have 'more leisure time' (p. 172). And this was much more than having more money:

> Americans work and play *en masse*. This is a powerful factor making for mass consumption to match mass production . . . Fashion moves *en masse* too . . .
>
> (Hutton, 1953, 187)

Hence Hutton was well aware of the importance of social habits to US economic success, but he did not think this more 'vulgarized' society appropriate for the English.

What for England, therefore? Hutton identifies the problem of repetitive work practices in modern industry but sees the solution in bringing back the 'satisfaction' of the craftsman in his work (p. 163). He does not predict that personal satisfaction might come from consumption outside work which the new work practices can make possible. He suggests 'peculiarly British goods' for a peculiarly British society (p. 231) in which, for instance, British people might choose to pay more for their household needs to preserve the small shops serving their traditional small packets of goods (p. 221). No thoughts of consumer modernity here! However, he does clearly realize the issues at stake, concluding with the question 'Who are we?' (p. 233), even if he goes on to provide an insular nationalist answer – 'a whole people, a family folk' (p. 234).

This difficulty of moving from the old British industrial modernity focusing on work to the new American modernity is clearly expressed by Harry Hopkins (1957) in his 'portrait at mid-century' based upon 'journeys into England' between 1952 and 1954. He too is concerned for repetitive work, in which the conveyor belt is 'king' (p. 67), and its relation to England's future identity. Visiting the Austin factory in Birmingham with its 16 miles of conveyor belts, he agrees that 'these great modern mass production factories are surely one of the outstanding phenomena of our times' (pp. 67–8). However, in a situation where workers just press buttons he worries about their minds (p. 68), which leads him to ask: 'Will the England of "automation" be an England we can recognise?' (p. 246). Like Hutton he is confident that little will change:

> And despite football pools, television, and the rest of it, the work a man does and the respect he receives while doing it remain the essential foundation of life. This is perhaps a Victorian notion, but I believe it remains true.
>
> (Hopkins, 1957, 69)

But again, like Hutton, he sees the social challenge and later revises his original thoughts:

> Human adjustment to change has lagged behind mechanical adjustment. In the docks, in the mines, everywhere, the old Rule by Fear has been banished. Yet new incentives, new social disciplines have still to grow up. The Victorian belief in work for its own sake has gone; a new philosophy has yet to develop to match the new fabric of our lives.
>
> (Hopkins, 1957, 248)

No pointers to the American solution – social discipline exchanged for more consumption – are even considered here. In fact Hopkins seems to think such Americanization would be the death knell of the country: 'A thousand Productivity Reports from across the Atlantic will not make a British worker think like an American worker – it would be the end of us if it could' (Hopkins, 1957, 249). Obviously hybridity had no place in these essentially nationalist discourses.

But, as we know through hindsight, Hutton and Hopkins got it very wrong. Consumer modernity soon overwhelmed England just as it did many other countries. Christopher Booker (1969) charts this 'profound change' in his book *The Neophiliacs*. He identifies 1956 as the key year when the country, like the 'other countries of the West', entered 'a period of upheaval'. Quite simply, society was transformed:

> Its major ingredient has been a material prosperity unlike anything known before. With it have come that host of social phenomena, which initially, in Britain and Western Europe, were loosely lumped together under the heading 'Americanisation' – a brash, standardised mass culture, centred on the enormously increased influence of television and advertising, a popular music more marked than ever by the hypnotic beat of jazz and the new prominence, as a distinct social force, given to teenagers and the young.
>
> (Booker, 1969, 35)

With the coming of commercial TV and Rock n' Roll nothing would be the same again. There was a ' "bubble of excitement" welling up in England over the years after 1956' (p. 79). By 1958 the word 'image' had came into common currency (from the American advertising industry) so that Britain's headlong rush into the new modernity was consummated by new imagery culminating in the 'swinging London' phenomenon of the mid-1960s (pp. 44–5).

But, of course, it was not America which was being reproduced but a hybrid, an 'Americanized England', one which was still England. Alongside emulation of things American there was also much anti-American feeling. But although intellectual elites might scorn the emptiness of the American way of life, it looked very attractive from the perspective of those (like me) living on council housing estates (Brogan, 1959, 16). Mass consumption proved seductive however much intellectuals from both right and left might be shocked by the 'philistine indifference' and 'bourgeois appetites' of the workers (Brogan, 1959, 18). These anti-Americans were the old guard, their positions of influence threatened by the new modernity. Coming into prominence was a ' "New Class" concerned with the creation of "images" – pop singers, photographers, pop artists, interior decorators, writers, designers or magazine editors' (Booker, 1969, 22). Although they looked upon America as 'the source of a new and unexpected inspiration, as a romantic land with an up to date culture, a hotbed of new sensibility' (quoted in Booker, 1969, 38), their alternative name gives away their continuing Englishness – the 'New Aristocracy'. Booker's (1969, appendix B) list of 'leading figures making up the "New England", 1955–66' is revealing, mixing as it does both many rising from 'the urban lower class' and a strong Oxbridge contingent. From the latter were drawn the 'Chelsea Set' (p. 133), part of this upheaval but hardly the stuff of American culture *en masse*. However powerful Americanization might have seemed, it certainly did not eliminate traditional English fascination with class and royalty. From the Beatles to Anthony Armstrong Jones (the photographer who married Princess Margaret), who were both in Booker's list, this was truly an English Americanization.

Globalization as the End of Americanization

Superficially at least, it sometimes seems as if Americanization has grown and grown to become contemporary globalization. Certainly the most conspicuous icons of America commonly appear in journalistic depictions of globalization – for example, McWorld, Cocacolaization and Disnification. What our world-systems analysis has shown is that there has been no such continuity: current resonant Americanization is a far cry from its capacious predecessor. In fact we can argue that the Americanization of high hegemony was the opposite of contemporary globalization.

By definition, under conditions of high hegemony the USA had no economic rivals within the world-economy. Hence America came to be seen as the model for other countries to emulate: the American Dream became 'world dream'. And this dream was built upon a simple promise: it you behave like us, you too can experience the good life like us. Americanization was successful because this seductive offer proved to be irresistible to hundreds of millions of people outside the USA. This was the optimistic world which America created. But that optimism has long since vanished. In the more competitive world which has followed high hegemony, foreboding has overtaken optimism. Whereas Americanization was a promise, globalization, with its dominant neo-liberal face, has been created as a threat, a new social discipline with the potential to undermine the lives of all managers and workers. The American world dream is replaced by a new world nightmare where all production is carried out for wages at the lowest world-wide level.

Fortunately the nightmare cannot happen because more mass production ultimately requires more mass consumption. With most people forced down to subsistence wages, who will buy the goods? There may be some time before this simple lesson of Americanization is rediscovered, with much consequent world economic upheaval in the meantime, but globalization is no more sustainable than the promise of the original Americanization.

References

Aron, R. (1959) 'From France', in F. M. Joseph (ed.), *As Others See Us: The United States through Foreign Eyes*. Princeton, NJ: Princeton University Press.

Arrighi, G. (1990) 'The three hegemonies of historical capitalism', *Review*, 13, 365–408.

Barzini, L. (1959) 'From Italy', in F. M. Joseph (ed.), *As Others See Us: The United States through Foreign Eyes*. Princeton, NJ: Princeton University Press.

Booker, C. (1969) *The Neophiliacs: A Study of the Revolution in English Life in the Fifties and Sixties*. London: Collins.

Brogan, D. W. (1959) 'From England', in F. M. Joseph (ed.), *As Others See Us: The United States through Foreign Eyes*. Princeton, NJ: Princeton University Press.

Duigan, P. and Gann, L. H. (1992) *The Rebirth of the West: The Americanization of the Democratic World, 1945–1958*. Oxford: Blackwell.

Ellwood, D. W. (1992) *Rebuilding Europe: Western Europe, America and Postwar Reconstruction*. London: Longman.

Hopkins, H. (1957) *England is Rich: A Portait at Mid-century*. London: Harrap.

Hutton, G. (1953) *We Too Can Prosper: The Promise of Productivity*. London: George Allen and Unwin.

Kuisel, R. F. (1993) *Seducing the French: The Dilemma of Americanization*. Berkeley: University of California Press.

Luce, H. (1941) *The American Century*. New York: Farrar and Rinehart.

Lukacs, J. (1993) *The End of the Twentieth Century and the End of the Modern Age*. New York: Ticknor and Fields.

Marling, S. (1993) *American Affair: The Americanization of Britain*. London: Boxtree.

Pells, R. (1997) *Not Like Us*. New York: Basic Books.

Price, H. B. (1955) *The Marshall Plan and its Meaning*. Ithaca, NY: Cornell University Press.

Rybcynski, W. (1986) *Home: A Short History of an Idea*. London: Penguin.

Taylor, P. J. (1996) *The Way the Modern World Works: World Hegemony to World Impasse*. Chichester, UK: Wiley.

Taylor, P. J. (1998) *Modernities: A Geohistorical Interpretation*. Cambridge, UK: Polity.

Wallerstein, I. (1979) *The Capitalist World-Economy*. Cambridge, UK: Cambridge University Press.

Wallerstein, I. (1984) *The Politics of the World-Economy*. Cambridge, UK: Cambridge University Press.

Chapter 2

Locating the American Century: Themes for a Post-colonial Perspective

David Slater

'Empire became so intrinsically our American way of life that we rationalized and suppressed the nature of our means in the euphoria of our enjoyment of the ends.'
(W. A. Williams, 1980, p. ix)

'America as Destiny'

I want to argue in this introductory chapter that for any review of the projection of American power to be meaningful it is necessary to keep in mind the twin significance of vision and mission. When we return to Henry Luce's (1941) statement on 'The American Century', penned at a rather crucial moment in twentieth-century history, it is noticeable that when considering the problems faced by America as it entered on to the world scene, much emphasis was given to the need for an authentic vision of America as a world power, and a vision 'which will guide us to the authentic creation of the 20th Century – our Century' (Luce 1941, p. 35). This sense of vision was defined in relation to: the development of 'free economic enterprise', with America as the dynamic leader of world trade, and the key guarantor of the freedom of the seas; the dissemination through the world of its scientific leadership, including its technical and artistic skills, and the fulfilment of its role of being a Good Samaritan for the entire world, undertaking to feed the world's people. Finally, Luce underscored the centrality of those great American ideals – a love of freedom, a feeling for the equality of opportunity, a tradition of self-reliance and independence and cooperation. In addition, Americans were defined as the inheritors of all the great principles of Western civilization – justice, the love of truth and the ideal of charity. This invocation of continuity, of Western rootedness, of civilized principles was then followed by a notion of mission. It is now America's time to be the powerhouse from which these great ideals can spread throughout the world and effect the elevation of the life of mankind from the level of the beasts to a little lower than the angels.

The prevailing sense of mission present in Henry Luce's portrait of America as an emerging global power, was shared by one of his contemporaries – Walter Lippmann. Lippmann (1943), writer and influential columnist, subsequent (re)inventor of the term 'Cold War', concluded a short book on US foreign policy by asserting that American destiny resided in the fact that fate had brought America to the centre of Western civilization, the 'glory of our world'. For Lippmann (p. 252) the American idea had a religious import; it was seen as historic and providential that the formation of the 'first universal order since classical times should begin with the binding together of the dismembered parts of Western Christendom' – America's emerging role in the world was to heal the old schism between East and West in a new universalizing mission of culture and faith. The presence in both these interventions of an interweaving of universality with America was also vividly captured in President Franklin D. Roosevelt's Annual Message to the Congress in January 1941 when he identified the 'four essential human freedoms' – freedom of speech and expression; the freedom of every person to worship God in his own way; freedom from want, and freedom from fear. For each freedom Roosevelt added the key phrase – 'everywhere or anywhere in the world' (emphasis added).[1] Vision, mission, destiny and universality come together in these statements, the content and timing of which can be used to draw out three points which are more generally relevant to my argument.

First, the prominence of universality, the idea of the wider, world relevance of American ideals and capacities, is a theme that can be traced back into the nineteenth century just as it continues through into the later part of the twentieth century, up to 1989 and beyond. What I would also suggest here is that there is an intertwining of the universal with a notion of destiny that draws sustenance from the past whilst being emblematic of the future, reminding us perhaps of Hegel's prescient comment that not only is America the 'land of the future', but also the place where 'the burden of the World's History shall reveal itself' (Hegel 1956, p. 86).

Second, the immediate reality of war is clearly crucial to our understanding of the positions expressed by Luce, Lippmann and Roosevelt, and although their statements are obviously a reflection of the time of their enunciation, it can be argued that war in general has been intrinsic to the creation and development of America. The War of Independence, the Second War with Britain (1812–15), the US–Mexican War (1846–8), the Civil War (1861–5), The Spanish–American War (1898), the two world wars of the twentieth century, the Korean War (1950–3) and the Vietnam War constitute some key markers, but war is more deeply rooted than this. The American historian Sherry (1995), for example, has shown how, since the 1930s, America has lived in the continuing shadow of war – war as memory, as metaphor, as model and as menace. A nation that was born through war, that expanded through conquest and that was reconceived through civil war cannot be adequately located if only interpreted in the context of progress, modernity and freedom. Moreover, the territorial constitution of the United States of America in the nineteenth century also involved a war against the indigenous inhabitants of the North American continent – colonization and settlement, the 'clash of cultures', the spread of Western civilization and order, and the first encounter with the 'native other', the 'savage Indian'. The connection between war and the indigenous other leads me to the last introductory point I want to make.

So far America has been referred to in the singular, but in an era of multiculturalism and a profound questioning of the meanings of identity and difference, we may want to ask ourselves such questions as – 'whose America?', and 'whose destiny?' In a similar vein we can destabilize the notion of 'The American Century' by asking what difference would a 'Century of the Americas' make to our conceptual, thematic and political priorities? (see González-Casanova in this volume). This questioning refers to both the duality of a North and a South of the Americas as well as to the heterogeneity within the 'America' of the United States itself, both past and present.

Space and the Other – Rethinking the Place of the Frontier

In 1891, the US Census Bureau declared that the frontier no longer existed. Americans were settling the entire continent from the Atlantic to the Pacific, and this conquest of nature had been driven by the explosive formation of an industrial economy, so that by 1890, for example, United States manufacturing production had surpassed the combined total of England and Germany (Takaki 1993, p. 225). But the end of the frontier signified far more than economic expansion. Frederick Jackson Turner (1962, pp. 37–8), in his renowned 1893 essay on the frontier in American history, argued that the process of settlement and colonization brought to life intellectual traits of profound importance – the emergence of a 'dominant individualism', a 'masterful grasp of material things', a 'practical, inventive turn of mind', a 'restless, nervous energy', which all reflected the specificity of the American intellect. In addition, and symptomatically, Turner (p. 3) posited that the frontier constituted 'the meeting point between savagery and civilization', and the 'line of the most rapid and effective Americanization' (ibid.). Of a related persuasion, Theodore Roosevelt (1889, pp. 26–7), in his four-volume examination of the frontier, entitled *The Winning of the West*, talked of the frontier farmers and 'warlike borderers', who, 'being spurred ever onwards by the fierce desires of their eager hearts . . . made in the wilderness homes for their children and by so doing wrought out the destinies of a continental nation'. Their adversaries, the native Indians, were considered to be the 'most formidable savage foes ever encountered by colonists of European stock' (p. 17). The dominant representations of the indigenous peoples throughout the nineteenth century, 'their savagery', or 'child-like nature', or subsequently, when no longer deemed a threat, their 'exotic nature', to be romanticized in expressions of nostalgia for a vanquished people, recur in other encounters with non-white peoples and cultures. The point that needs to be made here is that the frontier, territorial expansionism, colonization and settlement, Roosevelt's 'oncoming white flood', provide a founding context for any discussion of destiny, mission and geopolitical location.

By the beginning of the twentieth century, as the waves of colonization and settlement had passed their peak, the earlier Jeffersonian objective of separating the Indians from their land, of incorporating and assimilating the Indian into an advancing and superior civilization, had taken its toll. War and subsequent negotiated arrangements resulted in Native America being constricted to about 2.5 per cent of its original land base within the 48 contiguous states of the union (Rickard 1998, p. 58), and the violent appropriation of land and subsequent confinement of native Indian communities to limited reservations gave another, darker expression to the meaning of the frontier (Niess 1990, pp. 10–13; Takaki 1993, pp. 228–45; Zinn 1980, pp. 124–46). This reality may be contrasted with the traditional story of an untamed wilderness inhabited only by a few primitive Indians, who at times resisted savagely, but who eventually and inevitably melted away when confronted by civilization, technology and progress – a story that is critically countered by Durham (1993), a member of the American Indian Movement.[2] The presence within the dominant narrative of a sense of predestination, a future foreordained was previously expressed not by an American but by a learned Frenchman, writing in the earlier part of the nineteenth century. For Tocqueville (1990, p. 25), the 'implacable prejudices', 'uncontrolled passions' and 'savage virtues' condemned the Indians to 'inevitable destruction'; and, he went on, the Indians 'seem to have been placed by Providence amid the riches of the New World only to enjoy them for a season; they were there merely to wait till others came'.[3] This idea that a purportedly inferior people was predestined to wait for the advent and organization of another more advanced people, found a wider echo in the twentieth-century annals of modernization theory: Walt Rostow (1960, pp. 109–10), for instance, in considering colonialism, noted that colonies were often initially established to fill a vacuum, that is 'to organize a traditional society incapable of self-organization (or unwilling to organize itself) for modern import and export activity', and furthermore, he went on, 'in the four centuries preceding 1900 . . . the native societies of America, Asia, Africa and the Middle East were . . . structured and motivated neither to do business with Western Europe nor to protect themselves against Western European arms; and so they were taken over and organized'.

The nineteenth-century expansion of the frontier and white America's violent encounter with its Indian other came to form a deeply significant element of the nation's collective memory, and not only figured in the production of films about how 'the West was Won', but also found expression in twentieth-century warfare and foreign policy. In the 1960s, for example, during the Vietnam War, American troops described Vietnam as 'Indian country', and President Kennedy's ambassador to Vietnam justified military escalation by citing the necessity of moving the 'Indians' away from the 'fort' so that the 'settlers' could plant 'corn' (Slotkin 1998, p. 3). More recently too (as Campbell points out in this volume), the Indian metaphor has been deployed in the Bosnian conflict.

The expansion of the frontier, and the territorial constitution of the United States as we know it today, had another dimension which was also highly relevant to the later projection of power and influence in the Third World or South, and especially in the context of US–Latin American relations. On the eve of the US–Mexican War of 1846–8, and in the wake of the annexation of Texas from Mexico, a pivotal cause of the conflict, notions of 'Manifest Destiny' came to circulate in the worlds of journalism and politics.[4] John L. O'Sullivan, the editor of the *Democratic Review*, and the originator of the term, had already written in 1839 of a boundless future for America, asserting that 'in its magnificent domain of space and time, the nation of many nations is destined to manifest to mankind the excellence of divine principles'. But it was six years later in 1845, in relation to continuing opposition to the annexation of Texas into the Union, that O'Sullivan wrote of 'our manifest destiny to overspread the continent allotted by Providence for the free development of our yearly multiplying millions' (both quotations in Pratt 1927, pp. 797–8). The doctrine of 'manifest destiny' embraced a belief in American Anglo-Saxon superiority, and it was deployed to justify war and the appropriation of approximately 50 per cent of Mexico's original territory. Furthermore, as with accompanying characterizations of the native Indians' purported lack of efficient utilization of their natural resources, it was observed by President Polk, at the end of the War in 1848, that the territories ceded by Mexico had remained and would have continued to remain of 'little value to her or to any nation, whilst as part of our Union they will be productive of vast benefits to the United States, to the commercial world, and the general interests of mankind' (quoted in Gantenbein 1950, p. 560). Overall, and expressive of the dominant trend, the Mexican, like the Indian, was represented in denigrating racial stereotypes which were also replicated in later twentieth-century narratives (see Horsman 1981, Slotkin 1993, and Weinberg 1963).

The coalescence of destiny and the frontier, although not retaining its original territorial grounding, has remained a salient feature of the twentieth century. The frontier, both as metaphor and purpose, was invoked by President Kennedy in 1960, when he spoke of standing on 'the edge of a new frontier' at a time when the United States was faced by new circumstances and new threats – by the growing power of a rival creed – Communism – and by the proliferation of political turbulence in the countries of the Third World. Kennedy's message harnessed images of rugged individualism, restless energy and the desire for freedom, and it also formed the basis of new projects, such as 'The Alliance for Progress' in Latin America, rekindling the memory of Franklin D. Roosevelt's 'Good Neighbor Policy'. Continuity also pervades President Reagan's statement in the 1980s, during the resurgence of the Cold War, when he affirmed his belief in the nation of the future, assuring his audience that we can meet our destiny, and that destiny is to build a land that will be for all mankind – 'a shining city on a hill'.[5] In the twentieth century, the frontier was no longer rooted in the territorial formation of the nation – rather it was imagined in relation to America's destiny in the world. The mythical construction of new frontiers was interwoven with ideas of mission and destiny. The projection of geopolitical power and the development of a strategy to establish stability and order were both predicated on a deeply-rooted belief in America's pivotal role in the world. Naturally the economic, military and cultural capacities for global leadership had to be put in place, but it was the underlying and driving political will that was essential to the success of America's mission.

The Geopolitics of Order and the Changing Significance of Containment

An extension of spatial power, or the establishment of a new spatio-political order, a *Grossraum*, or 'Grand Area', as the German political theorist Schmitt suggested in his analysis of the Monroe Doctrine of 1823 (Ulmen 1987), needs a principle of legitimacy. In this case, such a principle rested on the belief that the United States must lead the New World of the Americas, as a separate domain from Europe. Moreover, in the Americas and subsequently globally, the construction of a geopolitical identity included the positing of difference as both inferiority and as a potential threat to security and order. At the beginning of the present century Theodore Roosevelt emphasized the importance of order and the 'proper policing of the world' in a context he described in terms of the increasing interdependence and complexity of international political and economic relations. In what became known as the Roosevelt Corollary (1904) to the Monroe Doctrine, it was contended that 'chronic wrongdoing, or an impotence which results in a general loosening of the ties of civilized society, may in America, as elsewhere, ultimately require intervention by some civilized nation', which in the Western Hemisphere may require the United States in acute cases of disorder to exercise 'an international police power'(quoted in Gantenbein 1950, p. 362).[6] But interventionism to preserve or restore order was the subject of much debate, especially in the context of the possible integration of what were regarded as inferior peoples within the Union.[7] For example, the assumption that the Cubans were incapable of governing themselves played a crucial role in preventing the recognition of a fully independent Cuba. This situation was reflected in the fact that the US Congress required Cuba to incorporate the Platt Amendment into both its constitution and the 1903 permanent treaty between the two countries. Cuba's sovereignty had been severely circumscribed by a number of conditions that were added on to an army appropriation bill by Senator Platt, and these conditions became law in the United States in March 1901, being known as the Platt Amendment. Article three of the Amendment stated that the Cuban government agreed that the United States may exercise the right to intervene for the preservation of Cuban independence, and the maintenance of a government adequate for the protection of life, property and individual liberty (see Brockway 1957, p. 71).

Although territorial expansion was an intrinsic part of United States expansion during the nineteenth century, the Spanish–American War of 1898 brought in its wake a qualitatively different form of expansion that entailed the acquisition or control over territories that were not contiguous. American leaders had to determine how these territories were to be governed; whether for instance they should become permanent colonies in an American Empire, as exemplified in the case of the Philippines, or rather more loosely connected through the granting of semi-sovereignty under American tutelage (a protectorate), as was the case with Cuba until 1934 when the Platt Amendment was abrogated by President F. D. Roosevelt. More generally, US interventionism in the early part of the twentieth century was closely connected to the chief concerns of the preservation of order and the diffusion of the American ideals of progress and civilization. In the occupations of Cuba (for instance, 1906–9), the Dominican Republic (1916–24), Haiti (1915–34), Honduras (1912–19 and 1924–5), and Nicaragua (1912–25 and 1926–33) these concerns were clearly present. The Taft administration, for example, continued what was called a 'moral protectorate' of Nicaragua and President Woodrow Wilson ordered the Marines to assume the entire functions of government in the Dominican Republic for five years to restore 'internal order' – as one academic at the turn of the century commented, 'disorganization and disorder will not be long permitted in a world grown as small as ours' (cited in Barnet 1980, p. 100).

With respect to containment, it needs to be emphasized that the Monroe Doctrine of 1823 represented the first codification of such a strategy in relation to the Western Hemisphere (Whitaker 1954). The objectives entailed staking out a geopolitical terrain for the Americas that would be free from the incursions or unwelcome interferences of the Old World. Thomas Jefferson's proposition that America would have a 'hemisphere to itself' (Niess 1990, p. 2) provided a basis for the first expression of a strategy of containment, being fuelled by Jefferson's belief in an 'Empire for Liberty'. Symptomatically, the independent nations of that other America – *Latinoamérica* – were not consulted, and Jeffersonian notions of liberty were critically received south of the Rio Grande, Simón Bolívar observing that the United States seemed 'destined by Providence to plague America with misery in the name of liberty'.[8] From the nineteenth century through the 'Good Neighbor Policy' of the 1930s and the Alliance for Progress of the 1960s to the 1994 signing of NAFTA (North American Free Trade Agreement), the ambiguity attached to the term 'America' (nation and hemisphere) was effectively deployed as part of the development of United States hegemony in the Americas/*Américas*. This ambiguity has been used in the process of justifying expansion whereby leaders of the United States have constructed a US identity and project to unify the Western hemisphere behind their own vision of an 'American future'. In the 1980s, for example, President Reagan, in an address to the Organization of American States, talked of this hemisphere being a 'special place with a special destiny', a chosen land that must be watched over to 'keep it secure from alien powers and colonial despotisms, so that man may renew himself here in freedom' (Kenworthy 1995, p. 21).

This hegemonic project for the hemisphere of the Americas combined a moral, cultural and political leadership with the capacity and will, when necessary, to use coercion, to play the role of enforcer, to deploy Theodore Roosevelt's 'international police power'. The development of US hegemony in the region of the *Américas* did not of course go unchallenged. The Cuban intellectual and revolutionary, José Martí, who defined the United States of America as the 'colossus of the North', argued in the early 1890s that whilst Hispanic America had almost entirely freed itself from its first metropolis, a new and much more powerful metropolis was overtaking it under the guise of economic penetration, using wherever necessary military means to extend its control (Martí 1961, pp. 19–29). During the first decades of the twentieth century, the intensities of Latin American nationalisms were closely associated with the spread of United States power in the Americas, and Franklin D. Roosevelt's 'Good Neighbor Policy' of the 1930s, which constituted a kind of 'New Deal' in foreign relations, was a response to the rapid emergence of nationalist forces in Latin America, carrying within it a recognition that the days of 'international police power' alone were in steady decline.[9]

The strategy of containment that was initiated with the Monroe Doctrine unfolded within a specific geopolitical arena, and as I have indicated above, the United States was primarily concerned with establishing and consolidating its position within a hemisphere that was deemed to require order, progress and the spread of civilization under US leadership. Threats to this order arose during the First World War, but far more acutely during the Second World War. However, the most compelling and sustained threat to US power in the Americas and more globally emerged in the post-World War Two era. Although the Bolshevik Revolution of 1917 had an important impact on political developments in Latin America before 1940, with the influence of socialist and Marxist ideas spreading through a wide range of countries in the South of the Americas, it was primarily in the era of super-power rivalry and the Cold Wars of the 1950s and 1980s that the real clash of political ideas and ideological conflict became fully manifest. In this extended period the strategy of containment was qualitatively different in the sense that Communism was constructed as a global threat to the *internal* stability of the nations of the developing world. It was perceived as a mobile, destabilizing political creed that had the potential to penetrate and infect the bodies politic of the countries of Latin America, Africa and Asia. These countries had to be protected from Communist subversion through modernizing and developing their social, economic and political structures, and where necessary order had to be restored through military intervention (Chomsky 1992, Kolko 1988, and Robinson 1996). Furthermore, in situations where democratically elected governments pursued radical policies, either in relation to land reform, or the nationalization of private property, or the formulation of an independent foreign policy, or an amalgam of these kinds of initiatives, US intervention was a distinct possibility, as in Guatemala in 1954, in Chile in 1973, in Grenada in 1983 and in Nicaragua through the 1980s.[10]

In these examples we can clearly see the relevance of the exercise of a coercive power, the capacity and determination to intervene in specific circumstances to effect a key political change in another sovereign but subordinated state (see Dodds in this volume). But it is important to keep in mind that the exercise of coercion was always legitimated within the broader context of moral, cultural, economic and geopolitical leadership, for which the United States always sought to promote consent. Significantly, the combined relevance of consent and coercion, of the need for both the will to lead and the will to enforce, was captured in a top-secret document on National Security (NSC 68) prepared by a special State and Defense Department study group in 1950. At a crucial moment in the reassessment of the world strategic situation, the document specifically laid out a series of political and psychological markers that were crucial to the overall development of a policy of containment towards the Soviet Union.

The document argued that the United States must lead in the construction of an effective political and economic system in the free world; it must strive to develop a healthy international community through vigorous sponsorship of the United Nations, the consolidation of the Inter-American system, the reha-bilitation of Western Europe and the promotion of our international economic activities. Values intrinsic to the United States, 'the essential tolerance of our world outlook, our generous and constructive impulses, and the absence of cov-etousness in our international relations' were seen as holding the promise of a 'dynamic manifestation to the rest of the world of the vitality of our system'. Therefore, the document went on, 'as we ourselves demonstrate power, confi-dence and a sense of moral and political direction, so those qualities will be evoked in Western Europe'; and 'in such a situation, we may also anticipate a general improvement in the political tone in Latin America, Asia and Africa' (NSC 68, pp. 401ff). If this was the vision of leadership, of the promotion of consent, it was also asserted that the maintenance of a strong military posture was essential to containment, which was defined as a policy of 'calculated and gradual coercion' (ibid.). The emphasis on coercion did not simply relate to the threat posed by the perceived expansionism of the Soviet Union; its deployment was incorporated into an overall and pervasive strategy of defending freedom and democracy in an increasingly turbulent world. Indeed, a decade and a half later, President Lyndon B. Johnson made it abundantly clear, in a context formed by military intervention in the Dominican Republic (1965), the war in Vietnam and a social crisis in the cities of the United States, that the domestic and the foreign were two sides of the same coin, that promoting democracy at home meant securing it abroad, that the United States was a great, liberal and pro-gressive democracy up to its frontiers, and 'we are the same beyond' – let us never imagine, he continued, 'that Americans can wear the same face in Denver and Des Moines and Seattle and Brooklyn and another in Paris and Mexico City and Karachi and Saigon' (Johnson 1969, p. 7).

This expression of the overlapping of the domestic and the foreign, of the inside and the outside, found an echo in many areas of political change. 'Containment', for example, as a strategy, was not only deployed in the inter-national arena (Gaddis 1982); internally, McCarthyism and the House Un-American Activities Committee which was funded by the Congress in the 1950s, generated an atmosphere in which the whole culture was permeated by anti-Communism, and fear of the 'red menace'. As Nadel (1995) explains, in his book on containment culture, surveillance was regarded as necessary but not sufficient since security at home and abroad required scrutiny not only of actions but of motives as well, and at the same time those motives and attitudes had to be channelled in an acceptable direction. This meant of course that accom-panying the development of military, economic and cultural capacities, in-fluencing public opinion and constructing consent around US policy was seen as particularly important. This perception became stronger after defeat in the Vietnam War, and during the 1980s, for example, the Reagan Administration established an Office of Public Diplomacy which had as a major function the dissemination through the media of anti-Sandinista propaganda (Chomsky 1992, pp. 76–9).

The protection of the 'free world' and the strategy of containment were characterized by a clear demarcation of friends and enemies, of allies and adversaries. One can suggest that US policy towards what were regarded as friendly states was oriented towards either accommodation or assimilation. With respect to Western Europe and Japan, the United States, whilst projecting itself as leader of the Western world, sought to establish and sustain relations of mutual benefit under US hegemony. As regards nation-states of the South, policies of benevolent assimilation, especially in relation to the less powerful nations, tended to be more strongly present, as reflected in the US role within the Organization of American States, and more recently in the context of NAFTA. Conversely, states that were regarded as hostile to US visions of freedom, progress and democracy, were confronted by policies of confinement and termination. The two were not always independent of each other. Hence, for example, in the case of Cuba, the Bay of Pigs invasion of 1961, which constituted an unsuccessful attempt to overthrow Castro's revolutionary government, was followed by a continuing blockade of the island and a confinement of its position in international relations. Similarly, with Nicaragua in the 1980s, confinement of position and termination of the Sandinista government were part and parcel of Washington's overall geopolitical strategy. The political identification of friends and enemies continues into the post-1989 period, but today the strategy of containment no longer refers to the global struggle against Communism.

The US invasion of Panama in December 1989, termed by the Bush Administration 'Operation Just Cause', represented both change and continuity in the context of interventionism. Change was reflected in the way the invasion was justified; no longer any reference to the Cold War and the fight against Communism, but instead the need to seize a corrupt dictator, facilitator of the entry of narcotics into the United States, and enemy in Washington's 'War on Drugs'. Continuity was to be found in the US transgression of Panamanian sovereignty, and the setting in place of a new government, post-intervention. Panama 1989 signalled the emergence of a new context for containment. The 'War on Drugs', with both its internal and external dimensions, has been associated with policies of eradicating the supply of cocaine by targetting the Andean areas of coca production (in Bolivia, Colombia and Peru), and in some cases calls have been made to bomb Medellín or even to invade Colombia. Containment in this domain is far more specific, far more geographically targetted, and references to Third World (dis)order, corruption and violence have been far more significant than the previous foregrounding of Communism as a dangerous and incursive ideology (Slater 1999).

In a different context, that of the geopolitical situation in the Gulf, the United States has had a strategy of 'dual containment' with respect to Iraq and Iran. This strategy, as recently indicated, has been less successful as regards Iran, and obviously the differences between the two countries makes such a strategy highly problematic.[11] US policy towards Iraq and Iran can be seen as part of a broader strategy aimed against so-called 'rogue states', a category which also includes Afghanistan, Cuba, Libya and more recently the Sudan. These are states which are seen as sponsoring 'terrorism', and as a result they have been subjected to sanctions, international isolation, and at specific moments direct missile attack, as most recently with Afghanistan and the Sudan. What can be seen evolving therefore, in the post-Cold War era, is the identification by the United States of a series of new threats to order and security, followed by the formulation of specific policies of targeted containment, which in the current period tend to be mediated through the United Nations (Morales 1994).

Overall, we can suggest therefore, that since the Monroe Doctrine there have been three general periods of containment: an initial period, until the Second World War, that was primarily concerned with the development of US hegemony in the Americas, and the restriction of European geopolitical influence; a second and major period of containment aimed at combating what were seen as the expansionary drives of the Soviet Union, and a third, post-Cold War phase characterized by a diversification of containment strategies as responses to a growing proliferation of perceived threats to world order and security. This contemporary phase is witness to strategies of containment which are contextualized globally but targetted nationally and regionally, according to their separate rationales.

Into the Twenty-first Century: Visions of 'America'

In 1993, President Clinton asserted that the new world toward which we are moving favours the United States – and he went on, 'we are better equipped than any other people on Earth by reason of our history, our culture and our disposition to change, to lead and to prosper' (quoted in Valladão 1998, p. 3). The United States as the 'indispensable nation' in a fast-changing, unstable and increasingly chaotic world is one vision of 'America' which will be strongly present in the next century. It is also a vision which can be bolstered by evidence which points to a unique combination of unrivalled military force, the largest and most dynamic economy on the planet, a culture with global scope and crucially the political will to remain hegemonic in the world of the future. Perhaps then going from the twentieth century to the twenty-first will be equivalent to moving from 'Imperial America' to 'World or Global America', where images produced in the United States will reflect a concept of the individual, separated from a particular culture, coping with the same challenges faced by other individuals on the planet, all being citizens of World-America – *pax americana* gone global. Alternatively, and in contrast to a vision of standardization, we might see, as several of the chapters in this volume intimate, a proliferation of hybridized identities, globally and within a multicultural United States.

At the outset of this chapter, I stressed the importance of vision, mission and destiny in our reflections on 'The American Century'. Referring to the wartime writings of Henry Luce and Walter Lippmann provided an introduction to a rather influential imagining of the place of 'America' in the world. Through thinking about some elements of war, spatiality, otherness, the frontier, order and containment, I have outlined one possible perspective that aims to problematize and question the dominant vision of American power and its projection in the world. It can be considered 'post-colonial' in the sense that the approach I am developing seeks to disturb those settled, singular visions of destiny, of freedom, progress and democracy by focusing on some of those other experiences and histories of subordinated peoples and states that are sometimes hidden from view. It is no more than the outline of another opening which might encourage us to juxtapose a 'Century of multicultural Americas' to 'American Century'. When thinking about alternative futures, alternatives to the normalizing and universalizing vision of 'America as destiny', we might want to think about those multiple forms of challenge, protest, mobilization, democratization and social re-imagining that continue to surface across the territories of the *Américas* (see Lima Costa in this volume and Alvarez, Dagnino and Escobar 1998). Moreover, we may want to highlight the potential within the manifold political cultures of the Americas for other ways of seeing beyond a singular vision of the future. In this way too we would be moving away from essentialist views of American politics and opening up the possibility of inventing multiple futures.

Notes

1   See Addresses and Messages of Franklin D. Roosevelt, Issued by the Senate of the United States of America, and Reprinted by His Majesty's Stationery Office, London, 1943, p. 71.

2   Durham (1993, p. 23) goes on to note that when Columbus first landed in the 'New World' there were estimated to be between 20 and 35 million people living in what is now the United States. The peoples were divided into nations, most of which had basically farming cultures, as opposed to the traditional image of Indian peoples being nomads and hunters. For a detailed and beautifully presented study of the history of the Native Nations of North America, see the Barbican Art Gallery's *Native Nations – Journeys in American Photography*, London, 1998, Booth-Clibborn Editions.

3   A parallel view of the posited inferiority of the Indian can be found in the scholarly writing of John Bach McMaster, described as the 'American pioneer of social history'. In the late 1880s he wrote that the Indian was a child of Nature, with an imagination that was singularly strong, but a reason that was singularly weak – for McMaster, the Indian was 'as superstitious as a Hottentot negro and as unreasonable as a child'. Quoted in Wish (1962, p. 223).

4   For an interesting discussion of the more informal 'agents of manifest destiny', not only influential politicians and businessmen, but the actual leaders of expeditions – the 'filibusters' (freebooters or pirates, the term having both a Spanish (*filibustero*) and Dutch (*vrijbuiter*) origin) who landed on foreign soil in the Caribbean and Central America to conquer a country with only a few armed men – see Brown (1980).

5   Quoted in Robert Hughes, 'The Decline of the American Empire', *The Guardian*, 18 Oct. 1996, p. 3. The reference to the 'city on the hill' has its origin with John Winthrop in the seventeenth century.

6   More pithily, Roosevelt suggested that America should adopt the homely adage – 'Speak softly and carry a big stick; you will go far' (Gantenbein, 1950, p. 361).

7   Paradoxically, it was often Americans from the Southern states who most vociferously attacked the policy of imperial annexation, not out of a sense of equality of human rights, but rather on the grounds of racial inequality – for a general discussion of the influence of racism in US imperialism, see Weston (1972).

8   Quoted in Aguilar (1968, opening page).

9   By 1940, however, the US Congress had promised to build up the strength of the armed forces of Latin American states, and this externally financed development of the military in Latin America provoked the Peruvian populist leader Haya de la Torre to refer to the Roosevelt Administration as 'the good neighbor of tyrants' – see Rosenberg (1982, p. 227).

10 In the Chilean case, which is topical at the moment of writing (October 1998), Henry Kissinger, National Security Adviser in 1970, (in)famously commented that 'I don't see why we need to stand by and watch a country go Communist due to the irresponsibility of its own people' – the remark was first printed in the *New York Times*, 11 Sept. 1974, p. 14. For a discussion, see Schoultz (1987, pp. 284–5).

11 See *The Guardian*, London, 13 June 1997, p. 19.

References

Aguilar, A. (1968), *Pan-Americanism – From Monroe to the Present*, Monthly Review Press, New York and London.

Alvarez, S. E., Dagnino, E. and Escobar, A. (eds) (1998), *Cultures of Politics, Politics of Cultures: Re-visioning Latin American Social Movements*, Westview Press, Boulder, Colo.

Barnet, R. J. (1980), *Intervention and Revolution: America's Confrontation with Insurgent Movements around the World*, Meridian Books, New York.

Brockway, T. P. (1957), *Basic Documents in United States Foreign Policy*, Van Nostrand, New York.

Brown, C. H. (1980), *Agents of Manifest Destiny: The Lives and Times of the Filibusters*, University of North Carolina Press, Chapel Hill.

Chomsky, N. (1992), *Deterring Democracy*, Hill and Wang, New York.

Durham, J. (1993), *A Certain Lack of Coherence: Writings on Art and Cultural Politics*, Kala Press, London.

Gaddis, J. L. (1982), *Strategies of Containment: A Critical Appraisal of Postwar American National Security Policy*, Oxford University Press, Oxford and New York.

Gantenbein, J. W. (ed.) (1950), *The Evolution of Our Latin American Policy – A Documentary Record*, Columbia University Press, New York.

Hegel, G. W. F. (1956), *The Philosophy of History*, Dover Publications, New York (first published in 1899).

Horsman, R. (1981), *Race and Manifest Destiny*, Harvard University Press, Harvard.

Johnson, L. B. (1969), 'Our Foreign Policy Must Always be an Extension of This Nation's Domestic Policy' (April 1966), in LaFeber, W. (ed.), *America in the Cold War: Twenty Years of Revolution and Response, 1947–1967*, John Wiley & Sons, New York and London.

Kenworthy, E. (1995), *America/Américas – Myth in the Making of US Policy Toward Latin America*, Pennsylvania State University Press, University Park, Pennsylvania.

Kolko, G. (1988), *Confronting the Third World: United States Foreign Policy 1945–1980*, Pantheon Books, New York.

Lippmann, W. (1943), *US Foreign Policy and US War Aims*, Overseas Editions, New York.

Luce, H. R. (1941), *The American Century*, Farrar & Rinehart, New York and Toronto.

Martí, J. (1961), *Obras Completas*, XIX, Estados Unidos y América Latina, Patronato del Libro Popular, Hávana.

Morales, W. Q. (1994), 'US Intervention and the New World Order: Lessons from Cold War and post-Cold War Cases', *Third World Quarterly*, 15, no. 1, pp. 77–101.

Nadel, A. (1995), *Containment Culture: American Narratives, Postmodernism and the Atomic Age*, Duke University Press, Durham and London.

Niess, F. (1990), *A Hemisphere to Itself: A History of US–Latin American Relations*. Zed Books, London and New Jersey.

NSC 68 (1978), 'United States Objectives and Programs for National Security, April 14, 1950', in Etzold, T. H. and Gaddis, J. L. (eds), *Containment: Documents on American Policy and Strategy, 1945–1950*, Columbia University Press, New York, pp. 385–442.

Pratt, J. W. (1927), 'The Origin of "Manifest Destiny"', *American Historical Review*, XXXII, July, pp. 795–8.

Rickard, J. (1998), 'The Occupation of Indigenous Space as "Photograph"', in Barbican Art Gallery, *Native Nations: Journeys in American Photography*, ed. and introduced by Jane Alison, London, pp. 57–71.

Robinson, W. I. (1996), *Promoting Polyarchy: Globalization, US Intervention and Hegemony*, Cambridge University Press, Cambridge.

Roosevelt, T. (1889), *The Winning of the West*, vol. I, G. P. Putnam's Sons, New York.

Rosenberg, E. S. (1982), *Spreading the American Dream: American Economic and Cultural Expansion, 1890–1945*, Hill and Wang, New York.

Rostow, W. W. (1960), *The Stages of Economic Growth: A Non-Communist Manifesto*, Cambridge University Press, Cambridge and New York.

Schoultz, L. (1987), *National Security and United States Policy toward Latin America*, Princeton University Press, Princeton NJ.

Sherry, M. S. (1995), *In the Shadow of War: The United States since the 1930s*, Yale University Press, New Haven and London.

Slater, D. (1999), 'Situating Geopolitical Representations: Inside/Outside and the Power of Imperial Interventions', in Massey, D., Allen, J. and Sarve, P. (eds), *Human Geography Today*, Polity Press, Cambridge.

Slotkin, R. (1998), *The Gunfighter Nation: The Myth of the Frontier in Twentieth Century America*, University of Oklahoma Press, Norman.

Takaki, R. (1993), *A Different Mirror: A History of Multicultural America*, Back Bay Books, Boston, New York and London.

Tocqueville, A. de (1990), *Democracy in America*, vol. I, Vintage Books, New York (this edn originally published 1945).

Turner, F. J. (1962), *The Frontier in American History*, Holt, Rinehart and Winston, New York and London (originally published in 1920).

Ulmen, G. L. (1987), 'American Imperialism and International Law: Carl Schmitt on the US in World Affairs', *Telos*, 72, Summer, pp. 43–71.

Valladão, A. G. A. (1998), *The Twenty-First Century will be American*, Verso, London and New York.

Weinberg, A. (1963), *Manifest Destiny*, Quadrangle Books, Chicago.

Weston, R. (1972), *Racism in US Imperialism*, University of South Carolina Press, Columbia, South Carolina.

Whitaker, A. P. (1954), *The Western Hemisphere Idea: Its Rise and Decline*, Cornell University Press, Ithaca and London.

Williams, W. A. (1980), *Empire As a Way of Life*, Oxford University Press, Oxford and New York.

Wish, H. (ed.) (1962), *American Historians: A Selection*, Oxford University Press, New York, pp. 215–53.

Zinn, H. (1980), *A People's History of the United States*, Longman, London.

Part II

Economic Capacities: American Know-how?

Chapter 3

Global Shift – the Role of United States Transnational Corporations

Peter Dicken

Introduction

The US-based corporation operating outside the United States itself – the *US transnational corporation* (TNC)[1] – is emblematic of the 'American Century'. US TNCs have established a presence in virtually all parts of the world, creating the most extensive corporate networks ever seen. They have been major agents in the transformation of the world economy and, especially, of the global shifts which have been occurring in its geography. They have been the bearers both of distinctive American ways of doing business (including how business operates and how labour relations are organized) and of American products (and culture). But there is more to it than this: TNCs, whatever their geographical origin, are creatures of *political economy* rather than merely economic actors. This is nowhere more clear than in the case of US TNCs. Such corporations have always been the focus of controversy, both inside and outside their home country. They have been seen, alternatively, either as free-wheeling capitalist agents, increasingly separate from US jurisdiction, or as agents of United States foreign policy and imperial expansion; an integral part of United States hegemony and the *Pax Americana*.

US TNCs have also had a disproportionate influence on the theoretical discourses surrounding TNCs and foreign direct investment (FDI).[2] Much of the theoretical explanation of TNCs is based, consciously or unconsciously, on the US model. Only with the rapid emergence of Japanese TNCs in the late 1970s and early 1980s was serious attention paid to alternative models of corporate structure and behaviour. Even then, a common tendency was to see Japanese TNCs as some kind of deviation from the US norm. Such a view has recently received renewed currency in the light of the East Asian financial crisis. After a period of US self-doubt and fear of losing out to the Japanese, a measure of US triumphalism – a renewed confidence in the superiority of the American way of doing business – has re-emerged. Two decades on, Rose's assertion that 'at the apex of its fame or infamy, the United States multinational shows incipient signs of decay' (1977: p. 111) looks distinctly off-target.

Within this broad context, the aim of this chapter is to examine the evolution and characteristics of United States TNCs within a globalizing world economy. The chapter is organized into three parts. In the first part I describe the evolution, geographical spread and sectoral and organizational characteristics of US TNCs in the world economy during the present century. In part two I outline some of the major arguments surrounding the effects of US TNCs on the host countries in which they operate. Finally, in part three, I present a brief vignette of US TNCs in Europe.

Emergence of the Colossus: United States TNCs in the Global Economy

The transformation of the world economy which has occurred during the twentieth century (Dicken 1998) has been dominated primarily by the United States. Despite the emergence of other countries – notably Japan and Germany – as major players, the United States remains the world's leading economic power. It is not only the world's largest producer of manufactured goods (27 per cent of the world total) but also exports more goods than any other country (17 per cent of the world total) and is the destination for the largest share of imports (9 per cent of the world total). Paradoxically, it is also the world's largest debtor and has the biggest external trade deficit. All of these characteristics are intimately bound up with the development and activities of US transnational corporations which still constitute the largest share of the world population of TNCs (Dicken 1998: ch. 2). In the mid-1990s, around 26 per cent of total world foreign direct investment originated from the US, more than twice as big a share as the next largest source countries, Japan and the UK.

Although the US share has fallen substantially (in 1960 almost 50 per cent of world FDI originated there) we must remember that world FDI as a whole has grown dramatically in the last few decades. In absolute terms, US TNCs are *more* significant today than they ever were in the past. They constitute the largest and most extensive network of international production and distribution facilities in the world (see figure 3.1). Indeed, it has been argued that if the overseas operations of US companies are taken into consideration and treated as US in origin (if not in location) then the massive US trade deficit becomes a *surplus* (Landefeld, Whichard and Lowe 1993).

Overtaking the Brits: the early development of United States TNCs

During the four decades between 1870 and 1914 the world economy experienced an unprecedented degree of growth and of openness to cross-border transactions of materials, products and capital (as well as of population). It was also a defining moment for the world economy because it was during this period that world economic leadership shifted from Britain to the United States. In 1870, Britain produced 32 per cent of world industrial output and the United

States produced 23 per cent. By 1914, the US share had soared to 36 per cent while Britain's share had fallen to 14 per cent (Hilgerdt 1945).

It was during this period, too, that substantial numbers of firms began to invest in production facilities overseas (Dunning 1993: ch. 5). Initially, much of this overseas investment exploited industrial and agricultural raw materials, but an increasingly significant proportion set up manufacturing facilities to serve overseas markets directly. Some of this manufacturing investment was undertaken to miminize transport costs but most was to 'jump over' national tariff barriers. But although the US economy had far outpaced that of Britain by 1914, Britain continued to account for the largest share of world foreign direct investment until after 1945. Thus, in 1914, 19 per cent of world FDI originated from the United States (Britain's share was 45 per cent); by 1938, the US share had grown to 28 per cent but Britain still generated a much larger share (39 per cent) (Dunning 1993: table 5.1).

However, these aggregate figures mask very important differences in both the sectoral composition and the geographical distribution of FDI from the United States compared with Britain and the other emerging industrial countries of Europe. Such distinctive *qualitative* differences reflected the specific nature of the United States' political economy, society and geography. In the late nineteenth and early twentieth centuries, US firms evolved in particular ways because of the distinctive institutional structure of the US business environment, the country's raw material base, and the geographical scale of its economy. In particular, these conditions created the circumstances for the growth of very large firms which had to find ways of operating on a continental scale to serve their growing domestic markets (Chandler 1962). Sectorally, although both US and British FDI in that early period shared some common features – both, for example, invested heavily overseas in food, chemicals, petroleum and engineering industries – there were significant differences which were to become reinforced in subsequent years. In particular, US overseas investments, from a very early stage, leaned towards the newer, more technologically sophisticated sectors in both producer and consumer goods.

Geographically, early US international investments were concentrated into two major geographical areas (Wilkins 1970). First, there was 'spillover' investment to geographically proximate countries, notably Canada, Mexico, the Caribbean, and Latin America. As table 3.1 shows, apart from Canada, virtually all of this investment by US companies was in mining, agriculture and petroleum. Second, there was substantial US investment in Europe – at that time the biggest market in the world – with Britain being an especially important destination (a characteristic which continues to the present day). US manufacturing investment – 88 per cent of the total – was overwhelmingly concentrated in Canada and Europe.

Take-off: the post-1945 surge in United States TNC activity

The period between the two World Wars was one of economic hiatus. The post-1918 growth phase was brought to an abrupt end by the 1929 Wall Street Crash.

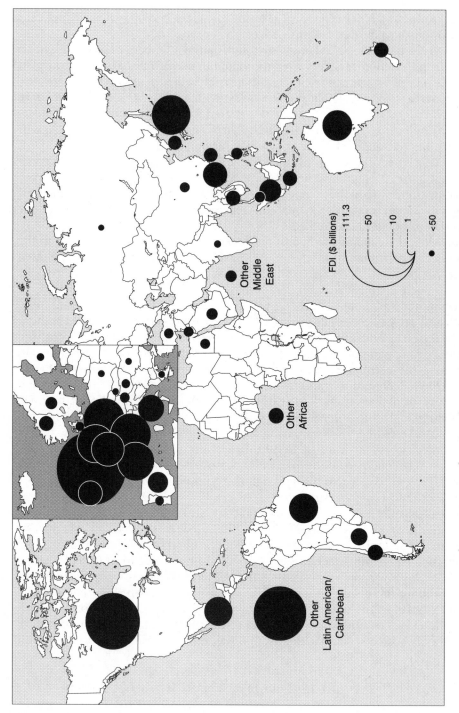

Figure 3.1 US foreign direct investment in the world economy, 1995
Source: Dicken, 1998, figure 2.17a.

Table 3.1 The geographical and sectoral distribution of United States foreign direct investment, 1914

Country/ region	Percentage of sector total				
	Total US FDI (%)	Mining	Petroleum	Agriculture	Manufacturing
Mexico	22.1	41.9	24.8	10.4	2.1
Canada	23.3	22.1	7.3	28.4	46.2
Caribbean	10.6	2.1	1.8	40.5	4.2
C. America	3.4	1.5	–	10.4	–
S. America	12.2	30.7	12.2	7.0	1.5
Europe	21.6	0.7	40.2	–	41.8
Asia	4.5	0.4	11.7	3.4	2.1
Africa	0.5	0.6	1.5	–	–
Oceania	0.6	–	0.6	–	2.1

Source: calculated from Wilkins 1970, table V.2.

Economic depression spread throughout the industrialized world; protectionism grew apace; the growth of FDI slackened, although the United States increased its share of the world total from 19 per cent in 1914 to 28 per cent in 1938 (over the same period, Britain's share fell from 45 per cent to 39 per cent). But it was in the period after World War Two that US FDI really took off and became, for a considerable time, the most significant driving force in the global economy. During the 1950s and 1960s, a whole wave of United States corporations established overseas operations to take advantage of the fact that, after World War Two, the United States was economically and technologically (as well as politically) the undisputed leader of the capitalist world-system. The 'technology gap' between the United States and the rest of the world was especially large, giving US firms a huge competitive advantage in both products and processes. Exploiting that gap through foreign direct investment was helped immeasurably by the strength and dominant global position of the US dollar and by the capital shortages being experienced by most other countries as they struggled to rebuild after the war. The United States was, after all, the only major country to emerge economically much stronger, rather than weaker, after 1945 (Kennedy 1987).

Apart from continuing to make major investments in petroleum and other raw materials, most US overseas investment was in technologically advanced sectors (such as pharmaceuticals, data processing, scientific instruments, synthetic fibres), which required a high level of research-and-development expenditure, and in mass produced branded consumer goods (such as automobiles, household appliances, tobacco products, convenience foods, household products), which required large production runs and heavy advertising expenditure. At the same time, US banks became increasingly internationalized. As the 1950s and 1960s wore on, the geographical structure of United States FDI changed substantially. The biggest relative shift involved a reversal in the importance of Canada and Europe. US investment in Canada grew much more slowly than in earlier periods whilst Europe became the most important single location for United States FDI. By 1979, almost 50 per cent of all US FDI was in Europe. Among developing regions, US manufacturing investment grew much more slowly in Latin America and much faster in East Asia (especially in Singapore, Hong Kong and Taiwan).

Inevitably, the remarkable growth rate of United States FDI between the 1950s and 1970s eventually slackened. In fact, in 1986, *The Economist* published an article headed 'American Multinationals: The Urge to Go Home' and predicted that American firms were 'turning inward'. But there has been little real evidence of such a tendency. Indeed, in six of the nine years between 1986 and 1994, the annual growth rate was still in double figures. As pointed out earlier, in absolute terms United States TNCs are more significant today that they were in the past. Of course, it is also true to say that they now face much greater competition from TNCs from other source countries, particularly from Japan and Europe. Nevertheless, United States TNCs remain the dominant foreign presence in many countries of the world, notably in Europe, Mexico and Latin America. We will look specifically at United States TNCs in Europe in part three of this chapter.

The geographical evolution of United States FDI for much of the past century can be explained, at least in part, by Vernon's conceptualization of the *product life cycle* (Vernon 1966, 1971). Although this particular theoretical construct is inadequate as a theoretical explanation of TNC activity in general (and even of more recent United States TNCs) it has considerable validity as a historically specific explanation of overseas manufacturing investment by United States corporations during the period of their most rapid growth. Vernon's starting point was the assumption that producers were more likely to be aware of the possibility of introducing new products in their home market than producers located outside that market. The kinds of new products introduced, therefore, would reflect the specific characteristics of the domestic market. In the United States case, the high average income level and high labour costs tended to encourage the development of new products which catered to high-income consumers and were labour-saving (both for consumer and producer goods).

In this first phase of the product cycle, Vernon argued that all production would be located in the United States and overseas demand served by exports. But, he argued, such a situation was unlikely to last indefinitely. US firms would eventually set up production facilities in the overseas market either because they saw an opportunity to reduce production and distribution costs or because of a threat to their market position. Such a threat might come from local competitors or from government attempts to reduce imports through tariff and other trade barriers. It follows from the nature of the product-cycle model that the first overseas production of the new product by US firms would tend to occur in other high-income markets. In the specific case of US investment this tended to be Western Europe and Canada. The newly established foreign plants would come to serve these former export markets and thus displace exports from the

United States, which would be redirected to other areas in which production had not begun. Eventually, the production cost advantages of the newer overseas plants would lead the firm to export to other, third-country markets and even, eventually, to the United States itself from these plants. Finally, as the product becomes completely standardized, production would be shifted to low-cost locations in developing countries.

It is perhaps not surprising that Vernon's product-cycle model fits the US experience during the early years of the twentieth century through to around the early 1970s rather well. It was, after all, based upon the empirical data collected by the Harvard Multinational Enterprise Project which was heavily biased towards US enterprises. As we have seen, the major geographical focus for US manufactures setting up plants abroad outside North America was, indeed, Europe; the biggest consumer market outside the United States. During the 1960s, too, US firms in certain industries – notably electronics – made the first substantial moves to locate certain types of production in low labour-cost locations (Grunwald and Flamm 1985; Dicken 1998). These were of two kinds. First, there was the development of a major concentration of US plants close to the United States itself, notably across the Mexican border, as part of the Border Industrialization Program initiated by the Mexican government in 1965, and in the Caribbean. Second, US electronics firms began to locate some of their assembly operations in East Asia. For example, the first offshore assembly plant in the semiconductor industry was set up by Fairchild in Hong Kong in 1962. In 1964, General Instruments transferred some of its microelectronics assembly to Taiwan. In 1966, Fairchild opened a plant in South Korea. By the early 1970s, in fact, every major US semiconductor manufacturer had established offshore assembly facilities, a process greatly encouraged by the offshore assembly provisions operated by the US government.

However, US transnational activity has become far more complex in subsequent years. In part this is because so many US TNCs have built extensive and sophisticated international networks of production in which new products/processes may be introduced virtually anywhere in the network. Hence, the notion of a simple unidirectional sequence starting in the United States and progressively working outwards in the manner of the product-cycle model has very limited validity, although it may well still apply to US firms making their first overseas investments.

The *sectoral* distribution of United States FDI has changed substantially in recent years. In 1985, 27 per cent was in the primary sector (predominantly in oil), 41 per cent in manufacturing and 31 per cent in services. In manufacturing, five industries (food, beverages, tobacco; chemicals; mechanical equipment; electric and electronic equipment; motor vehicles) accounted for three-quarters of the total. In services, nine-tenths was in finance, insurance and business services and in wholesale and retail trade. By the mid-1990s, the picture had changed in some important respects. The share of the primary sector had fallen to 12 per cent, that of manufacturing to 35 per cent, while the service sector had grown to 53 per cent. Within manufacturing, the major increases were in food, drink and tobacco and in chemicals set against a major decline in the share of mechanical equipment. The shares of FDI in motor vehicles and in electric and electronic equipment increased only slightly. The biggest relative shift was in the services sector where FDI in finance, insurance and business services had grown from 54 to 69 per cent of the sector total.

Geographically, 50 per cent of United States FDI was located in Europe in the mid-1990s. More than one-third of the European total was in one country: the United Kingdom. In the Americas, there was a marked contrast between Canada (whose share of United States FDI declined from 20 per cent in 1985 to 12 per cent in 1994) and Latin America, whose share grew from 12 per cent to 18 per cent. East and South East Asia contained only around 14 per cent of United States FDI in 1994 (up from 11 per cent in 1985). Within Asia, the major foci in 1994 were Japan (43 per cent of the regional total compared with 38 per cent in 1985), Hong Kong (15 per cent compared with 13 per cent) and Singapore (12 per cent compared with 8 per cent). It is notable that South Korea, Taiwan and China together contained only 11 per cent of United States FDI in Asia in 1994, up from 7 per cent in 1985. The spread of United States FDI outside Europe, Latin America and Asia was very thin indeed. The whole of the continent of Africa, for example, contained less than 1 per cent of the total.

Friend or Foe? The Impact of United States TNCs on Host Countires

The transnationalization of United States corporations during the twentieth century was an integral part of the increasing economic – and political – world dominance by the United States itself. Consequently, perceptions of their impact on the global economy in general, and on host countries in particular, are inseparable from this wider context. United States TNCs as a species have, indeed, been the focus of immense controversy both inside and outside the United States. The peak of such controversy occurred in the 1970s through the widely publicized involvement of some individual US corporations – such as Lockheed and ITT – in illegal political activities. In the latter case, for example, ITT was deeply involved with elements within the CIA to obstruct the democratic process in Chile by preventing the election to President of Salvador Allende. This story and others involving ITT are graphically recounted by Anthony Sampson in *The Sovereign State: The Secret History of ITT* (1973).

But it is dangerous to extrapolate general tendencies from individual cases, no matter how important these may be. Here, I will simply outline some of the general political and economic issues surrounding the possible impact of United States TNCs. Some of these are specific to the United States' position in the world; others apply to the question of TNC impact, whatever the country of origin (see Dicken 1998: ch. 8).

As essentially *political-economic actors*, TNCs tend to be perceived in two polarized ways. One view is that TNCs transcend political jurisdictions to create their own 'sovereign state' and, in so doing, threaten the sovereignty of nation-states and reduce their abilities to pursue autonomous economic policies.

Whereas the notion of national sovereignty is tied to specific geographical territory, that of 'firm sovereignty' is essentially *placeless*. The American academic, and former Secretary for Labor in the first Clinton administration, Robert Reich, expressed this view in graphic terms. He argued that because of technological developments in transport and communications, in production processes, and in the organization of production, 'the very idea of an American economy is becoming meaningless, as are the notions of an American corporation, American capital, American products, and American technology' (1991: p. 8). Similar views underlie much of the literature in what might be called the 'global reach' school (Barnet and Muller 1974). As argued elsewhere, this notion of 'placeless' corporation is greatly overdrawn (Dicken 1998: pp. 193–9) and difficult to sustain empirically. This is not to say that TNCs do not pose major problems for national policy-making – and US TNCs, because of their scale and ubiquity, certainly do – but rather that the demise of the nation-state in the face of TNCs has been much exaggerated.

A second, quite different, perception of the political-economic power of United States TNCs sees them as being, essentially, *agents* of United States economic – and even political – policy. According to this view, United States TNCs played a major part in establishing and maintaining the hegemonic position of the *Pax Americana*. Although, except in specific cases, it is difficult to sustain the notion of a United States government–TNC conspiracy there can be little doubt that, *de facto*, US hegemonic power was enhanced by the growth and spread of United States TNCs. As Robert Gilpin points out, as a result of the shift in world economic power from British financiers and bankers to United States industrialists,

> American corporations also became instruments of American global hegemony. Although government officials did not foresee that American corporations would expand so as to establish a significant American presence in and impact on all the economies of the non-Communist world, American policies did encourage and protect corporate expansionism after World War II. American tax policy, the insistence that American subsidiaries in the Common Market be treated as 'European' corporations, and, much later, the creation of the Overseas Private Investment Corporation (1969) to insure foreign investments are but three examples of measures designed to foster corporate expansionism. If British economic hegemony was based on the City of London, America's was based largely on her multinational corporations.
>
> (Gilpin 1975: p. 139)

As I implied earlier, TNCs – like all business organizations – are emphatically not placeless. They are deeply influenced in the way they evolve as organizations by the nature and characteristics of the place (i.e. country) in which they originate. In so far as countries differ in their role as 'containers' of distinctive cognitive, cultural, social, political, economic and institutional characteristics they will help to 'produce' different types of business organization. This does not mean, of course, that all firms from the same country will be identical. That would be to ignore the role of agency – of individual entrepreneurs, of a firm's own history, and the like. In the case of TNCs such a narrowly deterministic view would also deny the importance of the influence of the various host environments in which they operate. Firms adapt to local circumstances to a greater or lesser degree. Nevertheless, there is abundant evidence to suggest that 'nationality matters' in helping to create distinctive organizational forms. Thus, although not all United States firms are the same, even in the same industry – Ford and GM provide a case in point – they tend to be more alike than not. (The same applies, of course, to TNCs from other countries.)

Consequently, United States TNCs have tended to develop certain distinctive characteristics. Certainly during the major expansionary phase from the early twentieth century through to the 1970s, they tended to be predominantly very large firms which had already established a strong position in the large domestic market. They tended to be in the more technology-intensive and/or mass-production consumer goods industries in which, again, the nature of the United States domestic economy was important. They also tended to adopt a distinctive organizational form. The rigidly structured, strongly hierarchical, multi-divisional form, which developed specifically in the United States in the early part of the century (Chandler 1962), was generally transferred from the domestic to the international context (Dicken 1998: pp. 202–5). Although there were certainly individual exceptions, most United States TNCs, until quite recently, adopted such a structure. This had very important implications for the ways in which the overseas affiliates of United States TNCs interacted with the national and local economies in which they were established. A high degree of centralized control from the US headquarters, using highly formalized management planning systems, was the norm. Local affiliates, therefore, tended to have relatively little decision-making autonomy (Bartlett and Ghoshal 1989). The 'balance sheet' ruled, driven by the distinctive pressures of the United States financial system which emphasizes short-term profit maximization and shareholder value (Pauly and Reich 1997).

Undoubtedly, many United States TNCs have evolved different, in some cases more flexible, forms of international organization in recent years. Some of the newer generation of firms – like Nike and Microsoft, for example – have never operated in such a manner (for the case of Nike, see Donaghu and Barff 1990; Korzeniewicz 1994). In part this reflects a broader shift in organizational fashion; in part it also reflects the influence of strong and charismatic individuals. Such individuals have always played a key role in United States business. But Henry Ford, Harold Geneen (head of ITT at its peak), Bill Gates, Phil Knight (founder of Nike) are still all products of the United States system. The same point about embeddedness also applies to individuals. Though not deterministic, national context is an enormously powerful influence in creating a distinctively 'American way'.

As United States firms set up operations abroad they inevitably took with them these various characteristics of 'Americanness'. In some cases, this not surprisingly resulted in a serious culture clash; in others (most notably in Britain) the marriage was more benign. Of course, the larger the US presence in any host country the greater the potential friction. The problems have been especially apparent in three groups of countries: developing countries, where the need for access to scarce capital and technology are tempered by fear of economic domination by the US super-power; some older industrialized countries, like France, where a United States presence is seen as a direct threat to national culture;[3] in the two countries in closest geographical proximity to the United States: Canada and Mexico.

As the old saying goes, sleeping next to an elephant is never comfortable, especially when the elephant turns over. Both Canada and Mexico exemplify, in the most graphic manner, the tensions inherent in the relationship with the Untied States and, especially, those connected with the activities of United States TNCs. It is not surprising, then, that some of the most critical opposition to such firms has come from Canada. Not only is Canada next door to the United States but also – and not coincidentally – it has the highest level of penetration of United States companies anywhere in the world. Apart from the general fear of cultural domination, there has been a particular concern in Canada about technological dominance by the United States and about the creation, through the activities of US firms in Canada, of a *truncated and dependent* economy. Some twenty years ago, for example, Britton and Gilmour argued that Canada had been 'under-developed' because of this.

> Over the decades, an under-developed industrial structure, typical of satellite or hinterland economies has been generated . . . Canada is industrially backward . . . There is no avoiding the implication that the indifferent performance of high technology industry in Canada . . . reflects the high degree of foreign ownership of these activities.
>
> (Britton and Gilmour 1978: pp. 59, 80)

Part of the problem was that, at that time, Canada's economy was separate *de jure* from the United States. As a small domestic economy it tended to attract only the smaller, less innovative manufacturing branch plants of US firms oriented to the local Canadian market. Today, of course, Canada – along with Mexico – is part of NAFTA, which means that firms will increasingly make their location-investment decisions in relation to the entire North American market. As yet it is too early to see quite how NAFTA will work out. Fragmentary and anecdotal evidence suggests that United States firms are engaging in substantial restructuring which will alter the distribution and functions of their activities throughout the NAFTA area. This question of the behaviour of United States TNCs within a regional economic bloc leads us logically to a brief consideration of their activities in the most developed regional economic grouping of all: the European Union.

'The American Challenge': United States TNCs in Europe

Europe, particularly Britain, has been a focus for United States manufacturing firms from the beginnings of the internationalization of the US economy. The first foreign branch plant of any United States manufacturer was established in England in 1852 by Samuel Colt to produce firearms. However it, and another to manufacture vulcanized rubber in Edinburgh in 1856, failed within a few years. Arguably, the first truly transnational manufacturer from the United States was Singer, the sewing machine company. Singer built its first overseas factory in Glasgow in 1867, the first link in what was to become an extensive international production network. By the turn of the century, the penetration of the British and European economies by American firms had become so substantial that, in 1902, McKenzie was moved to write a book entitled *The American Invaders*. It contains graphic illustrations of the 'invasion', written in a style that foreshadowed the airport bookstall genre.

> America has invaded Europe not with armed men, but with manufactured products. Its leaders have been captains of industry and skilled financiers, whose conquests are having a profound effect on the everyday lives of the masses from Madrid to St Petersburg . . . The real invasion goes on unceasingly and with little noise or fuss in 500 industries at once. From shaving soap to electric motors, and from tools to telephones, the American is clearing the field . . . The American houses are finding our field so profitable that they are doing their utmost to become English as quickly as possible . . . So they are erecting works here, rushing over American machinery, and in some cases picked American workmen, and are floating English companies . . . *The many factories built by American firms in this country are today supplying not only England but also a large part of the world outside of America with goods which otherwise would have come from the U.S.*
> (McKenzie 1902: pp. 1, 2, 84, 240; emphasis added)

This pattern, very much in line with the product-cycle model, intensified during the following decades. For fairly obvious reasons Britain continued to be a major point of entry for United States firms establishing operations in Europe for the first time (interestingly, the Japanese followed a similar mode of entry from the 1970s). As a result, virtually all the leading United States firms had set up European operations by the 1960s. Indeed, so embedded had many of them become that the average consumer had no idea that such household names as Heinz, Kelloggs, Colgate, Mars, and the like were anything other than British. As United States manufacturing firms moved into Europe, so too did United States banks and other service firms.

As we have seen, United States international investment accelerated dramatically in the 1950s and 1960s as US firms capitalized on their technological strengths and the dominant position of the dollar. Initially, in the European case, the attraction was the weakness of local competition in the aftermath of war. But a far more important incentive soon emerged: the formation of the European Economic Community in 1957. Brought up in a unified continental economy at home, US firms were quick to see the potential of a more unified European market, even though it was to be nearly 40 years before the Single European Market came near to completion. The importance of the EEC to US investment can be seen by the fact that, as a non-member until 1973, Britain's role as a point of entry to Europe declined markedly. The FDI data show that the gap between the growth of United States manufacturing investment in the UK and in the EEC widened in the period after the Community was formed in 1957 and then narrowed again after Britain joined in 1973.

One sector in which the Bitish attraction was not diminished was banking and finance. US banks were also internationalizing apace in parallel with that of US manufacturers. The attraction of a location in Britain – that is, London – was the development of the Eurodollar market there in the 1960s. This provided a huge financial market which was, in effect, outside the regulatory control of the US authorities. 'Once it was appreciated that Eurodollars . . . were free of US political control, it did not take bankers long to recognise that the dollar balances were also free of US banking laws governing the holding of required reserves and controls upon the payment of interest' (Lewis and Davis 1987: pp. 225, 228).

Perhaps not surprisingly, the same fear of United States economic domination voiced by McKenzie in 1902 emerged again in Europe in the heady days of 1968:

> Fifteen years from now it is quite possible that the world's third greatest industrial power, just after the United States and Russia, will not be Europe but American industry in Europe.
>
> (Servan-Schreiber 1968: p. 3)

We can now see, of course, that Servan-Schreiber's prediction was wrong. But although 'American industry in Europe' is not 'the world's third greatest industrial power' it is immensely important both globally and, especially, regionally. Today, fully 50 per cent of total United States FDI is located in Europe. Despite the rapid growth of Japanese investment in Europe in the past two decades this is minuscule compared with the accumulated stock of US investment. Japanese investment may be more visible (every new investment being reported in the press) but it is much, much smaller. US firms occupy significant positions in most European countries, particularly in the UK and Germany, which, together, contain half of all United States FDI in Europe. Britain is still by far the most significant concentration: more than one-third of the European total.

It is difficult to overstate the cumulative impact of United States firms in Europe over the course of this century. They have been the bearers of innovative business practices and technologies in many industries. Long before we became obsessed by 'things Japanese' the way in which many industries were organized within Europe was being transformed by the import of United States management styles and methods (not to speak of the products themselves with their associated images of American culture). Geographically, United States firms have proved to be very adept at responding to the increasing integration and enlargement of the European economy. This is partly because of their domestic experience of operating over vast geographical distances in a unified market, and partly because few US firms had a legacy of large numbers of plants serving individual national markets in Europe. Ford, for example, took the first steps to create a pan-European structure as long ago as 1967 and to rationalize its operations on a Europe-wide basis. Whereas many large European firms, such as Unilever or Philips, had evolved a geographical structure consisting of plants in every major European country each of which served the local market, their US competitors did not have such a legacy. Hence, for example, Procter and Gamble had been able to create a Europe-wide structure far more easily than its arch-competitor, Unilever.

The moves made by the European commission in the late 1980s towards the completion of the Single European Market by 1992 had a strongly stimulating effect on United States investment in Europe. There was an especially large surge in acquisitions of European companies by US firms anxious to gain an initial or a stronger foothold in the single market. According to Hamill (1993: pp. 96–7), 'US companies have been the largest acquirers in Europe in value terms. In 1989, they paid almost 41 billion ECU for European companies – over 40 per cent more than the top European cross-border acquirer.' Paradoxically, although the rationale for completing the Single European Market was the desire to enhance the competitiveness of European firms by creating a larger and more integrated market in goods and factor inputs, it may well be US (and some other non-European) firms which will gain the early benefits. As a UNCTC study pointed out, 'it is possible that the strongest EC-wide firms could be from outside the region, notably from the United States and possibly Japan. Local content levels of United States transnational corporations in the EC are high enough and their operations are rationalized to the extent that their behaviour is, in many cases, indistinguishable from that of European-owned companies' (UNCTC 1991: pp. 34, 35).

Similar predictions are now being made about the beneficiaries of European Monetary Union. 'The imminent arrival of the euro has stirred a sense of expectation in US boardrooms – and a belief that the country's multinationals could be big winners as Europe's economies and product markets converge . . . the single currency is attracting as much interest in corporate America as the build-up to the single market programme in 1992 . . . The single currency has aroused hopes that Europe will become an easier, more "American" market in which to operate' (*Financial Times*, 2 June 1998). Only time will tell if the expectations of US firms (and the fears of European firms and politicians) are realized. What is clear, however, is that the 'American challenge' is far from dead.

Notes

1 The term 'transnational' is preferred to the longer-established term 'multinational' corporation for reasons explained elsewhere (see Dicken 1998). However, in much of the literature on the overseas expansion of US corporations the term 'multinational' is more common, especially in studies published in the 1960s and 1970s.
2 Foreign direct investment is the conventional aggregate statistical measure of TNC activity. Unlike portfolio investment, which is made for purely financial reasons, foreign direct investment is undertaken to establish *control* over production and distribution outside the investing firm's country of origin.

3 At the time of writing there was a furore in France over the threatened 'takeover' of the Eiffel Tower by American interests.

References

Barnet, R. J. and Muller, R. E. (1974) *Global Reach: The Power of the Multinational Corporations*. New York: Simon & Schuster.

Bartlett, C. A. and Ghoshal, S. (1989) *Managing Across Borders: The Transnational Solution*, Boston: Harvard Business School Press.

Britton, J. N. H. and Gilmour, J. M. (1978) *The Weakest Link*. Ottawa: Science Council of Canada.

Chandler, A. D., Jr. (1962) *Strategy and Structure: The History of American Industrial Enterprise*. Cambridge, MA: MIT Press.

Dicken, P. (1998) *Global Shift: The Transformation of the World Economy*. 3rd edn. London: Paul Chapman; New York: Guilford.

Donaghu, M. T. and Barff, R. (1990) 'Nike Just Did It: International Subcontracting and Flexibility in Athletic Footwear Production', *Regional Studies*, 24, pp. 537–52.

Dunning, J. H. (1993) *Multinational Enterprises and the Global Economy*. Reading, MA: Addison-Wesley.

'Big Boys in US Lick their Lips', *Financial Times*, 2 June 1998.

Gilpin, R. (1975) *US Power and the Multinational Corporation: The Political Economy of Foreign Direct Investment*. New York: Basic Books.

Grunwald, J. and Flamm, K. (1985) *The Global Factory: Foreign Assembly in International Trade*. Washington, DC: The Brookings Institution.

Hamill, J. (1993) 'Cross-border Mergers, Acquisitions and Strategic Alliances', in P. Bailey, A. Parisotto and G. Renshaw (eds), *Multinationals and Employment: The Global Economy of the 1990s*. Geneva: International Labour Office, ch. 4.

Hilgerdt, F. (1945) *Industrialization and Foreign Trade*. Geneva: League of Nations.

Kennedy, P. (1987) *The Rise and Fall of the Great Powers*. New York: Random House.

Korzeniewicz, M. (1994) 'Commodity Chains and Marketing Strategies: Nike and the Global Athletic Footwear Industry', in G. Gereffi and M. Korzeniewicz (eds), *Commodity Chains and Global Capitalism*. Westport, CT: Praeger, ch. 12.

Landefeld, J. S., Whichard, O. G., and Lowe, J. H. (1993) 'Alternative Frameworks for US International Transactions', *Survey of Current Business*, December, pp. 50–61.

Lewis, M. K. and Davis, J. T. (1987) *Domestic and International Banking*. Oxford: Philip Allan.

McKenzie, F. A. (1902) *The American Invaders*. London: G. Richard.

Pauly, L. W. and Reich, S. (1997) 'National Structures and Multinational Corporate Behavior: Enduring Differences in the Age of Globalization', *International Organization*, 51, pp. 1–30.

Reich, R. B. (1991) *The Work of Nations*. New York: Alfred Knopf.

Rose, S. (1977) 'Why the Multinational Tide is Ebbing', *Fortune*, Aug., pp. 111–20.

Sampson, A. (1973) *The Sovereign State: The Secret History of ITT*. London: Hodder and Stoughton.

Servan-Schreiber, J.-J. (1968) *The American Challenge*. London: Hamish Hamilton.

UNCTC (1991) *World Investment Report, 1991*. New York: United Nations.

Vernon, R. (1966) 'International Investment and International Trade in the Product Cycle', *Quarterly Journal of Economics*, 80, pp. 190–207.

Vernon, R. (1971) *Sovereignty at Bay: The Multinational Spread of US Enterprises*. New York: Basic Books.

Wilkins, M. (1970) *The Emergence of Multinational Enterprise: American Business Abroad from the Colonial Era to 1914*. Cambridge, MA: Harvard University Press.

Chapter 4

The Dynamics of US Managerialism and American Corporations

Michael Taylor

The US corporation has been pivotal in shaping the nature and functioning of twentieth-century capitalism. The pincipal contribution it has made to the social and economic circumstances of the twentieth century has been through its creation and refinement of dynamic, adaptive managerialism – the divorce of ownership from control in corporate business enterprises as a mechanism to enhance wealth creation and the accumulation of capital. It is the evolution, adaptation and international impact of US managerialism that is the focus of this chapter.

Through the processes of managerialism, power and control over corporations as wealth creators is centralized and vested in salaried managers. Ownership is dispersed and potentially disempowered through fragmentation, and responsibility for the success or failure of managerially directed accumulation strategies is devolved to the lower echelons and sub-units of any particular enterprise. These mechanisms of wealth creation through strategic management were forged and refined in the particular circumstances of industrial growth in nineteenth-century USA, and especially in the white heat of the Second Industrial Revolution of the 1890s. What was created was not a simple linear process of corporate development, in which environmental circumstances stimulate strategic reactions which, in turn, require the creation of appropriate organizational structures to maintain profitability, that can then be replicated. mimicked and diffused from place to place.[1] Instead, a recursive sequence of interactions was initiated in the USA, dynamically linking business enterprises, institutions, labour and regulation in ways which were subsequently extended, transformed and mutated, eventually to embrace capital accumulation on a global scale.[2]

The purpose of this chapter is to chart this sequence of recursive change. The central argument is that it was the needs of geography that initiated US managerialism and the excesses of collusion that brought the imposition of 'system forming' regulation[3] in the shape of anti-trust (competition) law. Innovation, not just in production but especially in management and marketing, gave first-mover corporations a competitive edge and an interest in shifting overseas.

Enhanced organizational capacities created by the adoption of multi-divisional structures allowed further concentration and the geographical expansion of those large US corporations to continue during the economically difficult inter-war years. Subsequently, it was R&D (research and development), organizational capabilities and the initiation of global 'system forming' regulation, through the World Bank, the IMF and GATT for example, that further reinforced the hegemony of those same US corporations between 1945 and the 1970s. What has happened since the 1970s has been more complex as the processes of globalization have embroiled US corporations in intense competition. They have turned multinationals into mutual invaders and have begun to make corporations stateless (Dicken, 1998; Erdilek, 1985). Now, organization, innovation and regulation appear to be interacting differently and less predictably. Within the corporation, R&D and marketing are now being separated from their traditional links with production, and a knowledge-based soft capitalism appears to be emerging. How much this complexity is real and how much it is simply the apparent confusion of the present thrown into relief by the seemingly stark clarity of the past is, however, difficult to know.

What is presented here is a broad-brush account of the changing nature of US corporations. The details of that change have been provided by others.[4] The narrative of the chapter is developed in four parts. First, the nineteenth-century origins of the US corporation are examined and their first moves overseas. Second, the impact of regulation and recession in the inter-war years is reviewed, with expansion mixed with disenchantment. Third, the decades of massive growth, US hegemony and the full flowering of managerialism between 1945 and 1970 are explored, together with the intense debate on US foreign investment that this sparked off. Finally, the manoeuvrings of US corporations as just one set of players in a competitive, borderless, globalized world are discussed as they attempt to become nimble, flexible knowledge brokers in an era of complexity.

The Creation of the Modern US Industrial Corporation

The modern US corporation and managerial capitalism were created by the conditions of the Second Industrial Revolution in the decades prior to the First World War. Chandler (1990) has attributed the emergence of this type of enterprise in the US to four principal causes. The first was geography; the great physical space of the USA with a scattered population. The second was fast domestic market growth, while the third was the opportunity to exploit economies of scale and scope in more industries than anywhere else in the world. Fourth was the relative self-containment of the US market, relieving US corporations of the need to cope with different national economic and social arrangements and the associated costs.

The new transport and communications technologies, the railroad and the telegraph, were vital to the emergence of the modern US corporation. However, it was not so much the space-shrinking technologies themselves that spawned these corporations as the managerial hierarchies and administrative structures that facilitated their efficient use. Here modern management was pioneered in line-of-staff task hierarchies so that traffic and trains could be controlled through newly devised accounting and information systems. The railroads were high fixed-cost businesses demanding continuous capital utilization. To counter competition an oligopoly emerged, as systems first became linked, through standardized equipment, organizational procedures and cartels to maintain prices, and then built parallel track systems to control traffic.

Commodity dealers, wholesalers and new mass retailers (department stores, mail order houses and chain stores) were amongst the first to capitalize on the transportation innovations at the end of the nineteenth century to gain economies of scale and scope through mass buying. Here too it was organizational innovation that brought success. Hierarchical buying systems were set up with a centralized buying office at the company headquarters, buyers coordinating flows of merchandise from the purchasing office to the sales operations, and purchasing offices and depots being set up in commercial and manufacturing centres across the USA and even in Europe.

The Second Industrial Revolution brought about a swift transformation of production in the 1880s and 1890s in the US. Though the new transport and communication technologies were essential to this transformation, it was investment rather than just innovation that was at the core of commercial success in production: investment in production to realize economies of scale and scope; investment in production-specific marketing networks; and investment in the recruitment and training of salaried managers to act as corporate strategists, monitors and coordinators (Chandler, 1990). In its turn, this technical and organizational transformation brought yet more technical and organizational change. The three forms of investment saw branded and packaged foodstuffs (beer and canned goods) and consumer chemicals (soaps, paints and drugs) sold to retailers by teams of travellers. That same pattern of investment underpinned commercial success in the mass produced light machinery production of sewing machines, office machinery and cameras, and in the production of standardized machinery such as boilers, pumps and printing presses. The same first-mover investments brought success to Edison, Thomson and Westinghouse in electrical equipment, and the same story has been charted for US corporations producing metals and industrial chemicals.

The form of the US corporation that was to persist into the 1940s was finally forged in the pre-1914 era through the intense interplay of competition, organizational innovation, and regulation, involving merger and rationalization. In very generalized terms, intense competition led to the formation of cartels by corporations that needed to protect their massive capital investments in the new technologies of the newly created industries. These horizontal combinations brought a storm of political protest and anti-monopoly feeling, culminating first in the Sherman Antitrust Act of 1890, which was later to be reinforced in the Federal Trade Commission and the Clayton Acts of 1914. However, organizational innovation continued, and the provisions of a New Jersey general incorporation law which allowed the formation of holding companies to hold the stock of existing corporations chartered in other states, opened the doors to further administrative centralization and the rationalization of production, purchasing, marketing and distribution.

What followed was a merger movement that reached its greatest intensity in the late 1890s, motivated variously by the desire for, 'control through legally enforceable contractual arrangements, gains from promotion and financing, personal aggrandizement, . . . [and] . . . market power through technological and administrative efficiencies' (Chandler, 1990, p. 78). The successful mergers created centralized operating enterprises which integrated volume production and distribution. What emerged were the large, capital-intensive, integrated firms of modern managerial capitalism funded by investment banks and managed by graduates of the newly established management schools.[5] Now 'inside' directors, principally the full-time senior managers, controlled the instruments of power, especially technology and knowledge, while the part-time 'outside' directors, representing families, banks and large investors, had a weaker grasp of issues, less information and less experience. The modern managerial corporation had arrived.

US corporations had also started to make an impact abroad in the decades before the First World War. They did so first by joining foreign cartels or by using the methods and procedures they had honed at home. US corporations began to invest abroad in response to a world-wide rise in protectionism (Jones, 1996). That protectionism began in the US in the Civil War, rose to an average of 57 per cent on protected commodities in 1897, and to even higher levels by 1920. Throughout Europe (except in the UK, the Netherlands and Denmark) protectionism grew after the severe recession of the 1870s, and it was used in Australia, India and Latin America to foster infant industries and manufacturing. US corporations invested primarily in Canada, Mexico, Central and South America and Europe, with their activities concentrated in order of importance in mining, manufacturing, agriculture and petroleum. The amount of investment was relatively small, but what is important is the fact that most of the 'leading corporations in key industries . . . were by 1914 involved in some kind of foreign business' (Wilkins, 1974, p. 214). First-mover advantages at home were translated into first-mover advantages abroad, and it was the new industries of the Second Industrial Revolution that the new US managerial corporations took abroad. Their impact was also disproportionate to their size, and this was nowhere better expressed than in the UK in 1901 by Frederick McKenzie:

> The most serious aspect of the American industrial invasion lies in the fact that these newcomers have acquired control of almost every new industry created in the past fifteen years . . . What are the chief features of London life? They are, I take it, the telephone, the portable camera, the phonograph, the electric street car, the automobile, the typewriter, passenger lifts in houses and the multiplication of machine tools. In every one of these, save the petroleum automobile, the American maker is supreme; in several he is the monopolist.
>
> (cited in Wilkins, 1974, p. 217)

The US car makers were yet to come to Britain.

The Interwar Years

Following the turmoil and economic disruption of the First World War, the inter-war years saw US corporations working in a radically changed and rapidly changing environment. In the decades before 1914, a borderless world had begun to emerge in which US managerialist corporations had been key players. This brave new world was to suffer major setbacks in the 1920s and 1930s. At home, increasing protectionism encouraged competition from foreign firms, though recession was to take its toll. Abroad, regulation and recession coupled with militarism and revolution heralded a new wave of resentment and resistance to US investment. The managerial corporation continued to emerge and to adjust, but its competitive edge was increasingly under threat.

Through the 1920s, the large US corporations that had developed to mass produce and mass market standardized goods continued their pre-war expansion, sharpening their managerial skills and building on their first-mover advantages (Chandler, 1990). Their investments in production, distribution and management, bolstered by patents, acted as formidable barriers to entry. New, small competitors appeared only in niche markets. The main challenges came from two sources; from large foreign corporations that had equally substantial resources and skills, and from domestic competitors in related industries. These domestic competitors developed their portfolios of products through cumulative innovation focused on product development and commercialization rather than fundamental research. The research and development strengths of US corporations was yet to come. From the 1920s into the 1930s, therefore, US corporations came to compete with ranges of firms in ranges of oligopolistic markets. Corporations began to blur their boundaries and become diversified in a way that had not existed before. The multi-divisional firm had begun to emerge.

The role of the salaried manager continued to expand, although in some notable cases, where owners retained managerial control, the advantages of previous decades of success were squandered. The most prominent examples were of Henry Ford with his automobile company and Judge Elbert Gary with United States Steel. Both men overruled, alienated and drove out their senior managers who, in turn, established competing businesses (Chandler, 1990).

In the 1920s, US corporations continued to move aggressively overseas, using both market orientated and supply orientated strategies and concepts of mass production, standardization and scientific management (Wilkins, 1974). The producers of small, standardized machinery continued their pre-war overseas expansion. Food and beverage manufacturers shifted overseas through joint ventures, acquisitions and by building branch plants. Vehicle producers were followed overseas by their suppliers making bodies, bearings, tyres, wheels, batteries and so on. US corporations in the electrical industry invested outside the USA to stabilize markets, to avoid competition, enforce patents, diversify, increase exports and appease host country sensitivities. Particularly large market-related investments were made overseas at this time by US oil companies.

Some US corporations invested overseas in the 1920s for reasons of supply: for cheap power, for oil (fearing future exhaustion in the US), for minerals, and for bulk agricultural products. Only occasionally did they invest to secure cheap labour. Agriculture, mining and oil ventures were invariably in developing countries, and their company towns and workers' quarters stood out as alien intrusions in those places. What is more, dealing with developing country hosts and colonial administrations was complex and so, for the first time, these US corporations also began to enlist State Department help to successfully negotiate these foreign regulatory mazes.

The late 1920s again saw the emergence of public anxieties in many countries about the impact of US corporate investment in their economies. They feared US monopolies and perceived this to be the goal of US mining corporations. Some 1,300 US companies offered strong competition in Europe in 1929 in a wide range of sectors including the vehicle, electrical and tyre industries, office machinery and oil. However, legal controls in the UK, Germany, France, Switzerland, Italy and the Netherlands tried to prevent further penetration of their economies by controlling 'foreign' (principally US) purchases of voting stock. In Spain, hostility to US oil companies led to the nationalization of that country's oil industry. Further resentment came from US corporations pushing into European countries' colonial spheres; into Latin America, the Middle East and the Far East. However, although there was rivalry in these places, there was also collaboration through associations and joint ventures in the UK, Canada, Chile, Australia, China and parts of East Africa, for example.

The 1930s saw US corporations operating in a new negotiated environment created by recession, nationalism, militarism, fascism and political and economic unrest. It was an era of corporate disenchantment. At home, the Fordney–McCumber and Smoot–Hawley Acts raised tariffs and sparked off retaliatory protection in Europe in the form of empire preferences and the creation of trading blocs. It also brought heightened foreign competition within the US domestic economy. Heightening their disenchantment, Roosevelt's New Deal was seen as anti-monopoly and anti-big business but pro-labour and pro-public ownership. Abroad, there was the spectacular spread of international collusion through cartels in manufacturing and primary commodities (Jones, 1996).[6] US corporations were in the thick of these changes. 'Not only did [US] corporations deal with the governments in each particular country, but many entered into international agreements to stabilise and to raise prices' (Wilkins, 1974, p. 205).

The war years intensified US corporate disenchantment. Their overseas investment activity was attacked by the US government as compromising US technology and capacities and for putting American interests into foreign hands. They also saw their overseas investments destroyed, confiscated, nationalized and expropriated, creating the view that overseas investments should only be for the acquisition of resources to supply the domestic US economy. When hostilities ceased, US mining companies still faced 'creeping expropriation' (Wilkins, 1974), especially in developing countries, through host country fiscal policies, social legislation, unionization, local purchasing requirements, trade restraints and adverse dollar conversion rates, all of which ate into their profits. Nevertheless, US overseas investment grew through the 1940s in an environment where collusion and cartels were the main vehicle of international business.

The Hegemony of US Corporations

After the Second World War, US corporations were in a uniquely powerful position. In this period, the managerial corporation reached its fullest development and came to dominate old and new industries alike. US corporations grew through their control of technology, their management and marketing skills, and their control of capital. What emerged were sharply defined corporations with command and control structures run by tiny managerial elites, separated from the structure of ownership, supported by multiple layers of middle management, and employing mass production techniques to produce standardized products for mass consumption. Three sets of factors created giant US corporations. First, there was little foreign competition. Europe and Asia had experienced extensive destruction during the Second World War, while US output had boomed. Second, the pace of technological change had accelerated during the war. The new electronics and associated technologies that emerged from the war (including telecommunications, computers, satellites and jet aviation) created new industries in which US corporations held paramount positions. As this impact on US corporations was described by Chandler (1990), 'The globalization of communication encouraged the internationalization of markets, while . . . domestic markets boomed' (p. 608). Third, in this period of *Pax Americana*, a new international regulatory regime began to emerge in the shape of the United Nations and its agencies, the World Bank, the IMF and the GATT, all cast very much in the US image and dominated by their financial input (Roberts, 1998). The cosy cartels and unwritten agreements between competing corporations began to melt away and tariffs and trade barriers began to fall, initially opening the world to US investment, but inviting international competition.

The managerial corporation grew through two mechanisms: diversification and overseas expansion. Diversification into related industries became the preferred route for US corporate expansion by the late 1960s, with firms building much more on R&D than they had in the inter-war years. However, although US corporations undertook R&D themselves, they also built on the innovative capacities of universities and research institutes and the 'externalized R&D' available to them in small firms. Their expertise was in bringing innovations profitably to market.

US corporations grew abroad in the post-war decades mainly by acquisition. Indeed, takeover and merger had turned effectively into a feeding frenzy by the late 1960s. The oil companies led the overseas charge with new tanker technology and exploration technology, spurred on by the closure of the Suez canal. It was corporations in manufacturing, oil and mining that dominated US foreign investment. However firms in the wholesale and retail sectors began to move overseas, as too did business service firms following US personnel and corporate clients. The growth was spectacular. In the 20 years following the Second World War 85 per cent of all new foreign direct investment was from the USA. The USA became the largest global investor by 1980, relegating Europe to second place (Jones, 1996, pp. 46–7).

As the size and global reach of US corporations reached unprecedented dimensions, major problems of internal coordination and control began to emerge. The international division now proved ineffective and multi-divisional structures began to emerge in the shape of world-wide product divisions and geographical divisions. Some corporations combined both structures in embryonic multiple-reporting matrix systems of management.[7]

The sheer size of US foreign direct investment created intense fears both within the USA and abroad. In the USA, it was argued that the overseas investments of US corporations created a raft of problems. It was argued vehemently that they destroyed American jobs, reduced tax revenues, affected the balance of payment, impaired US capital resources, gave away US technology, gave corporations foreign obligations that could conflict with US domestic loyalties, and could involve the US government in unwanted foreign situations. Arrayed against these arguments was an equally lengthy list of counter-views suggesting that the foreign investments of US corporations enlarged American markets and US capital resources, gave the US control over strategic raw materials, was a source of new technology for the US itself, created competition, and provided the US government with an instrument of diplomacy and foreign policy.[8]

Other countries, especially those of Europe, viewed US overseas investment with deep suspicion, and many of the arguments made in the US were rehearsed with the added colour of particular local circumstances. Countries' broad concerns were that US corporate investment would retard their own economic growth and destroy national sovereignty through:

- the repatriation of profits and earnings, through transfer pricing mechanisms in general (Plasschaert, 1979) and through the channelling of corporate finance flows through tax havens;
- the hollowing out of their economic structures, the creation of branches to serve host markets rather than export platforms to create export-led growth, and the intensification of dependency relationships;
- the distortion of their labour markets, by corporations creating generally unskilled assembly-line jobs (as part of the new international division of labour), putting only expatriates into key technical and managerial posts and so minimizing mechanisms of technology transfer;
- the transfer to them of essentially obsolete technology while at the same time using their corporate networks to acquire and repatriate host country technology.

There was also concern that the investment would undermine and destroy national cultures – a process of McDonaldization. Equally there was concern that these mechanisms would lead to some host countries being used as pollution havens where, because of weak environmental regulation, obsolete and polluting technologies would be sent to maintain output and extract residual profits. US corporations were, in this sense, the harbingers of the impacts of globalization that are so loudly criticized today (Taylor, 1996; Leonard, 1988). However, the relationships between multinationals and governments are not ones of narrow dependency but of complex interdependency. Governments have many sanctions they can impose on corporations, and they can promote their own multinationals. Multinationals also have good reason to avoid the attention of governments (Thrift and Taylor, 1989).

US Corporations and Global Uncertainties

But, the hegemony of US corporations was not to last. Globally, the growth of foreign direct investment was spectacular, rising from US$500 billion in 1980 to over $2,000 billion in 1993 (Jones, 1996). After 1983, it grew five times faster than world trade and ten times faster than world output. US corporations were now one set of players among many. European economic reconstruction was rapid in the post-war years and brought with it strong competition. European corporations also adopted US managerial techniques and hired US managers – in effect fighting fire with fire – which gave them the capacity to begin to compete in US domestic markets. Post-war transfers of technology from the USA to Japan helped to move Japanese industry into competitive mass production, but with distinctive national characteristics, and the tiger economies and newly industrializing countries of East and Southeast Asia began to emerge as a potent source of competition. Third World multinationals now appeared on the scene as a significant competitive force. Economic deregulation through international agreement also destroyed the cosy collusive environment of the 1950s and 1960s when profits had been easy to make.

Add to these gradual economic changes the intense shorter-term pressures of recession in the mid-1970s and again in the early 1980s, the demise of the Bretton Woods currency controls, the OPEC oil price hikes, and balance of payments problems in the USA, and there is a recipe for corporate crisis on a wide scale (Thrift, 1985). In the US context short-termism was further exacerbated by a shift in ownership patterns. Long-term institutional investors – the banks and insurance companies – were replaced by the portfolio managers of pension and mutual funds with much shorter time horizons. '[P]ortfolio managers became the new owners of American business' (Chandler, 1990, p. 625). The corporate investment certainties of the post-war decades were swept away and economic warfare[9] developed. That warfare and the involvement of US corporations in it is the story of the 1980s and 1990s.

With the end of economic certainty came the reemergence of the borderless world (Jones, 1996) and the growth of globally integrated corporations. US corporations now work in a complex, turbulent globalized environment within which rigid managerialist structures created to exploit economies of scale are insufficiently nimble and flexible. These structures were creatures of their time. They left shopfloor workers alienated, ill-used and undervalued, and customers unsatisfied. All corporations, and US corporations in particular, were left with bureaucratic bloat and inappropriate, even self-destructive, managerial cultures that made slow decisions in fast markets.[10] Head offices became separated from divisional managers, and monitoring and control depended on impersonal, poor-quality statistics devoid of personal contact. In short, internal communication began to break down. In the past 20 years, US corporations, together with US business schools and consultants, have been at the forefront in devising strategies to cope with this complexity and intensified competition.

Some corporations tried to build on their managerial capacities seeing this as a prime resource. Integral to this strategy was the use of conglomerate merger to move into new markets, seeking growth potential rather than synergies with core activities. Buying and selling corporations became a business in its own right with business service firms having developed to ease the process. Indeed, many corporations have now become increasingly fiscal, appraising their investments like banks and using financial derivatives to hedge their investment exposure.

A key corporate strategy to cope with complexity has been to reduce their size. They have slashed their workforces and rationalized their plants, sub-contracting out on an international basis increasingly large sections of their production and administration (Dicken, 1998). US corporations laid off some 450,000 employees in 1992 and a further 600,000 in 1993. They are now cutting the bureaucracies that collect and collate information, with GM for example having cut its head office from 13,000 to 2,000 employees. Handy (1991) maintains that slimmed down head offices now retain just four reserve powers: the right to control finance, the right to control information, the right to appoint key personnel, and the right to invade.

Associated with this strategy of downsizing and subcontracting, US corporations have also become increasingly involved in strategic alliances and joint ventures in order to move into new areas of production and new markets with minimal risk until a venture has proved commercially viable (Alvstram, 1995; Gaebe, 1997). In this way the boundaries between corporations have now become blurred. This blurring is increasing with close relationships being forged between corporations so that they can run more efficiently and share sensitive information. For example, airlines are linking with hotel chains and computer firms with counterparts in telecommunications.

Coupled with reduced size and an increasingly fiscal orientation, a key strategy adopted by many US corporations has been to integrate production, management and marketing on a global scale. Multi-domestic strategies ran into trouble in the 1980s and have been replaced by globalization strategies, creating global corporations. Global integration allows firms to increase their oligopolistic power by exploiting their scale and experience. It allows them to exploit the discrepancy between relatively efficient markets for goods and relatively inefficient markets for factors of production. It also allows them to exploit national differences in tax rates and tax structures, and to take advantage of transfer pricing opportunities. They can also use the strategy to avoid hostile government action (Dicken, 1998; Doz, 1986). There is, however, a down side to global integration that arises externally from exchange rate issues and government and labour interventions, and internally from the organizational complexity and demands of integration. It is with these issues of integration that corporations continue to grapple.

Major geographical consequences have arisen from these strategies of globalization (Thrift and Taylor, 1989). First, corporations have a more international workforce, with more workers and production overseas, prompting charges that corporations are becoming stateless – a hotly debated topic. Second, there has been a sifting of management and marketing jobs which are now increasingly concentrated in 'world cities', a process that has been greatly deepened and elaborated with the creation of regional head offices and the decentralization of corporate R&D activities. Third, there has been the locational shifting of production and clerical jobs, creating a new international division of labour.

As economic complexity has increased, so capital and managerial capacities are no longer seen as corporations' prime resources. Now, knowledge is the prime resource. Led by US management thinking, corporations are developing flexible structures to capitalize on knowledge and information. Corporations like GM, Ford, IBM, AT&T and Hewlett Packard have split themselves into profit centres and shifted power to the regional level, though some like GM seem to be unable to be unable to free themselves from rigid, hierarchic and intensely bureaucratic structures (*The Economist*, 1998). Others, like Xerox and Nike, have shifted to management systems based on complex integration. The aim is to win the hearts and minds of their employees. What they are attempting is simultaneously to become local insiders in many places so that they can become, 'dealers in locally-rooted insider knowledge and arbitrageurs between centres of excellence' (*The Economist*, 1995, p. 12). Corporations need to be locally sensitive with an internal network culture, a far cry from the command and control structures of the 1950s and 1960s.

Now, because image is all-important, the criticisms of corporations that have long been made are striking home. With de-industrialization at home in the US and production having been shifted to low labour cost countries, it is no surprise that Levi Strauss and Nike are now publicly sensitive to labour exploitation by their overseas subcontractors, with Nike having issued a public apology in the US. Neither is it a surprise, in the wake of environmental disasters like the lethal chemical releases at Bhopal and the role of CFCs in global environmental degradation, that Digital, Compaq and IBM are pushing for higher environmental standards, all in response to US domestic concerns. Most major corporations now have environmental policies, though to what effect is questionable. The corporations are equally sensitive to charges of cultural hegemony. Consequently, outside the US, Coca-Cola and McDonalds use local distribution systems, and Coca-Cola customizes its local products in Japan, for example, in order to be locally sensitive. The global media corporations, however, still seem impervious to these pressures.

The knowledge-based corporations are putting great effort into building internal network cultures to keep their images vibrant, marketable and locally sensitive. To build these network cultures, corporations are spending huge amounts of time on communicating, moving managers around the world so they can know and understand their colleagues, running training programmes, conferences, get-togethers and other bonding mechanisms to create face-to-face relationships, build trust and keep staff focused on a company vision. This is the management talk of soft capitalism (Thrift, 1998).

However, while the new knowledge-based corporation might be emerging, concentrating their efforts principally on R&D, product development and marketing – becoming designers and exploiters of brands – manufacturing itself is being anonymized. Production is being shifted to contract manufacturers – not small subcontractors, but multinational companies larger than the corporations themselves, with global production networks, who make to order under strictly controlled and imposed contracts. Contracts, not trust, govern these transactions, and the contract manufacturers engage in no R&D or product development themselves,[11] but make goods for image makers to brand, label and badge. This form of organization is now appearing in electronics manufacturing, and in sectors such as apparel and footwear, toys, data processing, home furnishings and lighting, semiconductor fabrication, food processing, vehicle parts, brewing and pharmaceuticals, for example. How far it will spread it is too early to say.

But there remains a very hard edge to soft capitalism. Harsh employment realities remain the same and jobs are still vulnerable – now as much in the developed as in the developing world. These realities are only too apparent in the late 1990s with the 'down-sizing' of head offices, the 'de-layering' of middle management and the shake-out in the semiconductor and automobile industries, for example. The emerging corporate duality of high-profile global branders and image makers and anonymous contract manufacturers might be interpreted as the latest corporate attempts to externalize and thus avoid public criticism over job losses, labour exploitation and environmental exploitation, assigning them instead to the anonymous realm of 'global forces'.

Conclusion: The Universal Accumulator

The US has given the world corporate managerialism in the shape of the large corporation that has dominated the world economic scene for the past century. From the very beginning it has excited managers and wealth creators and has struck fear into the hearts of the people and governments that have had to work with it. It has evolved and transformed through the decades, building on advantages of scale and scope, technological change and R&D, managerial capacities, and the control of knowledge. It has been copied and imitated around the world, and the US itself has been re-invaded by its foreign emulators. The large corporation, in a new global mantle, remains the principal engine of capitalist accumulation.

The conditions for the emergence of the US corporation were set in the geography, the size and the rate of growth of the US economy in the nineteenth century. These conditions provided the impetus for business owners to develop enterprises capable of realizing economies of scale and scope through investment in new production techniques, marketing and managerial personnel. The railroad companies were in the vanguard of these developments. In the borderless world of the late nineteenth century, US corporations used their first-mover advantages in the US as a springboard to move overseas, with Singer, the sewing machine manufacturer, becoming arguably more global than any present-day corporation. But, the growth of these managerial corporations brought protest at home from the small business lobby and protest abroad as their commercial hegemony grew. The reaction was increased regulation in the USA in the shape of anti-trust laws and increased protectionism elsewhere, especially in the wake of the 1870s recession.

After the First World War, the US managerial corporation continued to grow and evolve, finally decoupling ownership from control and concentrating power in the hands of professional managers. More corporations than ever moved overseas: seeking bigger markets; following other US corporations to supply components or services; and seeking primary inputs from agriculture, mining and oil. Cheap labour was not a particular draw card but, increasingly, US corporations sought US government help in their overseas dealings. However, anti-trust laws, protectionism, recession, militarism and widespread foreign hostility dampened the corporations' expansionary zeal and increasingly saw them seek sanctuary in cartels.

In the wake of the Second World War, the US managerial corporation reached its heyday, left without competition because of the massive destruction suffered in Europe and the East to capitalize on the massive technological advances stimulated by war, the *pax Americana*, and the demand created by the Cold War. As these giant bureaucracies fought to organize and cope with their own success, they created a storm of protest at home and abroad, as communities became concerned for their jobs, incomes, tax revenues, investment levels and capital formation, cultural values, and their physical environments. Home and host countries alike felt they were the losers and the corporations the winners. Certainly labour exploitation in the developing world expanded at the hands of the corporations as part of the new international division of labour, and the environment took a great toll, globally through massively increased emissions of carbon and other pollutants, and locally through disasters like Bhopal or the destruction of forests in the Philippines for pineapple plantations.

Since the 1980s, global economic circumstances have changed. The hegemony of the US corporation has been challenged. Now, there are more TNCs, from more countries, in more sectors and with a greater fiscal orientation than ever before. But, again, US managerialism is leading the way. One set are becoming increasingly knowledge-based and dependent on branding and marketing. These are the corporations that are increasingly sensitive to criticism over labour exploitation, environmental damage and cultural colonization; sensitive to the extent of apologizing to the American public but not necessarily sensitive

to the point of changing their commercial practices. Another group of corporations are becoming anonymous contract manufacturers – the makers, the
fabricators, the users of labour and the generators of pollutants and emissions.
The question is whether this is the next stage in the evolution of managerial
capitalism?

Notes

1   These views were developed as the structural contingency model of organization
    elaborated by Thompson (1967), and Lawrence and Lorsch (1967). For a summary
    of the model in a geographical context see McDermott and Taylor (1982). The
    model has been strongly criticized, for example by Hedlund and Rolander (1990).
2   This is an institutionalist, evolutionary view of organization, elaborated in various
    contexts by Nelson and Winter (1982) and Hodgson (1988).
3   Christopherson (1993) distinguished 'system forming' regulation from 'system
    guiding' regulation. The first form comprises codes and rules set up by government
    to establish the base conditions for economic activity in a place. The second form
    involves continuous negotiation and interaction between governments and eco-
    nomic actors, and continuous adjustment of behaviour.
4   The extensive writings of Alfred Chandler and Myra Wilkins have informed sig-
    nificant sections of this chapter. Of particular significance are: Chandler's *Structure
    and Strategy* (1962), *The Visible Hand* (1977), *Scale and Scope* (1990), and 'The
    Evolution of Modern, Global Competition' (1986); Wilkin's *The Emergence of
    Multinational Enterprise* (1970) and *The Maturing of Multinational Enterprise*
    (1974). An extensive summary of the evolution of international business is pre-
    sented in Jones (1996).
5   The interrelationship between the activities of US business schools and the devel-
    opment of corporate managerialism has been made strongly by Chandler for the
    borderless era before 1914. The argument has recently been remade in the new bor-
    derless era by Thrift (1998) as a fundamental of soft capitalism.
6   See, for example, the extensive historical study of the international electrical indus-
    try in the inter-war years and its impact in Brazil in Newfarmer and Topik (1982).
7   The evolution of organizational structures is reviewed in Bartlett and Ghoshal
    (1989), Dicken, Forsgren and Malmberg (1994) and Dicken (1998).
8   The arguments are elaborated more fully in Wilkins (1974), Barnett and Müller
    (1978) and Bergsten, Horst and Moran (1978).
9   Economic 'warfare' was described by Thrift and Taylor (1989) as arising from
    reduced demand on world markets, increased competition, and greater numbers of
    multinationals from more countries operating in a greater number of economic
    sectors.
10 Schoenberger (1997) has assessed the impact of top management culture (includ-
    ing 'trained incapacity') on the successes and failures of corporations to cope with
    and adapt to the new competition of the 1980s and 1990s that has challenged the
    standardized mass production systems that have been the very foundation of US
    corporate managerialism. The case studies of Lockheed and Xerox detail the dilem-
    mas involved.
11 Sturgeon (1998) argues that an organizational split is emerging in the corporate
    world between those involved in innovation and those involved in production.

Brand-name firms are controlling product definition, design and marketing in-house, while manufacturing is being shifted to 'turnkey production networks' creating a 'cadre of specialized merchant suppliers'.

References

Alvstram, C. (1995) 'Spatial Dimensions of Alliances and Other Strategic Manoeuvres', in Conti, S., Malecki, S. and Oinas, P. (eds), *The Industrial Enterprise and Its Environment: Spatial Perspectives*, Avebury, Aldershot, 43–55.

Barnett, R. and Müller, R. (1978) *Global Reach: The Power of the Multinational*, Jonathan Cape, London.

Bartlett, C. and Ghoshal, S. (1989) *Managing Across Borders: The Transnational Solution*, Harvard Business School Press, Boston, MA.

Bergsten, C., Horst, T. and Moran, T. (1978) *American Multinationals and American Interests*, The Brookings Institute, Washington, DC.

Chandler, A. (1962) *Structure and Strategy*, Harvard University Press, Cambridge, MA.

Chandler, A. (1977) *The Visible Hand*, Harvard University Press, Cambridge, MA.

Chandler, A. (1986) 'The Evolution of Modern Global Competition', in Porter, M. (ed.), *Competition in Global Industries*, Harvard Business School Press, Boston, MA.

Chandler, A. (1990) *Scale and Scope*, Harvard University Press, Cambridge, MA.

Christopherson, S. (1993) 'Market Rules and Territorial Outcomes: The Case of the United States', *International Journal of Urban and Regional Research*, 17, 272–84.

Dicken, P. (1998) *Global Shift: Transforming the World Economy*, 3rd edn, Paul Chapman Publishing, London.

Dicken, P., Forsgren, M. and Malmberg, A. (1994) 'The Local Embeddedness of Transnational Corporations', in Amin, A. and Thrift, N. (eds), *Globalization, Institutions and Regional Development in Europe*, Oxford University Press, Oxford.

Doz, Y. (1986) *Strategic Management in Multinational Corporations*, Pergamon, Oxford.

The Economist (1995) *A Survey of Multinationals: Big Is Back*, vol. 335, no. 7920, June 24–30.

The Economist (1998) 'The Decline and Fall of General Motors', vol. 349, no. 8089, Oct. 10–16, 92–7.

Erdilek, A. (ed.) (1985) *Multinationals as Mutual Invaders: Intra-Industry Direct Foreign Investment*, Croom Helm, Beckenham.

Gaebe, W. (1997) 'Strategic Alliances in Global Competition: Securing and Gaining the Competitive Edge', in Taylor, M. and Conti, S. (eds), *Interdependent and Uneven Development: Global-Local Perspectives*, Ashgate, Aldershot, 101–16.

Handy, C. (1991) *The Empty Raincoat*, Arrow, London.

Hedlund, G. and Rolander, D. (1990) 'Action in Heterarchies: New Approaches to Managing the NNC', in Bartlett, C., Dox, Y. and Hedlund, G. (eds), *Managing the Global Firm*, Routledge, London and New York.

Hodgson, G. (1988) *Economics and Institutions: A Manifesto for a Modern Institutional Economics*, Polity Press, Cambridge.

Jones, G. (1996) *The Evolution of International Business: An Introduction*, Routledge, London and New York.

Lawrence, P. and Lorsch, J. (1967) *Organization and Environment: Managing Differentiation and Integration*, Harvard Graduate School of Business, Boston.

Leonard, H. (1988) *Pollution and the Struggle for the World Product*, Cambridge University Press, Cambridge.

McDermott, P. and Taylor, M. (1982) *Industrial Organization and Location*, Cambridge University Press, Cambridge.

Nelson and Winter (1982) *The Evolutionary Theory of Economic Change*, Belknap Press, Cambridge, MA and London.

Newfarmer, R. and Topik, S. (1982) 'Testing Dependency Theory: A Case of Brazil's Electrical Industry', in Taylor, M. and Thrift, N. (eds), *The Geography of Multinationals*, Croom Helm, Beckenham, 33–60.

Plasschaert, S. (1979) *Transfer Pricing and Multinational Corporations: An Overview of Concepts, Mechanisms and Regulations*, Saxon House, Farnborough.

Roberts, S. (1998) 'Geo-governance in Trade and Finance and Political Geographies of Dissent', in Herod, A., Roberts, S. and Ó Tuathail, G. (eds), *An Unruly World: Globalization, Governance and Geography*, Routledge, London and New York, 116–34.

Schoenberger, E. (1997) *The Cultural Crisis of the Firm*, Blackwell, Cambridge, MA and Oxford.

Sturgeon, T. J. (1998) *Turnkey Production Networks: A New American Model of Industrial Organization?* Paper presented at the Association of American Geographers Conference, Boston, April.

Taylor, M. (1996) 'Industrialisation, Enterprise Power and Environmental Change: An Exploration of Concepts', *Environment and Planning A*, 28, 1035–51.

Thompson, J. (1967) *Organizations in Action*, McGraw-Hill, New York.

Thrift, N. (1985) 'All Change: The Geography of International Economic Disorder', in Taylor, P. and Johnston, R. (eds), *A World in Crisis? Geographical Perspectives*, Blackwell, Oxford, 12–67.

Thrift, N. (1998) 'The Rise of Soft Capitalism', in Herod, A., Roberts, S. and Ó Tuathail, G. (eds), *An Unruly World: Globalization, Governance and Geography*, Routledge, London and New York, 25–71.

Thrift, N. and Taylor, M. (1989) 'Battleships and Cruisers: The New Geography of Multinational Corporations', in Gregory, S. and Walford, R. (eds), *Horizons in Human Geography*, Macmillan, London, 279–97.

Wilkins, M. (1970) *The Emergence of Multinational Enterprise*, Harvard University Press, Cambridge, MA.

Wilkins, M. (1974) *The Maturing of Multinational Enterprise*, Harvard University Press, Cambridge, MA.

Chapter 5

Overseas Investment by US Service Enterprises

P. W. Daniels

Introduction

While overseas investment is not solely linked with the ambitions of large corporations, they exert an important influence on its volume and geography. Over half (278) of the largest 500 corporations in the world in 1996 were service companies (*Fortune*, 1997) and 32 per cent were owned and controlled from headquarters in the US.[1] With 28 per cent of the global revenues of the largest service corporations, almost 47 per cent of the employees and an average of 98,941 employees each, US service corporations are therefore not easily overlooked.[2] There are eight US service companies in the top 100: Citicorp (banking), Prudential of America, State Farm Insurance and American International Group (insurance), Wal-Mart Stores, Sears Roebuck and Kmart (general merchandisers), US Postal Service (mail services), and Pepsico (food services). Lower down are international household names such as McDonalds (food services), Chase Manhattan Corporation (banking), American Express (diversified financial services), Marriott International (hotels), AT&T (telecommunications) and United Airlines. Not all these US service corporations are engaged in overseas investment; the US domestic market, for example, is still large enough to sustain corporate ambitions (Wal-Mart or Kmart) or the service does not easily or logically require overseas markets (US Postal Service). For others, saturation of the domestic market or the tradability of the service combined with improved technology and the opening up of overseas markets to foreign suppliers has encouraged US companies to look for opportunities well beyond their national boundary. This has been encouraged since the mid-1990s by the General Agreement on Trade in Services (GATS), which is the first ever global agreement that binds countries to explicit rules governing investment and trade in services (American Foreign Services Association, 1995). It is administered by the World Trade Organization (WTO) Council for Trade in Services.

There are two principal ways in which US service companies can serve markets in another country: trade or foreign direct investment (FDI). As a general rule these should not be viewed as substitutes since most service firms do not exclusively pursue one or the other. The balance will depend on the needs of individual companies at particular times in their corporate development, the kinds of service that they provide and the requirements placed on them in different national markets. This chapter will focus on FDI by US service companies, but before looking more closely at some of its characteristics, it is worth digressing very briefly to consider US international trade in services. In 1994 global trade in services was valued in excess of $1 trillion or about 22 per cent of all trade in goods and services. The value of US service exports exceeds the value of imports and this helps to compensate for a trade deficit (excess of imports over exports) in goods (figure 5.1) (see also US Department of Commerce International Trade Administration, 1997). Some categories of services make only a very modest contribution to the overall trade surplus for US services, but of particular interest in the context of this chapter is the striking growth of the balance of trade in private services since the mid 1980s (figure 5.2). These include telecommunications, financial services, information services, commercial, professional and technical services and transportation (including port and freight services). In the seven years between 1987 and 1994 exports of private services doubled by $99 billion. A $60 billion surplus on trade in private services in 1994 helped to offset 36 per cent of the deficit in US goods trade. The important point is that, in contrast to the surplus from royalties and licence fees, the private services surplus reflects the way in which many of the companies involved must be engaged in some form of production overseas. This is the case even if they can also serve those markets at 'arm's length' from their offices and facilities in the US.

This is where FDI enters the picture. It is now an important factor in global economic development. World-wide FDI increased dramatically in the late 1980s (OECD, 1994; UNCTAD, 1995). This coincided with the growing international ambitions of service firms in one nation to serve markets or clients in another nation, integrating their operations across borders by trading intermediate goods and services (Dunning, 1981, 1993b; Cantwell, 1991; McCulloch, 1996). Indeed, the structural and dynamic characteristics of the world economy are increasingly a function of FDI and its close links with financial flows, technology transfer and international trade. This is most clear for activities such as downstream services (where FDI in dealer networks and after-sales services is often necessary to promote sales) and financial services (where the overseas activities of home-country clients often prompt FDI by the service provider).

Between 1982 and 1990 the US was the major destination for global FDI; almost 40 per cent of the inward flows between 1982 and 1987 went to the US (table 5.1). This was almost double its share of world outflows. This balance deteriorated between 1988 and 1990 but was followed by a dramatic reversal during 1991–3 when the US share of world inflows of FDI declined to just over 10 per cent but its share of world outflows reached more than 22 per cent. This is double the equivalent figure (11.2 per cent) for Japan which, apart from 1988–90, has lagged behind the US in outward FDI and is also very much second best in relation to inward FDI.

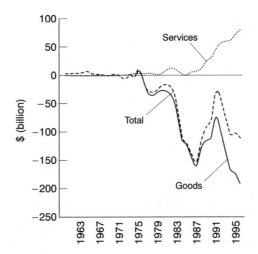

Figure 5.1 US trade balance in goods and services, 1960–96
Source: US Dept of Commerce International Trade Administration, 1997.

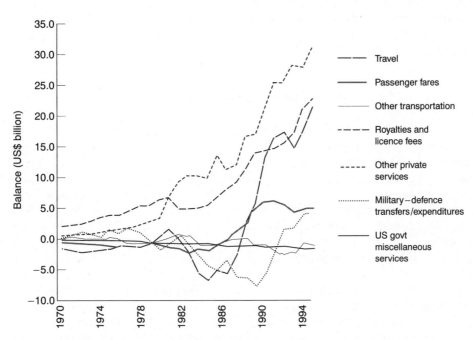

Figure 5.2 US services trade: by major category, 1970–96
Source: US Dept of Commerce International Trade Administration, 1997.

Table 5.1 World-wide inward and outward shares (%) of FDI flows, 1982–93

Region/country	1982–7		1988–90		1991–3	
	Inward	Outward	Inward	Outward	Inward	Outward
Developed countries	78.1	98.0	84.6	95.8	67.0	95.3
Developing countries	21.9	2.0	15.3	4.2	33.0	4.7
European Union	28.2	47.4	42.3	48.2	44.4	50.2
United States	39.9	19.8	31.3	11.6	10.2	22.4
Japan	0.7	13.4	–	20.3	0.9	11.2

Source: Derived from UNCTAD (1994) and IMF, after Economists Advisory Group (1998), extracted from tables 2.1 and 2.2.

All the evidence therefore points to the significant contribution of US service firms to the rapid recent expansion of outward FDI. Some of the consequences for the geography of economic development will now be considered, but before turning to this it is necessary to dwell briefly on some of the principal elements of FDI theory.

The Theory of FDI[3]

FDI arises from the behaviour of (service) firms seeking to create or to exploit opportunities for growth in markets outside their home country. Whether or not individual firms engage in production outside the home country will depend on whether there are firm-specific advantages from some form of representation in the target market. A combination of ownership-specific advantages, internalization opportunities and locational considerations will shape the firm-specific advantages (Dunning, 1981, 1993b; UNCTC, 1992). Firm-specific advantages refer primarily to the production skills and knowledges (tangible and intangible) that are assumed to be unique to firms of a particular nationality, ownership or, possibly, industry sector. If production techniques and knowledges are uniformly distributed across national borders, the location of production will simply depend on differences in factor costs, tariffs, transportation costs, and the size of markets. Such factors can make it advantageous to locate production in the market being served rather than exporting (see for example Buigues and Jacquemin, 1992; UN Transnational Corporations and Management Division, 1993).

The issue of the location of production also encompasses an element of business strategy, which is developed in relation to the firm's competitive environment. Firms typically pass first through an internationalization phase, where investments are located in response to local markets and supplies. This is followed by a regionalization and/or globalization phase, where the emphasis, for investment and divestment, is on rationalizing existing operations to secure efficient networks, often under strong competitive pressures (Dunning, 1993a).

The location of production that accompanies FDI by service multinationals (MNEs)[4] (as well as those in manufacturing) is very uneven. Globalization or the liberalization of world trade has not resulted in an even landscape of production, even if technological advances suggest that this is possible (Petri, 1994). Geography remains important in so far as, for example, national government support for domestic service enterprises, regulation of service markets or adoption of the terms agreed in bilateral or multilateral trade negotiations is variable. At the same time it is important not to overlook the importance of its home base to the success of a service MNE. It may be multinational but it is not stateless in the sense of being indifferent as between countries or not having a home nation (Reich, 1992; Ohmae, 1990). Successful service MNEs depend a great deal on the experience and successes that they have achieved to a high level in their home markets (Hymer, 1976; Porter, 1990). Over time, it is these home-built (and tested) advantages that are combined with those obtained from locating particular operations in other nations that provide the foundation for effective overseas investment. Thus, in spite of the increasing geographical reach of their operations, the majority of service MNEs continue to derive a large proportion of their competitive advantages from the location-bound assets of their home country (Stopford and Strange, 1991).

As to where FDI by US service MNEs is located, it is increasingly influenced by the availability of created assets (human skills and innovatory capacity) in the production process. This reflects the tendency for specialist, knowledge, and human resource intensive services to be at the forefront of overseas investment. Financial services, consulting, business and professional services are tradable and are largely embodied in the people who provide them. They can be provided by transferring staff from home to overseas locations or by a combination of transfers and local recruitment in the overseas location (Beaverstock, 1990). The availability and cost of complementary support assets, many of which are supplied either directly or indirectly by governments and located in close proximity to each other (city centres, science parks, industrial districts) may also determine where the investments are ultimately made (Eaton, Lipsey and Safarian, 1994). Put another way, the availability of clusters of educated and well-trained manpower, technological capacity and a good transportation and communications network have become the critical location attractions of the late twentieth century, especially for high technology or information/ knowledge intensive sectors such as many of the service industries. The size of a country is less important as a demand variable than its location in relation to other national markets. Indeed, it has been noted that the trend for firms to invest directly for market access will grow as product and service differentiation continues and as the service sector expands (Thomsen and Woolcock, 1993).

FDI by US Services

Using international accounts' data the US Department of Commerce (1997) provides statistics on US direct investment abroad.[5] It also publishes data on the assets, sales, net income and total employees of the non-bank foreign affiliates of US companies by country and by industry. Between 1994 and 1996 the sectoral distribution of FDI has remained quite stable (table 5.2). Services account for approximately 55 per cent of the investment annually with almost two-thirds arising from FDI by finance, insurance and real estate. Many of these services are following their multinational manufacturing or other service clients who expect levels of support equivalent to that which they receive in the US. Others are taking advantage of the expertise developed in the very competitive domestic market in the US, combined with the opportunities offered by information and communications technology (ICT), to gain control of new markets in under-supplied or less competitive markets overseas. In certain sectors such as professional services, this has been helped by service trade liberalization allowing direct representation or equal treatment, for example, in certain overseas markets. In others, such as financial services, the deregulation of financial markets in London and other European financial centres, or much more recently in Tokyo, has been instrumental in enabling the large and dynamic US firms to extend their operations.

A more detailed breakdown of the foreign affiliates of US companies by industry (excluding depository institutions, i.e. banks) shows that some 50 per cent of more than 21,000 affiliates are in the service sector (table 5.3). In 1995 they has assets (buildings, land, equipment) that amounted to more than 55 per cent of the total. With property often located in high-cost downtown locations in some of the world's major corporate control centres, this share is not surprising. However, the services share of total sales by affiliates is only 26.9 per cent and their share of all employment is even lower at 20.5 per cent. The foreign affiliates of US service firms are on average also much smaller than those of manufacturing firms (table 5.3). Thus, the relative importance of the overseas activities of US service firms depends on the measures that are used to assess their significance for the host countries. They score more highly on the less tangible impacts such as the value of their total FDI, than on tangible inputs such as the number of jobs or the size of the establishments located overseas.

Some 56 per cent of the FDI by US services in 1996 was made in Europe (figure 5.3). Almost two-thirds of the investment was attributable to finance, insurance and real estate services with almost 50 per cent located in the UK, followed by the Netherlands and Germany. With 22 per cent of the FDI balance going to Latin American and Other Western Hemisphere locations, financial and related services are again prominent, especially Bermuda, the Bahamas, Jamaica and Trinidad and Tobago. These Caribbean islands have been attractive locations for investment in back office facilities providing routine data processing or for offshore investment and banking (Roberts, 1994). Africa and the Middle East are very minor destinations for US services direct investment, while

Table 5.2 US direct investment abroad, by industry of foreign affiliate, 1994–6

Industry[1]	1994		1995		1996	
	Investment[2]	%	Investment	%	Investment	%
Petroleum	67,104	10.5	70,229	9.8	75,479	9.5
Manufacturing	211,431	33.0	250,253	34.9	272,564	34.2
Wholesale trade	62,608	9.8	67,222	9.4	72,462	9.1
Banking	26,693	4.2	28,123	3.9	32,504	4.1
FIRE[3]	213,175	33.3	228,744	31.9	257,213	32.3
Other services	26,734	4.2	32,769	4.6	36,673	4.6
Other industries	32,575	5.1	40,213	5.6	49,600	6.2
Total	640,320		717,553		796,495	

Notes: 1. Industry of foreign affiliate.
2. Direct investment position on a historical-cost basis (US$ billions). Historical cost is the only basis on which detailed estimates by country and industry are available. The figures for 1996 are preliminary.
3. Finance, insurance and real estate (except banking).
Source: Survey of Current Business, 1997.

Table 5.3 Selected data for non-bank foreign affiliates of US companies, by industry, 1995

Industry	Number of affiliates	Total assets (US$ mill.)	Sales (US$ mill.)	Employees (000s)	Employees per affiliate
Petroleum	1,520	272,087	428,030	230.9	152
Manufacturing	8,023	779,339	984,868	4,376.6	546
Wholesale trade	4,878	206,015	367,525	538.3	110
FIRE	2,742	1,229,643	108,441	191.0	70
Other services	2,671	114,995	100,035	779.8	292
Other industries	1,484	213,062	151,548	1,260.4	849
Total	21,318	2,815,141	2,140,447	7,377	346

Source: Survey of Current Business, 1997.

the Asia-Pacific region does rather better but perhaps attracts a smaller share of total investment than might be expected from its share of total global output. Most of the investment went to Australia, Singapore and Hong Kong rather than Japan, which until very recently has protected its domestic market for services from foreign suppliers.

A sectoral breakdown of US service FDI in 1996 by world region reveals that 55.6 per cent of more than 356 billion dollars invested by US service firms abroad went to Europe (figure 5.4). FDI by finance, insurance and real estate services amounted to 66 per cent of all the investments in Europe. This is less

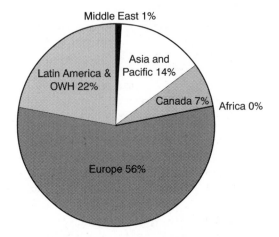

Figure 5.3 US services direct investment position abroad (historical cost basis), 1996
Source: US Dept of Commerce, 1997.

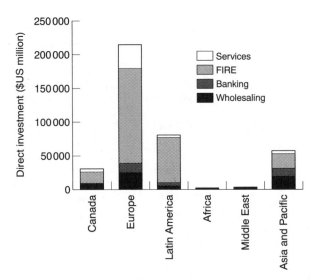

Figure 5.4 US direct investment abroad, 1996 (historical cost basis), regional distribution by sector
Note: Latin America includes Other Western Hemisphere (Central America, Caribbean).

Table 5.4 Comparative distribution of US and Japanese FDI stock in Europe, by sector

Sector	US		Japan	
	US$ mill.	%	US$ mill.	%
Services	**133,464**	**60.0**	**53,406**	**77.8**
Wholesale trade	24,875	18.6	8,329	15.6
FIRE	74,370	55.7	35,415	66.3
Other services	6,159	4.6	5,583	10.5
Other n/c industries	28,060	22.1	4,079	7.6
Manufacturing	**89,090**	**40.0**	**15,231**	**22.2**
Total (all sectors)	222,554	100	68,637	100

Note: Stocks of investment at Dec. 1991 (US), March 1992 (Japan).
Source: Extracted from EAG (1998), table 9.9, p. 111.

than the equivalent proportion for investment by FIRE services into the Latin America region (83 per cent), which includes the Caribbean which, as already mentioned, is an important destination for US offshore data processing and related facilities. Indeed, although well behind Europe in its share of FDI in 1996, Latin America is ahead of Asia and the Pacific that were the destination for just over 14 per cent of the total. FIRE services also only comprise 41 per cent of the investment.

It is interesting to compare the sectoral distribution of US and Japanese FDI stocks in Europe (table 5.4). The information relates to 1991 and 1992 respectively. Although the volume of overall investment and that specific to services by US firms is more than double that by Japanese firms, almost all the investment (78 per cent) by the latter is service-related. The popular perception is that the Japanese have been major investors in Europe, an impression that is probably reinforced by a preference for greenfield investments (new manufacturing plants for example) rather than the preference for merger and acquisitions in the case of US MNEs. This has given FDI by US service (and manufacturing) MNEs a much lower profile than might be expected from the longstanding historical and cultural ties with Europe. Their low visibility is also promoted by a preference to use reinvested profits or locally raised capital as a source of finance for FDI in Europe (Dunning, 1993b). In many ways much of the US services FDI in Europe has been part of a natural sequence that started with manufacturing to be followed by services in line with wider structural and regulatory shifts in the international economy (Burgenemeier and Mucchielli, 1991; Goldberg, 1992; Graham, 1991; Kogut, 1993). The size of the EU market, the removal of intra-EU barriers to trade, easier cross-border movement of human resources, and EU-wide recognition of professional qualifications obtained in one country and previously not recognized in another are just some of the

Table 5.5 Direct investment by US services (historical-cost basis) in selected EU Member States, 1994–6

Country	Year	Wholesale trade (%)	Banking (%)	FIRE (%)	Other services (%)	Total (%)
UK	1994	23.2	49.0	60.9	36.1	53.2
	1995	26.3	47.4	52.6	31.8	46.7
	1996	27.8	51.5	53.6	38.4	48.7
Netherlands	1994	13.6	1.7	12.5	15.5	13.5
	1995	12.5	1.4	17.7	12.9	16.4
	1996	14.8	1.3	18.5	10.9	17.0
Germany	1994	12.7	13.4	8.6	8.0	10.6
	1995	11.8	13.2	10.3	5.5	11.2
	1996	10.9	13.7	9.1	0.0[1]	9.5
EU (15)[2]	1994	21,250	8,856	106,329	14,701	153,130
	1995	24,399	9,798	113,332	20,532	170,056
	1996	26,460	10,212	127,498	22,218	188,384
	Change (%)	24.5	15.3	19.9	51.1	23

Notes: 1. Data suppressed to avoid disclosure of information on individual companies.
2. Millions of $US.
Source: US Department of Commerce, 1997.

factors that have encouraged US firms to target a large part of their overseas investments on Europe. These changes arise from the measures included in the Single Market programme: they have the effect of raising the locational advantages of the European market for non-European firms.

The UK is by far the favoured destination for US services FDI in Europe (table 5.5). This is a position, in relation to FDI by manufacturing and services, which it has held since it joined the EU. There has been a decline in its share from 53 per cent in 1994 to 49 per cent in 1996, which was a recovery from the 1995 position (48 per cent). It is only possible to speculate that the EU Single Market as well as further deregulation of markets for banking and financial services in countries such as France and Germany is diverting services investment to previously less attractive locations. It is important, however, to examine the data disaggregated into four categories of services identified by the US Department of Commerce (see table 5.5). This shows that whereas investment by US banks in 1994 in the UK was below its overall share of US services investment in the EU, it was actually higher than the overall figure for 1996. Although the latter is a preliminary estimate it suggests that the increased competition for inward investment (or retention of earlier rounds of investment) between European financial centres via deregulation of markets has yet to seriously challenge the attraction of the UK (specifically the City of London) for US banks. A combination of historical, cultural, linguistic and localization economies contribute to the continuing pull of the UK. But investment in the EU by US banks increased by only 15 per cent between 1994 and 1996 while that by other services (including business, professional and technical services), which in volume terms was already larger than banking in 1994 increased by more than 51 per cent (see table 5.5). While the UK's share has increased it has been well below its overall share of US service investment. The Netherlands and Germany have also underperformed, another indication that many of the US firms in this category are developing new markets in emerging/peripheral areas of the EU such as Spain and Portugal where the competition from domestic or other international suppliers is less well established. This may also be true for FIRE services. The UK remains the pre-eminent destination for US investment but its share has fallen sharply while Germany and the Netherlands have remained relatively stable. Investment is therefore trickling down to less saturated, more 'open' markets elsewhere in the EU.

In order to appreciate the scale and global reach of investment by US services it is useful to examine some case studies. The choice is of course large, but three contrasting examples will be briefly discussed. The first is McDonalds, a food service company which was ranked 426 in the *Fortune* global 500 in 1996 and is perhaps one of the best-known US brands around the world. The second is Lehman Brothers, a diversified financial services firm that represents a very different kind of overseas investment (and brand image) to McDonalds. The third is Arthur Andersen Consulting, a global management and technology consulting organization.

Case Study 1: McDonalds

McDonalds is very clear about the fact that their 'vision is to dominate the global food service industry', and it plans to expand its leadership position 'through convenience, superior value and excellent operations' (McDonalds WWW page, June 1998). It has approached this not only by extending its global reach after firmly establishing its position in the US market but also by very tightly defining the operational, presentational and experiential dimensions of the service that it provides.[6] While nuanced to meet particular local tastes or regulations (beer can be purchased with your meal in McDonalds in some countries but not in others, for example) there is a high degree of certainty on the part of the customer about what they can expect wherever they use McDonalds food service. The interior and exterior design of each restaurant also conform to a closely specified format that makes it instantly recognizable and 'familiar'. It could be argued that these features have underpinned the company's expansion outside the US that accelerated dramatically during the mid-1980s (table 5.6). Just under one in four of 9,811 restaurants were outside the US in 1987; by 1997 the number of restaurants had more than doubled to more than 23,000 and almost half (46 per cent) were at locations outside the US. McDonalds operated in 47 different countries in 1987, rising to 109 countries in 1997. Much

Table 5.6 McDonalds: indices of global reach, 1987–97

Index	US		Outside US		% Outside US	
	1987	1997	1987	1997	1987	1997
Restaurants (No.)	7,567	12,380	2,344	10,752	24	46
Sales (billion $US)	11	17	4	16	27	48
Revenues (billion $US)	3	5	1	7	25	58
Assets (billion $US)	6[1]	8	6[1]	10	50[1]	56

Note: 1. Figure is for 1992.
Source: Extracted from data available on McDonalds WWW page, June 1998.

of the country-based expansion has occurred between 1992 and 1997 when restaurants were opened for the first time in 45 different countries. These range from some of the new post-socialist states such as Belarus and the Ukraine to Pacific Islands such as New Caledonia and Tahiti and countries in the Middle East such as Qatar and Oman.

The expansion of McDonalds into markets outside the US has also been accompanied by a shift in sales and revenues, especially the latter. US sales grew by 4.9 per cent per annum (compound), while outside the US they increased by 16 per cent for the 10-year period ended 31 December 1997. Thus, 48 per cent of sales originated in overseas restaurants in 1997 but 58 per cent of revenues (table 5.6).

Case Study 2: Lehman Brothers

While on any one day McDonalds serves just less than one per cent of the world's population, Lehman Brothers has contact with a minute fraction of global clients by comparison but is no less important. It is a global investment bank providing state-of-the-art research, trading, distribution and financing services to MNEs, institutions, governments and high-net-worth investors worldwide. Henry, Emmanuel and Mayer Lehman founded the firm in Montgomery, Alabama in 1850. They provided commodities broking and trading services to clients in the US South and expanded sufficiently to be able to open a New York office in 1858. A seat was acquired on the New York Stock Exchange in 1887 and in 1929 the Lehman Corporation was created as a prominent investment company. Following further growth via the acquisition of two other US investment companies in 1975 and 1979, the Lehman Corporation was itself acquired by Shearson/American Express Company in 1984. In common with many other US service firms with international ambitions, a seat was acquired on the London Stock Exchange following the 'Big Bang' in 1986 and two years later in 1988 on the Tokyo Stock Exchange. The Lehman Brothers division was established in 1990 before it was spun off from American Express Company to become an independently owned public company, Lehman Brothers Holdings Inc.

With one-third of the organization engaged in international business, almost half of Lehman Brothers' revenues are now generated outside the US. The worldwide headquarters is in New York with regional headquarters in London and in Tokyo; Lehman Brothers participates in the global capital markets via a network of 39 offices. In addition to New York there are 15 other offices in the US, five in Latin America, eight in Europe and the Middle East, and eight in the Asia-Pacific. The firm stresses not only its ability to provide integrated client access to a wide range of global investment opportunities but the role of its professionals at various locations outside the US that are readily accessible to clients in those markets. Thus, its fixed income sales force that serves the investment and liquidity needs of major institutional investors comprises 325 professional staff in 14 locations world-wide. The equity sales group has 315 professional staff at 10 locations in the US, Europe and Asia, while the private client services group, that concentrates on the needs of private investors and mid-sized institutions, has 288 professionals in a world-wide network of 12 offices. There is also a centralized fixed income research group of 275 specialists providing in-depth expertise in areas such as global government securities. derivatives, emerging market debt, and real estate. Finally, there is an equity research department of 260 professionals that emphasizes its global economic, strategic and industry expertise including media and telecommunications, financial institutions, healthcare, power and technology.

Case Study 3: Andersen Consulting

Andersen Consulting was founded in 1989 as offshoot from Arthur Andersen, which is best known for its accounting and tax services and was founded in the US in 1913. The two branches together comprised Andersen Worldwide until Andersen Consulting initiated action in December 1997 to break off from the former. The latter has 363 offices in 78 countries and describes its work as being to acquire and share knowledge in ways that will allow clients to grow and profit. It has adopted a one-firm philosophy, enabling its employees (58,000 in 1997) to 'work together – across boundaries of competencies, functions, and geographies – to deliver to each client a multidisciplinary, complete solution' (Arthur Andersen, WWW page, June 1998).

Andersen Consulting has adopted a similar approach by creating the concept of business integration that involves combining services such as strategy and change management with the firm's competencies in technology solutions. Clients can expect a 'one-stop' service that links their people, processes and technologies to their strategies. While Andersen Consulting stresses this operating model is based on competencies rather than locations it has nevertheless extended its network of offices to 46 countries in 1997 with 1,126 partners forming part of a worldwide personnel count of 53,426. This compares with an employment total of 21,400 in 1989 when revenues amounted to $1.6 billion compared with $6.6 billion in 1997. An example of the emphasis on shared competencies within the firm is the implementation of Release 1.0 of Knowledge Exchange® which enables the partners and employees to electronically share knowledge globally. This means that clients also have access to business solutions that are fully informed by international rather than regional trends. This knowledge network is complemented by the establishment of specialist centres such as those for strategic technology in Palo Alto, California and Sophia Antipolis, France (1994) or the formation of global alliances such as that between Andersen Consulting and Dow Chemical (1996) to advance Dow's information technology organization. In 1996 the firm also set up ServiceNet as a business utility that enables companies to plug in and to quickly begin conducting electronic commerce and related activities via the Internet. All these and other initiatives represent a business philosophy that is promoted by Andersen Consulting which, as it extends its highly integrated network of partners and offices globally, exposes a growing list of clients to the same philosophy. Since the firm already serves in 1997 some 75 per cent of *Fortune* magazine's global 200 largest public companies, it is possible to appreciate the significance of Andersen Consulting's overseas expansion during the 1990s.

Conclusion

During this century the US has consistently been at the leading edge of the shift towards service industries that has been typical of the advanced economies. It is therefore not surprising that it is the home of some of the world's largest service companies. Since many of these expanded their geographical and market coverage within the US through their links with clients in manufacturing and other services, it was logical for them to follow these into overseas markets if they were to retain the business. As US manufacturing firms, in particular, set up production plants and other facilities outside the US in Europe and elsewhere, from the 1950s onwards US firms in advertising, marketing and accountancy were amongst the first in the world to trade outside their traditional national markets. This early experience accompanied by market saturation and/or increased competition at home made it increasingly likely that they would export to new markets or undertake foreign direct investment as a way of maintaining their growth trajectories. The international trade and the services FDI statistics summarized in this chapter illustrate the ability of US service firms to capitalize on their significant competitive advantage (experience, knowledge, networks, pricing, human resources, etc.) to make them major players in most corners of the global economy since the 1970s.

There are of course variations in the presence of US service firms by global region, There are many reasons but most significant is probably the extent of market opportunity and potential. Thus, Europe is a prominent target while Africa, the Middle East and Asia are much less visible in the investment and trade portfolios of US service firms. The three case studies help to amplify aspects of the bigger picture. In particular, they suggest that the ever-expanding global reach of US service firms is accompanied by a particular approach or philosophy of service provision. Whether it takes the rather different forms typified by McDonalds on the one hand and by Andersen Consulting on the other, the net effect is to create for the client or the place a service experience that clearly incorporates the values, culture and corporate identity of the supplier. There are some concessions to geography, but *where* the service is provided around the globe is less important than ensuring it meets predetermined criteria for presentation, quality, value-for-money, access to and use of best available knowledge, and so on.

There can be no question that overseas investment and trade by US service enterprises will continue to be significant. This applies not just to the activities of the financial, professional and business services that are already prominent but to new services and enterprises based on the information and technology and the borderless environment created by the Internet. With more than 100 million people connected to the Internet by the end of 1997 (and growing rapidly) US telecommunications and media services will be well placed to realize the global potential of the digital economy (or electronic commerce).

Notes

1   It is worth noting that the *Fortune* list only includes companies that must publish financial data and report all or part of their figures to a government agency. It therefore excludes some very large service firms that operate as partnerships rather than as registered companies. These are not required to publish details of their fee and other forms of income. Thus some major US accountancy, management consulting, advertising and various professional services companies are overlooked.

2   Note, however, that Japanese service corporations are more numerous (87) than US service corporations in the Global 500 but account for only 18 per cent of the employees and, on average, are much smaller (24,664 employees).

3   Foreign direct investment (FDI) is direct investment (the investment by one firm in another with a view to gaining control of that firm's operations) which takes place across the boundaries of nation states, i.e. where a firm from one country buys a controlling investment in a firm in another country, or where a firm from one country sets up a branch or subsidiary company in another country.

4   The term 'transnational corporation' (TNC) may also be encountered. An MNE has operations in several countries whereas a TNC simply implies operations in at least two countries. All MNEs are therefore TNCs but not all TNCs are MNEs. The term MNE will be used throughout this paper.

5   It is important to note that there are numerous country and sectoral gaps as well as variations in the way that countries record FDI flows (inward/outward). The use of flow rather than stock data for FDI can lead to inaccurate conclusions because annual flows can fluctuate widely. Reinvested profits have not been included until very recently; they can form a significant part of FDI, particularly for the more 'mature' investors such as US service firms). FDI data availability is far less comprehensive than for trade.

6   McDonalds operates as a franchising company with some 80 per cent of the restaurants operated in this way. It only franchises to individuals and on the basis of overall business experience, personal qualifications and background, an ability to effectively lead and manage people, good common business sense and a track record of success in whatever they are doing.

References

American Foreign Services Association (1995) *World Trade in Services* (Highlights from a Conference at the US Department of State, Washington, DC). Washington: American Foreign Services Association.

Beaverstock, J. V. (1990) New international labour markets: the case of managerial labour migration within large chartered accountancy firms, *Area*, 22, 151–8.

Buiges, P. and Jacquemin, A. (1992) Foreign direct investment and exports to the European Community, in Encarnation, D. J. and Mason, M. (eds), *Does Ownership Matter? Japanese Multinationals in Europe*. Oxford: Oxford University Press.

Burgenemeier, B. and Mucchielli, J. L. (eds) (1991) *Multinationals and Europe 1992: Strategies for the Future*. London: Routledge.

Cantwell, J. (1991) A survey of theories of international production, in Pitelis, C. N. and Sugden, R. (eds), *The Nature of the Transnational Firm*. London: Routledge.

Dunning, J. H. (1981) *International Production and the Multinational Enterprise*. London: Allen and Unwin.

Dunning, J. H. (1993a) *The Globalisation of Business: The Challenge of the 1990s*. London: Routledge.

Dunning, J. H. (1993b) *Multinational Enterprises and the Global Economy Workplan*. Berkshire: Addison-Wesley.

Eaton, B. C., Lipsey, R. G. and Safarian, A. E. (1994) The theory of multinational plant location in a regional trading area, in Eden, L. (ed.), *Multinationals in North America*. Calgary: University of Calgary Press.

Economists Advisory Group (1998) *The Single Market Review: Impact on Trade and Investment, Vol. 1: Foreign Direct Investment*. Luxembourg: Office for Official Publications of the European Communities.

Fortune (1997) *The Global 500*, 136 (4 Aug.), F-2–F-28.

Goldberg, M. A. (1992) The determinants of United States direct investment in the EEC, Comment, *American Economic Review*, 62, 692–9.

Graham, E. (1991) Strategic responses of US multinational forms to EC 1992, in Yannopoulos, G. N. (ed.), *1992: Europe and the United States*. Manchester: Manchester University Press.

Hymer, S. H. (1976) *The Operation of National Firms: A Study of Direct Foreign Investment*. Cambridge, MA: MIT Press.

Kogut, B. (1993) Foreign direct investment as a sequential process, in Dunning, J. H. (ed.), *The Theory of Transnational Corporations*. London: Routledge.

McCulloch, R. (1996) New perspectives on foreign direct investment, in Froot, K. (ed.), *Foreign Direct Investment*. Chicago: University of Chicago Press.

OECD (1994) *International Direct Investment Statistics Yearbook*. OECD: Paris.

Ohmae, K. (1990) *The Borderless World: Power and Strategy in the Interlinked Economy*. London: HarperCollins.

Petri, P. A. (1994) The regional clustering of FDI & trade, *Transnational Corporations*, 3, 1–24.

Porter, M. E. (1990) *The Competitive Advantage of Nations*. New York: Free Press.

Reich, R. (1992) *The Work of Nations: Preparing Ourselves for 21st Century Capitalism*. New York: Vintage Books.

Roberts, S. M. (1994) Fictitious capital, fictitious spaces: the geography of offshore financial flows, in Corbridge, S., Martin, R. and Thrift, N. J. (eds), *Money, Power and Space*. Oxford: Blackwell, 91–115.

Stopford, J. M. and Strange, S. (1991) *Rival States, Rival Firms: Competition for World Market Shares*. Cambridge: Cambridge University Press.

Thomsen, S. and Woolcock, S. (1993) *Direct Investment and European Integration: Competition among Firms and Governments*. London: Pinter.

UNCTAD (1995) *World Investment Report*. New York: UNCTAD.

UNCTC (1992) *The Determinants of Foreign Direct Investment: A Survey of the Evidence*. New York: United Nations.

UN Transnational Corporations and Management Division (1993) *The Transnationalisation of Service Industries: An Empirical Analysis of the Determinants of Foreign Direct Investment by Transnational Service Corporations*. New York: United Nations, Department of Economic and Social Development.

US Department of Commerce (1997) *Survey of Current Business*, no. 9, 75–148.

US Department of Commerce, International Trade Administration (1997) *Foreign Trade Highlights*. Washington, DC: US Department of Commerce.

Chapter 6

The Rise and Decline of US International Monetary Hegemony

Mark Holmes and Eric Pentecost

1 Introduction

The projection of US power in international monetary relations, that is, the United States ability to act, control or influence the international monetary arrangements, has gone through four phases during the twentieth century. In 1900 the UK economy was the dominant global economic power. Sterling was as good as gold and the international gold standard was managed by the Bank of England. The US fixed the dollar to gold and so cooperated with the international consensus, but had no role in the operation or management of the system. The second phase, between the wars, was when the UK no longer had the resources to lead the international financial community, but the US, although it possessed the economic power, was unable or unwilling to offer global leadership. The third phase, from 1945 until 1971, is that which corresponds to the peak of the projection of US economic power. Under the Bretton Woods system, the design of which was heavily influenced by the US, the dollar was fixed to gold and all other currencies were pegged to the dollar. The dollar was the principal international reserve asset and medium of exchange. The final phase since 1971 has seen some decline in US influence, with dollar–gold convertibility surrendered in August 1971, followed in March 1973 by floating exchange rates. More recently, the increasing economic power in Europe, especially after 1989, and the Far East make the German D-mark and the Japanese yen potential rivals to the US dollar.

There are two general characteristics that are necessary for a country to have hegemonic economic power. First, the country must have a stable political environment supported by an established institutional framework. In particular, a central bank with the necessary credibility and influence to manage the international financial system. Second, for the particular country's national currency to become adopted as international money, the country needs to have a global trading system, low inflation and low variability of inflation, with well-developed financial markets to facilitate the settlement of third-party debts. This chapter is organized around these two characteristics. Section 2 considers the

US's influence on the credibility and management of the international monetary system in two ways. First, it examines the US's reluctance to participate in international monetary cooperation after the Great War, despite being the dominant world economic power. Second, it considers the US influence on the role and structure of the international monetary institutions established after the Second World War. Section 3 considers the US dollar's changing role as an international money. Section 4 reviews the main arguments and speculates on the future role of the US and the US dollar in international monetary relations.

2   Institutional Factors in US Monetary Hegemony

In 1900 the prevailing international monetary system was that of the gold standard. The US dollar had been practically linked to gold since 1879, although this was made official in 1900. The international gold standard was managed by the Bank of England from London, with cooperation from the central banks of France and Germany. In contrast to the hegemonic stability theory (see Keohane, 1980), whereby a dominant economic power ensures international financial stability, Eichengreen (1992) argues that the key to the success of the gold standard lay in the areas of credibility and cooperation. In the countries at the heart of the system – Britain, France and Germany – the credibility of the official commitment to the gold standard was beyond reproof. Moreover the cooperation with France and Germany enabled the Bank of England to regularly intervene to prevent an outflow of gold from London, by raising interest rates to attract foreign capital. If sterling weakened, funds would flow toward Britain in anticipation of capital gains that would arise once the Bank of England moved to strengthen the exchange rate. Because the central bank's commitment to the existing parity was beyond question, capital flows responded quickly and in considerable volume. The United States was not a party to the cooperative arrangements supporting the gold standard system. The absence of a US central bank precluded American participation in these ventures, but so long as the US was not the leading user of gold reserves, her failure to participate did not threaten the system.

After the First World War the UK's ability to act as hegemonic stabilizer of the international monetary system came to an end. The evidence for this is the fact that Britain's attempt to return to the gold standard in 1925 at the pre-war parity lasted only five years and was a major factor behind the General Strike of 1926 and the subsequent economic depression. By 1920 the US was the dominant economic power. In December 1913 the US had established a central bank, and returned to the gold standard in 1919. Although the US had both the economic power and the institutional structure it refused to participate in international conferences aimed at building a framework for international economic cooperation. In particular, the US declined to participate in the meetings in Brussels in 1920 and Genoa in 1922, because they anticipated that European negotiators would argue that international monetary cooperation required war debt liquidation. In addition, the US did not support the proposals for an international organization such as an International Bank of Issue on the grounds that the extension of new loans would reduce the prospects for repayment of those already outstanding, most notably war debts. Moreover the US refused to re-inflate, despite huge and growing gold reserves, for fear of stimulating inflation. The result was that other countries had to deflate, to restore stability to prices and the exchange rate.

At the World Economic Conference in 1933 the US again declined to act as leader of the international financial community despite her by now overwhelming economic power. Roosevelt refused to consider any concerted action with respect to the exchanges and simply reaffirmed the US's own attachment to the gold standard. The US wanted an exchange rate of at least $5 to the pound, chiefly to raise the domestic prices of the primary products which American farmers and foresters sold in Britain. This was achieved by the UK temporarily importing gold, but sterling fell in 1938–9 and stood at only just over $4 at the outbreak of war. From this it is clear that no state was in charge over this period. There was no monetary hegemony and no reliable international lender who could be counted upon in a crisis. In Washington there were serious problems of a personal and intellectual sort, while relations between the government and the central bank were such that no one except the Secretary of the Treasury could possibly take much initiative, and he, in turn, was not inclined to be helpful. Although the authorities in London were more sympathetic to larger questions, especially after the 1933 conference, they were chiefly concerned with Britain's own recovery, with the problems of the overseas sterling area and, particularly after 1938, the husbanding of resources under threat of war. Thus in 1939, as in 1931, each nation was very much on its own (Kindleberger, 1973). The principal problem of the inter-war years was not so much that the US refused to take on the hegemonic stability role vacated by the UK, but that in declining this role, they also declined to encourage and participate in international cooperation.

The lack of international cooperation was recognized by the early 1940s, and as a result the international monetary institutions were comprehensively reformed during and immediately after the Second World War. The extent to which these US-led reforms were successful is open to debate. On the one hand the Bretton Woods framework can hardly be credited with having inaugurated a golden age of international monetary stability. The first step toward activating the new monetary order, Britain's abortive restoration of convertibility in 1947, was an unmitigated disaster that had to be reversed in short order. A round of major devaluation's followed in 1949–50. Nearly a decade passed before the nations of Europe finally succeeded in restoring currency convertibility in 1958. The Bretton Woods system then operated on an international scale for barely 13 years, collapsing the first time the system was confronted with a major disturbance, namely the US rise in inflation in the late 1960s.

On the other hand, the International Monetary Fund (IMF) and the World Bank offered venues in which countries could meet periodically to exchange information, discuss monetary strategies and remind one an other of the economic interdependencies in the modern world. The Bank of International Settlements (BIS), the EEC Commission, the OECD and the G-10 all provided additional opportunities to exchange opinions and information. These new developments appear even more important when it is acknowledged that the formulation of monetary policy became still more politicized after 1945. Increased international cooperation may have helped to offset to some degree the decline in the credibility of commitments to particular exchange rates. The hegemony of the Keynesian model endowed policy-makers with a common conceptual framework facilitating international cooperation.

After the Second World War the US provided a focal point for harmonizing policies internationally. The Federal Reserve expanded and contracted credit as required to help stabilize the US economy. Given the size of the US economy international economic transactions and the consequent synchronization of business-cycle fluctuations internationally, other central banks simply followed the Fed's lead (see Giovannini, 1989). In times of crisis cooperative support operations were required. The monthly meetings of the BIS were used to arrange three-month credit lines for sterling. When the dollar came under stress the countries provided collective support to the Fed through gold pooling, reserve swaps and other devices (see Tew, 1988: 109–19).

Ad hoc cooperation orchestrated by the US, however, proved to be inadequate to sustain Bretton Woods beyond 1971. Ironically the existence of a dominant economic power able and willing to veto such proposals led to their defeat. It was American officials who struck from the first draft of White's post-war plan for monetary reconstruction articles providing for international oversight of domestic monetary and fiscal policies (see Block, 1977: 47). White's initial conception may have been over-ambitious, yet given sufficient US leadership policy-makers might have been persuaded to sacrifice some autonomy over domestic monetary and fiscal policies for greater international cooperation. However, officials in the US, which as the world's dominant economic power and repository of the majority of the free world's gold reserves, was the nation that would have the most autonomy over domestic policy, were unwilling to concede the point.

The influence of the US over the structure of the post-war international monetary institutions did not end there. America's leverage was such that she was able to resist pressure for concessions on liquidity. The British plan, drawn up by Keynes, proposed making available $26 billion of credits to countries that found it necessary to run temporary balance of payments deficits. The US, anticipating it would be the major surplus country, and would be the one to extend the vast majority of the credits, sought to limit their total to $5 billion and the maximum US obligation to just $2 billion. The final compromise was $8.8 billion in total with a US ceiling of $2.75 billion. These credits were limited by the quantity of gold that member countries contributed to the IMF. Thus a casualty of American self-interest was Britain's scheme to regulate the global supply of international reserves through the provision of a synthetic asset. Keynes proposed that the new international clearing union, which evolved into the IMF, should have the power to create paper credits that its members would accept as in balance of payment settlement. It would be able to adjust the supply of credits to meet the liquidity needs of trade, but if all the credits ended up in American hands, this amounted to giving foreign countries a printing press to be used to create dollars for purchasing American goods.

There are circumstances in which a dominant power facilitates the conclusion of an agreement conducive to international stability – when it uses its influence to fashion institutions for cooperatively structuring and managing international relations. If their durability and resiliency gauge the adequacy of such institutions then it is hard to apply this argument to the Bretton Woods arrangements after the Second World War. It was precisely America's dominance of the post-war international economy that allowed US officials to resist international pressures to sacrifice narrowly defined domestic interests. American hegemony may have been the basis for the post-Second World War international money order, but it was also the source of contradictions that gave rise subsequently to international monetary instability. American dominance of the negotiations at Bretton Woods was responsible for the inadequacies of the Bretton Woods international adjustment mechanism. Owing to US opposition, no sanctions on surplus countries were instituted. Anticipating that the US would be the main surplus country after the war, American officials used their influence to eliminate provisions that might have forced them to revalue the dollar or pay tax on their international reserves.

The demise of the Bretton Woods exchange rate arrangements in the early 1970s led to the re-adoption of floating exchange rates. These arrangements did not give rise to the monetary chaos of the 1920s and 1930s, partly because of the existence of the international institutions, such as the IMF, that were set up in 1945. Although these institutions were not able to set down formal rules for policy cooperation, they remained a forum for discussion and debate, which with the advent of floating prevented any major lapse into protectionism and competitive devaluation as happened in the 1930s. Moreover, the period of US economic hegemony was coming to an end as economic power was increasingly becoming geographically more widespread. The rise of the European Union, with its inherent cooperation on economic matters, and the growth of the Japanese and South-east Asian tiger economies have tended not only to reduce US power, but also to be forces for international economic cooperation.

3   The Dollar as International Liquidity

In 1900 the prevailing international monetary system was that of the gold standard. The US officially tied the dollar to gold in 1990, although it had been practically linked to gold since 1879. The US was already a large economy in terms of international money, with some 15 per cent of global gold reserves, which rose to 22 per cent by the outbreak of the First World War. Despite these stocks of gold held in the US, over the period 1900–13 currencies were more popular than gold in settling international transactions. The inter-war period saw the US emerge as the most powerful economic nation, and after 1945 the

US dollar replaced the pound sterling as the dominant international currency with which international transactions in real and financial assets are conducted. This dominance confers on US authorities a significant degree of economic power over the rest of the world. This section discusses why a dominant international currency projects power, the factors which give a currency dominant international status, and likely developments in the future.

US power is derived from the dollar's international currency status because US monetary management can affect the value of the dollar. If goods and financial assets and liabilities are denominated on dollars, then fluctuations in the value of the dollar affect the market value of transactions as well as the value of stocks. This is particularly the case since the demise of the Bretton Woods system in 1971 and the reintroduction of floating exchange rates. A key example is the oil price that is quoted and traded in dollars per barrel. Oil importers must contend not only with the price of oil but also with their bilateral exchange rates with respect to the dollar. The positive oil price shocks of 1979–80 were accompanied by an appreciating dollar that served to magnify the adverse impact of higher oil prices on the rest of the world. However, the benefits of falling oil prices in the mid-1980s were enhanced by a depreciating dollar during 1986–7. Similarly, the value of financial assets and liabilities can change dramatically as a result of exchange rate developments. As with changes in the value of commodities, this can have far-reaching implications for the behaviour of the economy. From the perspective of the US, the benefits of having international currency status can be seen in terms of the ability to derive seigniorage revenues (these are essentially revenues obtained by the US from money creation) and good earnings for its financial institutions. However, there are factors that might actually undermine American power. Shifts in foreigners' preferences can lead to large capital flows that undermine domestic monetary control. Under floating rates, such shifts can lead to a volatile exchange rate. Thus, the formulation of US monetary policy takes into account any likely reaction on the part of foreign speculators.

We may now turn our attention to the factors that determine a currency's international status. The US dollar has retained its status as the dominant international currency since the end of the Second World War. During the Bretton Woods period, most currencies were pegged to the dollar, much trade and finance was conducted in dollars, and the international reserve currency was the dollar. However, the role of the dollar in relation to other major international currencies such as the D-mark and yen has varied over time. With the collapse of Bretton Woods in 1971, dollar exchange rates became more volatile, causing investors to switch to other currencies. Also, the relative importance of the US economy has declined as other countries have grown faster, and the deregulation of financial systems throughout the industrial world has expanded the number of currencies used to finance trade, invest and borrow. In Europe the move towards European Monetary Union (EMU) has promoted the D-mark as the dominant currency among EU members. However, evidence suggests that although the dollar's relative role has declined, it remains the most important international currency.

Table 6.1   Average annual inflation rates

	US	Germany	Japan	UK
1970–82	7.8	5.1	8.2	12.7
1983–96	3.5	2.3	1.4	4.6
1970–96	5.6	3.6	4.7	8.5

Note: Inflation rates are based on consumer price indices.
Source: *International Financial Statistics.*

Table 6.2   Inflation variability

	US	Germany	Japan	UK
1970–82	3.1	1.4	5.3	5.4
1983–96	1.0	1.3	1.1	2.2
1970–96	3.1	1.9	5.0	5.7

Note: These are standard deviations of the inflation data
reported in table 6.1.

The theory of international money identifies the factors that underlie the use of national monies as domestic financial instruments by non-residents and as offshore financial instruments by both residents and non-residents. An international currency fulfils three basic functions in the international monetary system – a medium of exchange, a unit of account, and a store of value. However, two further sets of conditions must be satisfied before a currency can be used internationally. First, there must be confidence in the value of the currency and in the political stability of the issuing country. This means that low levels of inflation and inflation variability contribute to the international demand for a currency. Second, a country should possess financial markets that are substantially free of controls. Well-developed financial markets are required to supply assets that are appropriate for international use. Markets should be broad, in terms of offering a range of financial instruments, and deep, in terms of offering well-developed secondary markets. A number of countries satisfy both of these criteria. During the 1960s, average annual inflation for the US, Germany and Japan was 2.4, 2.5 and 5.4 per cent respectively. Tables 6.1 and 6.2 show that since 1970, the US, Germany and Japan (and to a lesser extent, the UK) have experienced fairly low and stable rates of inflation. However, the tables also show that in more recent times the inflation performance of Germany and Japan has been superior to that of the US. Each of these countries now offer well-developed financial systems; thus the dominance of the dollar is no longer supported by Germany and Japan having financial systems that are relatively restricted in terms of capital movements and the range of available assets and liabilities.

These factors, however, do not completely explain why a particular currency emerges as the dominant currency. Private agents may choose a vehicle currency to denominate and execute foreign trade and capital transactions that do not involve direct dealings with the issuing country. Economies of scale prevail once a currency assumes vehicle status. The more it is used, the greater is the degree of familiarity with it, the lower are the costs of information and uncertainty, the greater is the possibility of finding a matching transaction and thereby the lower are transaction costs. Bearing these considerations in mind, the following stylized facts characterize invoicing behaviour in international trade. First, two-thirds of international trade between developed countries in manufactured products is invoiced in the exporter's currency or in dollars. Second, currency hedging by importers is uncommon. Third, invoicing in the exporter's currency is most frequent for differentiated manufactured products with long contractual lags. Fourth, according to Grassman's Law, trade between developed and less-developed countries tends to be denominated in the currency of the developed country, although the US dollar is used frequently. Fifth, trade in primary products is usually denominated in US dollars and, to a lesser extent, in sterling (Tavlas, 1997).

The above discussion explains the dominant status of the US dollar as an international currency. Dominance appears to be related to the US share of world exports, the proportion of US exports comprising specialized manufactured products, and the extent of US trade with developing countries. Its share of world exports is in turn influenced by the relative size of the US economy. Tables 6.3–6.5 provide a useful insight into the dominant yet changing role of the US dollar. Since 1960 the relative size of the US economy has declined from 30 to 26 per cent of world economic activity. This can be seen against the background of an increasing share for Japan. The US share of world trade has been steady while Japan has increased its share of activity from 5 to 8 per cent. To some extent, these experiences are reflected in table 6.5, a and b, which report the currency denomination of imports and exports for the US, Germany, Japan and the UK. For imports, the US dollar appears to be used less extensively in the 1990s as compared to 1987, whereas the evidence concerning exports is somewhat more mixed.

The second dimension to the dollar's dominance as an international currency is its role in denominating financial assets and liabilities, for example reserve assets, external debt and international bonds. A key development regarding the status of the dollar in international liquidity occurred in the late 1960s. Triffin (1960) highlighted a potential problem with the Bretton Woods system in terms of dollar liabilities increasing faster than the gold stock (Triffin's Dilemma). The US could essentially finance any balance of payments disequilibria by issuing its own currency and hence an increased foreign holding of dollars. This could lead to serious credibility problems. In response to this, the IMF introduced the concept of an artificially created asset called the special drawing right (SDR) in

Table 6.3 Percentage shares of world GNP

	US	Germany	Japan	UK
1960–9	30	5	6	5
1970–9	26	5	8	4
1980–9	26	5	9	4

Source: *Handbook of Economic Statistics*, Directorate of Intelligence, 1990.

Table 6.4 Percentage shares of world trade

	US	Germany	Japan	UK
1960–9	15	10	5	8
1970–9	13	10	7	6
1980–9	14	10	8	6

Note: Trade is measured as the value of imports plus the value of exports.
Source: *Handbook of Economic Statistics*, Directorate of Intelligence, 1990.

Table 6.5a Currency denomination of imports of selected industrial countries (%)

	US Dollar	DM	Yen	Sterling	Other
1980:					
US	85	4	1	2	9
Germany	33	43	2	3	20
Japan	93	1	2	1	2
UK	29	9	1	38	23
1992–6:					
US (1996)	89	3	3	0	5
Germany (1994)	18	53	2	2	25
Japan (1995)	70	3	23	0	4
UK (1992)	22	12	2	52	12

the late 1960s. SDRs, rather than dollars, could be created to finance the growing volume of international trade, thereby supplementing the lagging supply of gold.

Monetary authorities hold foreign exchange reserves to intervene in foreign exchange markets and to offset international shocks. The choice of reserve currency can be explained by a number of considerations. For industrial countries

Table 6.5b Currency denomination of exports of selected industrial countries (%)

	US Dollar	DM	Yen	Sterling	Other
1980:					
US	97	1	0	1	1
Germany	7	82	0	1	9
Japan	66	2	29	1	2
UK	17	3	0	76	5
1992–6:					
US (1996)	98	0	0	0	2
Germany (1994)	10	76	1	2	11
Japan (1995)	53	0	36	0	11
UK (1992)	22	5	1	62	11

Source: Page (1981), ECU Institute (1995) and various official sources.

one looks towards the nature of exchange rate agreements and policy objectives (regarding the extent of foreign exchange intervention), the distribution of trade flows, the currency composition of government borrowing on foreign capital markets, and profit and risk consideration. In the case of less-developed countries (LDCs), there is scope for portfolio selection criteria based on profit maximization and risk minimization. In addition, historical ties may account for the higher shares of the French franc and pound sterling in the foreign exchange reserves in some cases, and regional financial and trade agreements may also explain higher shares of 'unspecified' currencies in the portfolios of these countries. Overall, developed countries and LDCs who intervene in dollars hold a high proportion of their reserves in that currency in order to reduce transactions costs. Table 6.6 reports data on the official holdings of foreign exchange for 1975 and 1995. The most noticeable change is the marked decrease in the use of the dollar accompanied with large increases in the role of the D-mark and yen. This can be partly explained by a more volatile dollar and higher US inflation since the breakdown of Bretton Woods, and Germany and Japan now offering an increased range of sophisticated financial assets. In addition, fewer countries now peg their currencies to the dollar (54 per cent of all countries pegged to the dollar in 1980 compared with 44 per cent in 1996), thus less dollar reserves are required for stabilizing bilateral dollar exchange rates. These findings are also reflected in the data on external bond issues reported in table 6.7, and further evidence can be found in terms of external bank loans and eurocurrency deposits. In each case the dollar is accounting for a smaller proportion of external assets in financial portfolios across the world.

Power can also manifest itself through the use of the dollar in denominating liabilities. A powerful example is the LDC debt crisis. Following the positive oil price shocks of the 1970s, the surplus revenues of the Organization of Petroleum Exporting Countries (OPEC) were largely recycled into dollar

Table 6.6 Percentage share of national currencies in world official holdings of foreign exchange

	1975	1995
US dollar	80	62
DM	6	24
Yen	1	7
Sterling	4	4
Other	9	13

Note: End of year data. ECU–dollar swaps included as dollars.
Source: IMF, *Annual Report*, various issues.

Table 6.7 Currency composition of external bond issues (%)

	1981–4	1996
US dollar	63	43
DM	6	14
Yen	6	9
Sterling	3	9
Other	22	25

Sources: BIS, *International Banking and Financial Market Developments* and OECD, *Financial Market Trends*, various issues.

Table 6.8 Currency composition of external debt of LDCs (%)

	1970	1980	1995
US dollar	47	54	46
DM	9	8	7
Yen	2	7	14
Sterling	11	4	2
Other	31	28	31

Source: International Bank for Reconstruction and Development.

denominated loans for LDCs where variable rate private loans were based on US interest rates. Table 6.8 shows that dollar-denominated loans accounts for about half of LDC external debt. The worry for LDCs was that higher US interest rates were likely to lead to an appreciation of the dollar and consequently more onerous debt servicing obligations. Clearly, monetary policy changes in the US can have profound effects on the health of LDC economies that need to channel vital resources towards debt servicing. For instance, the high value of the dollar in the early 1980s contributed to the debt crisis where many LDCs were unable to meet obligations regarding debt servicing. Most notably, some forty countries were involved in annual rescheduling negotiations with commercial banks by the end of the 1980s (Gibson, 1996).

Overall, the evidence suggests that while the US dollar remains the dominant international currency, the German mark and, to a lesser extent, Japanese yen, are being increasingly used as international currencies. The decline of sterling as the dominant international currency since the Second World War was primarily influenced by sharp fluctuations in sterling's external value beginning in the late 1940s, the continuation of exchange and other controls after the war, and a declining share of world trade. The pound sterling is still used as a world currency largely because of the status of London as a free and sophisticated financial centre. Within the EU, the D-mark has become the dominant international currency on account of its strong reputation for keeping its value and the increasing sophistication of German financial markets. Also, Germany has become the leading monetary authority within the exchange rate mechanism (ERM), where the stance of German monetary policy is more relevant than the US stance in maintaining exchange parities (De Grauwe, 1996). As eleven members of the EU proceed towards European Monetary Union (EMU), a single currency will dispose of the need for a vehicle currency in trading between EMU members.

4 Conclusions

This chapter has considered two dimensions of US monetary hegemony in the twentieth century: the role of the US in the design and management of the international monetary system and the role of the dollar as an international medium of exchange and store of value. Since the end of the Second World War, the US dollar has assumed dominance in terms of the denomination of international trade and financial assets. Therefore variations in the value of the dollar influence not only the value of trade but also the value of financial assets and liabilities across the world. This in turn has strong implications for the health of the world economy. The initial rise of the dollar towards dominance was supported by the dollar's reputation for keeping its value and the openness of US financial markets. This was against a background of the Bretton Woods system that enforced fixed exchange rates and placed the dollar at the centre of international monetary relations. Since the demise of Bretton Woods and the move to floating exchange rates, Germany and Japan have offered currencies with reputations for keeping their values and financial markets that have become increasingly open and sophisticated. While the dollar remains the dominant form of international money, it is no longer at the centre of a fixed exchange rate regime, and the inflation performance of Germany and Japan have served to promote the D-mark and yen as alternative forms of international money. In addition, the exchange rate mechanism and moves towards monetary union within the

EU have promoted the central role of the D-mark within Europe. The dominant role of the US dollar may be further moderated if the Euro is backed by a central European power with a sound reputation for controlling inflation.

The other aspect of US power in the twentieth century is US influence on international monetary arrangements. In this aspect US hegemony has been less successful. According to the hegemonic stability hypothesis, the US led the international financial community over the 25 years following the end of the Second World War, but failed to exert its power in the inter-war years, when it was the only credible, hegemonic power. In terms of international economic cooperation the US not only failed to foster cooperation in the inter-war years, but also failed to embed international rules for cooperation in the institutions set up after the Second World War. It is the case, however, that the existence of the IMF has enhanced the operation of the flexible exchange rate regime since the early 1970s, and as a result there has been little movement towards protectionism, as there was in the 1930s. In a multipolar world the spirit of cooperation has become more evident with the relative decline of the US. The shape of international monetary arrangements in the twenty-first century does not lie solely in US hands: Europe, reasserting itself with the Euro, and the as yet unknown international status of the Far Eastern economies, both have an important role to play in conjunction with that the US, in the reform of the institutions and operation of the international monetary system.

References

Block, F. L. (1977), *The Origins of International Economic Disorder*. Berkeley: University of California Press.

De Grauwe, P. (1996), *International Money*. Oxford: Oxford University Press.

ECU Institute (1995), *International Currency Competition and the Future Role of the Single European Currency*. London: Kluwer Law.

Eichengreen, B. (1992), *Gold Fetters*. Oxford: Oxford University Press.

Frenkel, M. and J. Sondergaard (1998), 'The Effects of European Monetary Union on the Dollar and the Yen and International Reserve and Investment Currencies', *International Journal of Business*, 3, 15–32.

Gibson, H. (1996), *International Finance: Exchange Rates and Financial Flows in the International System*. London: Longman.

Giovannini, A. (1989), 'How Fixed Exchange Rate Systems Work: The Gold Standard, Bretton Woods and the EMS', in M. Miller, B. Eichengreen and R. Portes (eds), *Blueprints for Exchange Rate Management*. New York: Academic Press, pp. 13–42.

Hakkio, C. (1993), 'The Dollar's International Role', *Contemporary Policy Issues*, 11, 62–75.

Hume, D. (1752), 'On the Balance of Trade', in *Essays, Moral, Political and Literary*, vol. 1. London: Longman Green, 1898, pp. 330–45.

Kenen, P. (1983), *The Role of the Dollar as an International Currency*, Group of Thirty Occasional Papers, 13, New York.

Keohane, R. (1980), 'The Theory of Hegemonic Stability and Changes in International Economic Regimes', in Ole R. Holsti et al. (eds), *Change in the International System*. Boulder: Westview Press, pp. 131–62.

Kindleberger, C. P. (1973), *The World in Depression, 1929–39*. Berkeley: University of California Press.

Page, S. (1981), 'The Choice of Invoicing Currency in Merchandise Trade', *National Institute Economic Review*, 85, 60–82.

Scammell, W. M. (1987), *The Stability of the International Monetary System*. Basingstoke: Macmillan.

Tavlas, G. (1991), *On the Use of International Currencies: The Case of the Deutsche Mark*. Essays in International Finance. Princeton: Princeton University Press.

Tavlas, G. (1997), 'The International Use of the US Dollar: An Optimum Currency Area Perspective', *World Economy*, 20, 709–47.

Tew, B. (1988), *The Evolution of the International Monetary System*. London: Hutchinson.

Triffin, R. (1960), *Gold and the Dollar Crisis*. New Haven: Yale University Press.

Chapter 7

The 'New' Developmentalism: Political Liberalism and the Washington Consensus

Kate Manzo

The term Washington Consensus has been used by scholars such as Gills and Philip (1996) and Wade (1996) as shorthand for a shared paradigm (or model) of development promoted by United States development agencies (such as USAID) and the twin architects of structural adjustment, namely the International Monetary Fund (IMF) and World Bank. According to Bodenheimer (1971: 8), 'it is generally agreed that one of the main characteristics of a paradigm is a consensus as to the basic assumptions, the main theoretical and empirical problems, and the procedures for investigating them.' A paradigm is therefore a body of ideas; an ideology that can be contested. But a paradigm is not necessarily recognized as ideology by those party to the consensus. This point is also illustrated by Bodenheimer (1971: 8), who goes on to note that the field of development studies in the 1960s was governed by

> a paradigm-surrogate, a strikingly pervasive consensus on fundamentals, whose core is liberal democratic theory, as modified by the particular conditions of twentieth century America – a consensus which is all the less frequently recognized or challenged precisely because it is generally taken for granted.

The relationship between liberal economic theory (specifically, neoclassical economics or neoliberalism) and development policy has been well-studied already (see for example Brohman, 1995a; Escobar, 1995; and Wade, 1996). This chapter asks whether the core of the so-called Washington Consensus is also liberal *political* theory. The question is not one of politics versus economics, for they are shown to be intertwined. It is about novelty (or change) versus repetition and continuity.

Is there anything particularly novel about a parcel of ideas marketed by the World Bank, in particular, as a recipe for successful development and insurance against state collapse? Novelty is suggested by several factors, namely: the World

Bank's embrace of the concept of governance in 1989;[1] the attempt at compromise in the early 1990s between 'the well-established World Bank view and the newly-powerful Japanese view' of development (Wade, 1996); the idea of a 'New Policy Agenda' for development assistance in the 1990s (Robinson, quoted in Hulme and Edwards, 1997: 5); and suggestions that neoliberalism has been revised (see for example Kiely, 1998). On the other hand, development institutions are arguably still impelled 'by beliefs organised around the twin poles of neo-liberal economics and liberal democratic theory' (Hulme and Edwards, 1997: 5) – neither of which is novel. And despite the struggles within the World Bank between Japanese and American ideologies of development, the Bank managed to emerge 'with its traditional paradigm largely unscathed' (Wade, 1996: 5).

To address the question of novelty, the chapter analyses the Washington Consensus in relation to an established paradigm of political development – one referred to by many as developmentalism (see especially Bodenheimer, 1971; see also Smith, 1985 and Manzo, 1991). This analysis demonstrates, firstly, that the Washington Consensus is markedly political. Political factors feature prominently in discussions of economic development, and vice versa. The twin pillars of liberalism (i.e. the economic and the political) are not separate but interdependent; woven together into a neoclassical political economy. At the same time, the Washington Consensus is revealed as a political ideology with political consequences. Whether or not they serve as 'an important instrument by which the US state seeks to project a powerful external reach' (Wade, 1996: 4), Washington-based institutions are significant actors in world politics. They set the terms of development debate as well as formulate policies for the recipients of development assistance to implement. As Wade (1996: 5) has noted, 'the World Bank enjoys a unique position as a generator of ideas about economic development'; so much so that 'debates on development issues tend to be framed in terms of "pro or anti" World Bank positions'. The ideas circulated by such authoritative institutions have material consequences and therefore warrant continued critical analysis.

The second finding of the paper is that the Washington Consensus is novel in only one respect. That is the way that established (sometimes incompatible) ideas modified by American conditions in the twentieth century have been repackaged and resold. Liberal ideas have been presented as reasonable explanations for recent events and lessons in adaptation to 'the demands of a globalizing world economy' (The World Bank, 1997: 1). But though marketed as an objective message of experience, the Washington Consensus is not a new paradigm – born only of dispassionate appraisal of global events and trends. World development does not simply speak for itself; the message it emits is filtered through interpretive frameworks provided by established modes of thought. The criteria used by the World Bank and IMF to rank countries along an ascending scale from the collapsed to the miraculous – indeed the very understanding of politics those institutions employ – derive from an American liberal tradition and not from objective observation of conditions elsewhere.

The Washington Consensus:
Modernization Theory or Developmentalism?

Dubbed the 'model of the modern' by O'Brien (1972: 353–7), modernization theory was integral to American development assistance programmes and doctrines until the early 1960s. A liberal consensus was able to form around the wisdom of modernization theory because its 'basic premises: namely, those of the liberal tradition', went largely unquestioned (Packenham, 1973: 303). Political development 'was quintessentially about *state-building*' in modernization theory, because the elements of change it sought in so-called new states were differentiation, capacity, equality and democracy (Manzo, 1991: 13; emphasis in the original).

Though neoliberal arguments for shrinking the state are arguably incompatible with a state-building logic, a number of studies of more recent development doctrines and practices have found their roots in modernization theory. Gills and Philip (1996: 591) have noted that 'the Washington consensus *may* be little more in political terms than an updating of old-fashioned (Almondian) modernisation theory' (emphasis added). Other authors have been more definitive. In questioning whether neoliberalism is 'an innovative new development strategy', Brohman (1995b: 121) accuses it of universalism, Eurocentrism and ideological bias. His argument is that 'many of the most serious contradictions of modernisation theory seem to be reappearing in the neoliberal development framework. These shortcomings make neoliberalism susceptible to many of the same criticisms that have plagued modernisation theory and, eventually, contributed to its demise' (Brohman, 1995b: 121).

A rethinking of modernization and dependency theory undertaken by Scott (1996) also raised questions about the Washington Consensus. It included analysis of two publications providing 'a comprehensive articulation of the overall philosophy guiding structural adjustment programs' in the 1980s.[2] In a chapter entitled 'Gender and the World Bank: Modernization Theory in Practice', Scott (1996: 71) argued that 'like modernization theorists, the World Bank stresses that state capacity and penetration are most effective when private enterprise and initiative are supported.' Subsequent examination of 'the likely consequences of structural adjustment and increased dependence on the operation of the contemporary African state' (Scott, 1996: 71) revealed to Scott (p. 78) the 'liberal dimension' of World Bank philosophy:

> The [World] Bank . . . treats women's role in economic development as a problem of access and opportunity. The implicit liberal assumption is that women are the same as men but are barred from equal opportunity by old-fashioned customs and laws that can easily be swept away. Furthermore, questions about the state's role in reproduction, child-care, and housework are not asked because the liberal imperative is to move women beyond the household: women should be more like men.

The aforementioned studies are complementary to this one, and not contradicted in what follows. But Scott's identification of embedded liberalism only goes to show that the Washington Consensus and its consequences cannot be reduced to the conceptual and political parameters of modernization theory. As I have argued elsewhere (Manzo, 1991), modernization theory was only one branch of the larger paradigm of developmentalism. The focus of what follows is therefore not modernization theory *per se*, but the American liberal tradition in all its guises.

The Politics of Developmentalism: Liberal Theory and Its Consequences

Two investigations into the liberal aspects of earlier theories of development are particularly relevant to questions about the Washington Consensus. One is the work of Bodenheimer (1971), whose critique of developmentalist ideology was designed to 'make explicit the implicit' and uncover the liberal roots of theories of development and American foreign policy as applied to Latin America. Bodenheimer (1971: 7–10) showed that developmentalism 'consists of four analytically distinct but integrally interrelated themes, which are manifested both on the substantive level in development theories about Latin America, and on the epistemological level as general presuppositions of political science'. The four themes (explained later in the chapter) were identified in brief as *cumulation, stability, end of ideology* and *diffusion*.

The other investigation is that of Packenham (1973). *Liberal America and the Third World* identified 'three main types of doctrines about political development in the American aid programs' of the 1950s and 1960s. One of these was the 'explicit democratic approach', which linked the export of American-style democracy to 'the norm of gradualism, to the idea of reform and evolution within a "moving equilibrium"' (Packenham, 1973: 189). This doctrine co-existed with the 'economic approach' (which saw economic development as the bedrock of social and political development) and the 'Cold War approach' (whose highest political ideals were anti-Communism, gradual change and stability). Despite their differences, these doctrines were all rooted by Packenham (1973: 111–60) in four implicit or 'inarticulate' liberal assumptions, namely: change and development are easy (as well as desirable); all good things go together; radicalism and revolution are bad; and distributing power is more important than accumulating power.

The Washington Consensus is revealed in what follows through critical analysis of a number of flagship publications of the IMF (especially its annual reports) and World Bank. The World Bank classifies some of its own writings as 'development theory' (see for example the World Bank, 1995: 107), and it is with publications of that ilk – those that provide 'the conceptual framework and evidence' meant to justify structural adjustment (Wade, 1996: 5) – that the chapter is concerned. The *World Development Report 1997* receives particular attention in what follows. Written well after Japan's aforementioned challenge to the World Bank ended, this is basically an expanded version of those sections of the *World Development Report 1991* devoted to the role of the state in development.

Wade's (1996: 26) point, that the *World Development Report 1991* 'seeks to persuade by ignoring serious alternative explanations' for global events and conditions, is pertinent to this investigation. What is *not* said by either the IMF or the World Bank – especially about the power relations that govern structural adjustment programmes – is considered in subsequent analysis. For when those relations are understood, the explanation offered for global conditions in the 1990s by the World Bank appears misleadingly simple. However 'chronically weak' and lacking in 'governance capacity' they may be, so-called developing states exist in a global context; they are not answerable only to their own 'prominent economic interests' (The World Bank, 1994: xix). At least some of the responsibility for assorted crises and difficulties must therefore lie with the architects of structural adjustment.

The politics of data: cumulative knowledge and global supervision

The 'cumulative notion of knowledge and development' starts from the premise 'that knowledge it built up through the patient, piecemeal accumulation of new observations, which has reached its triumphant culmination in the modern data bank' (Bodenheimer, 1971: 11). An immediate problem identified by Bodenheimer is that indiscriminate collection of data cannot sift significant information from irrelevant material, thereby obscuring knowledge instead of exposing it. Data-driven research is not necessarily open-ended and welcoming of innovation either, because existing stocks of knowledge define the research problems to be addressed and discourage genuinely new approaches: 'Like common law, received research can become an arbiter of future endeavors, putting the burden of proof upon those who would break precedent' (Bodenheimer, 1971: 12).

The World Bank's own 'accumulation of new observations' is published annually in its *World Development Reports* – one of which speaks of 'a gradual accumulation of knowledge and insight' (The World Bank, 1991: 4). A set of country-based World Development Indicators follows without much explanation the lengthy defence of structural adjustment that opens each *Report*. The Bank (1991: 198) has admitted that though 'the statistics are drawn from the sources thought to be most authoritative', the actual data 'are subject to considerable margins of error'. None the less, the contribution of quantitative data to knowledge production is taken for granted. The Indicators are said to 'provide information on the main features of social and economic development' (The World Bank, 1991: 198). They apparently enable 'a fairly clear understanding' of some aspects of development (ibid.: 4). And they seemingly represent 'a wealth of cross-country experience' available from 'international agencies' (The World Bank, 1997: 14–15).

Cumulation is a particularly insidious element of the Washington Consensus because of its use as an instrument of political discipline and control – not by the government of the United States but by the IMF in its role as global supervisor. Surveillance 'lies at the heart of the Fund's responsibilities'; it is designed to ensure 'the effective functioning of the international monetary system' (The International Monetary Fund, 1996: 4, 41). Surveillance 'involves examining the policies of individual countries, encouraging countries to adopt policies that enhance the functioning of the international monetary system, and . . . assessing the consequences of individual country policies for the operation of the global system' (ibid.: 41).

Annual 'consultations' are a main mechanism of surveillance. 'Data issues' feature prominently in these, because 'the Fund has put additional emphasis on members' data provision for surveillance purposes. A core set of indicators has been identified, which all members are encouraged to report monthly to the Fund.' Lest this pose too great a burden, 'the Fund is providing technical assistance to members to improve their ability to compile and report economic data.' For some countries, 'annual consultations have been supplemented with interim staff visits. In several cases, Fund management has followed up on the Board's annual consultation discussions with letters to country authorities on important policy issues' (ibid.: 42).

The belief that more and more data accumulated over time will amount eventually to scientific knowledge 'has appeared in the form of the continuum theory, according to which development and "modernization" proceed in a continuous, linear progression from "traditionalism" to "modernity"' (Bodenheimer, 1971: 13). Bodenheimer is not the only one to question whether piecemeal reform must amount eventually to fundamental and irreversible change. Even the World Bank (1991: 4) has admitted that 'progress has not moved along a straight line from darkness to light. Instead there have been successes and failures.' Yet aspects of the continuum theory continue to reappear, especially in attempts to explain 'contrasting developments' around the world (The World Bank, 1991: 1). As the following section demonstrates, obstacles to development are still derived 'from prior conceptions as to the "prerequisites" or "typical characteristics" of development,' and the extent to which development 'is being conditioned by an international environment dominated by the developed nations' (Bodenheimer, 1971: 14) remains unexplored.

The politics of order: the role of the state in development

One branch of developmentalism revived in the Washington Consensus is the so-called political order school of development theory made famous in the late 1960s by Huntington (1965 and 1968). An early critic (O'Brien, 1972: 363) noted that 'order is seen to rest on "effective" political institutions, which may or may not be formally democratic in character: in any case this question has been of declining significance. Order is imposed from above on the mass.' O'Brien (1972) also argued that participatory democracy was eroded as a political ideal when associated with three forms of disorder: a) political instability (such as military coups) in so-called new states; b) the 'counter-insurgency' objectives of the United States government (the Vietnam war, in particular); and c) the preoccupation with law and order in the United States itself as a consequence of street demonstrations around various issues (such as civil rights and Vietnam).

The question driving the *World Development Report 1997* is why some states have collapsed, others been gripped by financial crises, and still others (until recently) been the beneficiaries of economic 'miracle'. Explicitly, the *Report* asks '*why* and *how* some states do better than others at sustaining development, eradicating poverty, and responding to change' (The World Bank, 1997: 1; emphasis in original).

Evidence of crises and collapse seems to discredit the liberal political assumption that change and development are easy (Packenham, 1973: 112–23). As a previous quote from Scott (1996: 78) implied, 'old-fashioned customs and laws' cannot always 'easily be swept away'. Structural adjustment (a type of change normally signified in economic parlance by competition, deregulation, privatization, liberalization and globalization) is certainly a difficult process. The 'conventional wisdom' that structural adjustment causes political unrest has been refined by research showing that 'the Fund's high-conditionality adjustment programmes may be conducive to generating political protest when there are high levels of urbanisation and economic development, as well as in the presence of a democratic political regime' (Auvinen, 1996: 395). Even the World Bank (1994: xix) has acknowledged that 'structural adjustment imposes high costs on diverse segments of society'.

Yet the World Bank continues to suggest that development is relatively simple – as long as political obstacles to progress are removed and its prerequisites understood. Questions about the feasibility (let alone desirability) of structural adjustment are not entertained. Instead, 'the state is in the spotlight. Far-reaching developments in the global economy have us revisiting basic questions about government: what its role should be, what it can and cannot do, and how best to do it' (The World Bank, 1997: 1).

Some 'very poor outcomes' are apparently the fault of governments – those that embark on 'fanciful schemes,' allow 'endemic' corruption to go unchecked, and fail to maintain investor confidence (The World Bank, 1997: 1–2). If it is true that 'good policies and more capable state institutions . . . produce much faster economic development' (ibid.: 13), then the old 'economic approach' to change has been turned on its head again by the World Bank. The World Bank (1994: xix) has implied before that 'strong institutions' are needed so that 'developing country governments can deploy enough governance capacity to overcome political resistance, design and implement appropriate adjustment policies, and sustain the course of economic reform.' In social science parlance, economic development is the dependent variable and not the independent variable; it is the consequence of 'institutional development' and not its cause.

Like the political order school in the past, what the World Bank (1997: 1) describes as 'central to economic and social development' is not a *democratic* state but an *effective* state. The establishment and maintenance of democratic accountability and popular participation in policy-making are not listed among the 'five fundamental tasks' that 'lie at the core of every government's mission' (The World Bank, 1997: 4). An 'effective state,' according to the World Bank

(1997: 4–6), does the following: it establishes law and order; maintains 'a nondistortionary policy environment, including macroeconomic stability' (i.e. manages privatization, regulation, and industrial policy); invests in social services and infrastructure; protects the vulnerable (through a social security system that falls well short of 'even pared-down versions' of the European welfare state); and protects the environment (itself considered an economic resource to be used and managed efficiently).

A noteworthy ambivalence toward democracy is apparent in the Washington Consensus. Where liberal theory once took for granted that core state institutions are democratically accountable – that 'distributing power is more important than accumulating power' (Packenham, 1973: 151–60) – separation of powers is a mixed blessing for the World Bank (1997: 7):

> Over the long term, building accountability generally calls for formal mechanisms of restraint, anchored in core state institutions. Power can be divided, whether among the judicial, legislative, and executive branches of government or among central, provincial, and local authorities. The broader the separation of powers, the greater the number of veto points that can check arbitrary state action. But multiple veto points are a double-edged sword: they can make it as hard to change the harmful rules as the beneficial ones.

Between the lines of the above quote is a warning; that unrestricted access to the state constitutes an obstacle to development. The political marginalization of 'ethnic minorities or women, or disfavored geographic areas' *is* considered a problem, but only because 'such groups are fertile ground for violence and instability' (The World Bank, 1997: 4). The argument that 'the Bank is less concerned with the achievement of equality and more with the conditions that will allow free-market forces' to flourish (Scott, 1996: 71) is therefore still valid. As in the late 1960s, 'the masses of the underdeveloped world, with high material and other aspirations, and low levels of allegiance to governments which fail to meet these aspirations, are in this context seen as a great potential danger' (O'Brien, 1972: 364–5).

The World Bank's entire rethinking of the state – including its offer of 'ways to narrow the growing gap between the demands on states and their capability to meet those demands' (The World Bank, 1997: 3) – recalls the political development theory of Samuel Huntington. Huntington (1965: 415) was dismissive of the 'literature on modernization', treating it as largely irrelevant to the pressing task of explaining disorder and instability: 'More relevant in many cases would be models of corrupt or degenerating societies, highlighting the decay of political organization and the increasing dominance of disruptive social forces.' Then as now, political problems encountered by 'changing societies' were attributed to weak public institutions unable to cope with rapid political mobilization (see in general Huntington, 1965 and 1968).

An 'absolute priority accorded to the achievement of political order' has been described as more authoritarian than liberal: 'The label "liberal" might appropriately be applied to most of the influential scholars of the early 1960s

(Almond, Deutsch, Apter) but not to those who dominate the latter half of the decade (Pye, Huntington)' (O'Brien, 1972: 369). This misses the point that the term 'political development' has always encompassed more than democratization – even in liberal thought. Those who would find liberalism at work in the political order school do not necessarily 'use the epithet "liberal" indiscriminately, as a political custard pie' (ibid.).

Equilibrium, stability and order are as much economic as political ideals; considered essential in neoliberal economics so that markets can operate freely 'under the watchful eye of the government as referee' (Wade, 1996). Smith (1985: 537) has noted that an 'unfortunate tendency throughout much of developmentalism' is 'to exaggerate the congruence of elements within a given social organization (a preference for static equilibrium models which often classified contradiction and change as "dysfunctional").' The 'third liberal premise' of US development doctrines and theories in the 1960s was that 'radical politics, including intense conflict, disorder, violence, and revolution, are unnecessary for economic and political development and therefore are always bad' (Packenham, 1973: 132).

In the developmentalist model, instability is particularly bad when it creates uncertainty and poses an obstacle to prediction:

> Closely linked to the cumulative view of development . . . is an underlying preoccupation with stable and orderly change and a corresponding fixation upon instability as the main hindrance to development . . . The premium is on development which maintains the system in equilibrium and promotes social integration . . . These tendencies are not only specific to substantive structural functionalism and systems theory, but are also echoed at a more general epistemological level in the preoccupation with *prediction* . . . For if stable patterns or regularities in the development process can be identified, policy makers can be provided, as it were, with a map for guiding Latin American development toward their desired objectives.
>   (Bodenheimer, 1971: 16–18; emphasis in original)

The World Bank (1997: 4) insists that when states fail 'to ensure law and order, protect property, and apply rules and policies *predictably*', their credibility in the eyes of 'local entrepreneurs' is diminished and 'growth and investment suffer as a consequence' (emphasis added). As for the IMF, equilibrium in balance of payments was one of its founding objectives. Like the World Bank, it now promotes the view that political stability and order are vital prerequisites of economic development.

Political stability is prized by the IMF for two reasons. First, stable institutions are considered necessary for economic growth: 'We [the IMF] concentrate our technical assistance on a few areas: designing macroeconomic policy, central banking, establishing treasuries and budgets. The better the institutions, the better the chances for promoting growth' (Camdessus, 1994: 52). And second, structural adjustment apparently requires the fertile soil of civil order and political certainty: 'Sound macroeconomic policies and a vibrant private sector were unlikely to take root in conditions of strife and uncertain security. [IMF] Directors hoped that several countries would soon be able to re-establish the conditions necessary for civil order and growth' (The International Monetary Fund, 1996: 69).

In *Governance: The World Bank's Experience*, the World Bank (1994: xix) uses the phrase 'successful economic reform' as a synonym for structural adjustment. If structural adjustment does impose high costs on certain social groups and classes then it must be a political as well as economic process. This is recognized implicitly in the World Bank's own political advocacy in the past – its insistence that states undergoing structural adjustment must 'avoid capture by prominent economic interests' and 'deploy enough governance capacity to overcome political resistance' (1994: xix). And yet the following quote from the World Bank (1997: 7, 13–14) does not treat politics as integral to development itself. Instead it gives the impression that politics is an obstacle to development that can eventually be overcome:

> Countries struggle to build the institutions for an effective public sector. One reason the task is so difficult is political. Strong interests may develop, for example, to maintain an inequitable and inefficient status quo . . . Comprehensive reform along these lines will take a great deal of time and effort in many developing countries, and the agenda varies considerably from region to region. Reform will also encounter considerable political opposition . . . Reforming the state requires cooperation from all groups in society. Compensation of groups adversely affected by reform (which may not always be the poorest in society) can help secure their support.

The idea that development can one day leave politics behind – so that governance becomes no more than public administration – is only made possible by a particular understanding of politics and the state, as the following section demonstrates.

The politics of apolitical development: political consensus and the end of ideology

Another school of thought reflected in the Washington Consensus is the so-called new political economy of the late 1960s. Rather than stability, the 'new' political economy promoted the liberal political values that underwrite American pluralism, namely consensus, bargaining, cooperation and compromise among individuals and interest groups. This renewed celebration of the 'liberal democratic consensus society' (Bodenheimer, 1971: 21) was hardly novel. More inventive was the extension of economic theory to political systems and processes.

Enthusiasts of the 'new' political economy celebrated 'the distinct outlines of economics and economic behavior in the basic model of politics' (Mitchell, 1969: 105). The effect of understanding politics 'as essentially an *exchange* phenomenon not totally different from economic exchange' (ibid.) is to depoliticize the state. Instead of 'a place of *struggle* for power or influence', the 'political system' becomes 'an essentially *cooperative* division of labor within which various forms and degrees of competition may take place' (ibid.: 105, 119; emphasis in original). The state is thus stripped of politics; analogous not even to a civil service but to a marketplace or public corporation.

Nelson (1995: 113) has pointed out that the World Bank goes 'to some lengths to represent its own activity as above politics'. An obvious reason is that 'the Bank's constitution requires it to be "apolitical"' (Wade, 1996: 16). However, what Leftwich (1994: 363) has labelled 'the technicist fallacy . . . that the effective administration or "management" of development is essentially a technical or practical matter', also serves a political agenda:

> The World Bank is neither apolitical nor a neutral servant of the wishes of its member states, but an organization with a measure of autonomy, considerable power over many of its borrowing members, and a variety of ways to exercise that power . . . Development plans with profound political and distributional implications can be promoted forcefully if their basis is agreed to be apolitical and scientific. And when real economic choices are sharply constrained by debt and the absence of financial and institutional alternatives, emphasizing borrowers' formal sovereignty softens the hard realities of power relations in their dealings with the Bank.
>
> (Nelson, 1995: 112–13)

The technicist fallacy is consistent with the *end of ideology* thesis prominent in the United States in the 1950s (see Bell, 1962). Reflected in the 'new' political economy and more recently in the work of Fukuyama (1992), the thesis celebrates achievement of a 'political consensus or "agreement on fundamentals"' and predicts an eventual demise of the 'currently prevailing politics of conflict and often of violence' (Bodenheimer, 1971: 18–19). Around the world, the Marxist idea of permanent struggle among social groups and classes is supposed to be discredited and replaced with 'the familiar pluralistic model of competing pressure groups whose respective interests can be accommodated by bargaining and compromise' (ibid.).

Ideological politics has apparently disappeared already from international relations; hastened by 'the collapse of command-and-control economies in the former Soviet Union and Central and Eastern Europe' (The World Bank, 1997: 1). An international birthday party was held in Madrid in 1994 to celebrate fifty years of the Bretton Woods regime (i.e. the triple institutions of the General Agreement on Tariffs and Trade – known by its acronym GATT – the IMF and the World Bank). According to the IMF (1995: 37), that conference was marked by 'a strong consensus on the policy strategies that all countries needed in order to reach the [stated] objective' of 'high quality growth'. The subliminal message of that statement is that the Washington Consensus is now universal; no longer confined within the United States but shared with neoliberal forces everywhere.

If a political consensus on structural adjustment policies is supposed to be universal, then it sits uneasily with prevailing concerns about political disorder and instability. Even the IMF cannot wish away political struggles around its programmes:

> When you go to structural action like trade liberalization or privatization, you have powerful vested interests against you. So it's immensely important to have leaders who can convince their people that these changes are essential for growth. One of the most important elements for which we are mobilizing our forces is institution-building and governance. In many countries, 30 years after independence, the very structure of the state must be reconceived.
>
> (Camdessus, 1994: 52)

Yet the IMF continues to emphasize 'the importance of explaining what we're doing to people who can form part of the national *consensus*' (Camdessus, 1994: 52; emphasis added). Consistent with the *end of ideology* thesis, 'notions of "class struggle" or conflict based in power relations' are not entertained at all. Instead there are only 'conflicts of interest . . . reconcilable by bargaining and compromise' (Bodenheimer, 1971: 20).

According to Bodenheimer (1971: 20), 'this type of analysis is based upon a concept of politics and "politicization" which makes ideological conflict and violence a form of "antipolitics" or "pathology" in societies which have not yet achieved consensus.' This recalls the term 'anti-politics machine'; coined by Ferguson (1994) to describe the political ideal of a depoliticized and bureaucratic developmental state. A state that eradicates conflict from governing institutions and 'promotes the competition of ideas within a framework of underlying agreement' (Bodenheimer, 1971: 21) is not actually apolitical. It is an agent of the same model of politics promoted by the 'new' political economists.

Other than convincing its people that structural adjustment is essential, the World Bank's 'effective' state is supposed to reform itself. It is to become more like a competitive marketplace or monopoly corporation (an 'anti-politics machine') than 'a place of *struggle* for power or influence' (Mitchell, 1969: 119). According to the World Bank (1997: 9), 'governments can improve their capability and effectiveness by introducing much greater competition in a variety of areas: in hiring and promotion, in policymaking, and in the way services are delivered.' Within their own borders, state governments are supposed to boost competition within the civil service and ensure more competition in the provision of public goods and services (The World Bank, 1997: 9). The other side of this competitive coin is what the World Bank (1997: 12) calls globalization:

> 'Globalization' is not yet truly global – it has yet to touch a large chunk of the world economy. Roughly half of the developing world's people have been left out of the much-discussed rise in the volume of international trade and capital flows since the early 1980s . . . The cost of not opening up will be a widening gap in living standards between those countries that have integrated and those that remain outside. For lagging countries the route to higher incomes will lie in pursuing sound domestic policies and building the capability of the state . . . In that sense, globalization begins at home.

The politics of diffusion: expertise and external assistance

The liberal assumption that 'all good things go together' (Packenham, 1973: 123–9) is apparent in the above thesis about globalization, i.e. in the belief that successful reform requires progress on many fronts. It has recurred before, amid the 'seven lessons in reform' taught to 'policymakers everywhere' (The World Bank, 1991: 152). Components of structural adjustment like tax reform and tariff reduction have been described as 'complementary actions' that must occur simultaneously because *'partial attempts often fail'* (The World Bank, 1991: 152; emphasis in original).

Universal lessons beg a 'crucial question' for Gills and Philip (1996: 587), which is 'whether development can indeed be taught. All liberal international institutions in the development issue area rest on the unquestioned assumption that it can be.' Such lessons also recall the *diffusion* hypothesis identified by Bodenheimer (1971: 22). This contained 'two major assertions about development: a) that development occurs largely through the spread of certain cultural patterns and material benefits from the developed to the underdeveloped areas; and b) that within each underdeveloped nation a similar diffusion occurs from the modern to the traditional sectors.'

In developmentalism, international diffusion entails inputs of capital and technology as well as 'the values and institutions which made America and Western Europe "great"' (Bodenheimer, 1971: 27). That second aspect explains why 'the most frequently heard, and the bitterest, charge against the developmentalist paradigm is that it was "unilinear" or "ethnocentric" in its concept of change; that is, it projected a relatively inflexible path or continuum of development in which social and political forms would tend to converge, so that the developmental path of the West might well serve as a model from which to shed light on transformations occurring in the South' (Smith, 1985: 537).

A Fund and a Bank are evidently agents of capital diffusion. But the second aspect of international diffusion identified by Bodenheimer is equally pertinent to analysis of the Washington Consensus. It is not only that liberal values are encoded within discussions of reform. It is also that the World Bank (1997: 12), at least, is quite explicit in using the Western state as a model of universal development:

> The history of state reform in today's established industrial countries offers hope – and gives pause – to today's developing countries. Until the last century many of the problems that now appear to have reduced the effectiveness of the state in the developing world were in plain evidence in Europe, North America, and Japan. But the problems were addressed, and modern states with professional systems emerged. This gives us hope. But it also gives us pause, because institutional strengthening takes time.

To speed up change, the 'developing world' is supposed to stiffen its resolve and then turn to those 'international agencies' equipped with valuable lessons: 'If the history of development assistance teaches anything . . . it is that external support can achieve little where the domestic will to reform is lacking' (The World Bank, 1997: 15). External supporters can 'encourage and help sustain reform' through the international diffusion of: a) technical advice; b) a 'wealth of cross-country experience'; and c) financial assistance (The World Bank, 1997: 14). An 'effective political leadership [that] is skilled at devising strategies for building consensus or compensating losers' (ibid.: 28) should take care of the 'domestic' side of the diffusion equation.

If 'experience' is the World Bank's stock in trade, then 'technical advice' is the province of the IMF:

> The Fund offers members a broad range of technical assistance and in-country training in areas of macro-economic management covering monetary, fiscal, and related statistical, legal, and information-technology areas . . . The Fund provides technical assistance in the field through missions by Fund staff and the assignment of short- and long-term advisors. At headquarters, and occasionally in member countries, courses and seminars are given by the IMF Institute and the technical assistance departments.
>
> (The International Monetary Fund, 1996: 147)

Financial assistance is designed to 'help countries endure the early, painful period of reform until the benefits kick in' (The World Bank, 1997: 15). Should financial pay-offs fail to numb the pain and reform appear threatened, then international agencies 'can provide a mechanism for countries to make external commitments, making it more difficult to backtrack on reforms' (ibid.). Such a mechanism (though unspoken) is presumably financial. Debt relief under the HIPC (Heavily Indebted Poor Countries) initiative, for example, 'is conditional on six years of faithfully obeying demands from the Fund and Bank' (The Jubilee 2000 Coalition, 1998).

Conclusion

The clear implication of the IMF and World Bank's own writings is that structural adjustment is a set of far-reaching policies with inevitable political consequences. Many of the states undergoing structural adjustment will be caught between a political rock and a hard place. Either they inflict pain and invite political resistance, or they renege on their external commitments and risk being disciplined by their international 'teachers'. In this context, explanations for 'vicious cycles of poverty and underdevelopment' that cite only 'the chronic ineffectiveness of the state' (The World Bank, 1997: 15) are surely misguided. They fail to give due attention to the complex web of relationships within which so-called developing states are embedded.

This chapter has shown that the Washington Consensus is not a novel paradigm. As in post-war United States doctrines of political development, there is a pervasive consensus on fundamentals within the IMF and World Bank that reflects liberal political theory as well as neoclassical economics. Given that the intellectual background of many IMF and World Bank staff is in economics rather than politics (on this see Wade, 1996), it may seem odd to take these institutions seriously as theorists of political development. But the continued circulation of developmentalist assumptions, problems and policies highlights the political character of supposedly apolitical development.

Two general conclusions about the Washington Consensus are suggested by the preceding analysis. Firstly – as Wade (1996) has emphasized – the World Bank, in particular, remains an important vehicle for the promotion of American values and interests. Despite its name and international composition, the World Bank is less a microcosm of global intellectual currents and more a carrier of traditional American development ideology. However they are marketed, publications like the *World Development Reports* 'are really *advocacy* statements' on behalf of developmentalism (Wade, 1996: 35; emphasis in original). They are a political ideology that can be contested, and not objective science to be accepted at face value.

Secondly, the reassertion of the Huntingtonian ethos of institutional capacity and political order (in conjunction with the economism of the 'new' political economy) raises an important question about the character of American-style democracy. This chapter has shown that order and democracy are not necessarily zero-sum ideals, in the sense that the assertion of the one does not presuppose the erosion of the other. The *end of ideology* thesis, for example, treated political order as the logical consequence of a pluralistic model of politics marked by bargaining, consensus and the absence of permanent class-based conflict. But when stability is threatened by political unrest, is the first priority to strengthen the state in order to restore order or to examine the underlying causes of political struggles and resistance to change? The answer suggested by the Washington Consensus (as by the political order school in the past) is the former, which is why the language of political effectiveness is not to be treated as a synonym for participatory democracy.

Acknowledgements

The author would like to thank David Campbell, Richard Dodgson, the editors of this volume, and the participants at a conference on 'Globalization, State and Violence' (Sussex University, April 1998) for helpful comments on earlier versions of this chapter.

Notes

1   According to the World Bank (1994: xiv), 'governance was a rarely used term in development circles until employed in the World Bank's 1989 report, "Sub-Saharan Africa: From Crisis to Sustainable Growth." '
2   The two publications were: The World Bank's 1981 report, *Accelerated Development in Sub-Saharan Africa: An Agenda for Action*; and the previously-mentioned *Sub-Saharan Africa: From Crisis to Sustainable Growth*.

References

Auvinen, Juha Y. (1996) 'IMF Intervention and Political Protest in the Third World: A Conventional Wisdom Refined', *Third World Quarterly* 17, 3: 377–400.

Bell, Daniel (1962) *The End of Ideology: On the Exhaustion of Political Ideas in the Fifties* (New York: Free Press).

Bodenheimer, Susanne J. (1971) 'The Ideology of Developmentalism: The American Paradigm-Surrogate for Latin American Studies', *Comparative Politics Series Volume II* (Beverly Hills: Sage).

Brohman, John (1995a) 'Economism and Critical Silences in Development Studies: A Theoretical Critique of Neoliberalism', *Third World Quarterly* 16, 2: 297–318.

Brohman, John (1995b) 'Universalism, Eurocentrism, and Ideological Bias in Development Studies: From Modernisation to Neoliberalism', *Third World Quarterly* 16, 1: 121–40.

Camdessus, Michel (1994) 'The IMF's Next Task', *Newsweek* (15 Aug.): 52.

Escobar, Arturo (1995) *Encountering Development: The Making and Unmaking of the Third World* (Princeton: Princeton University Press).

Ferguson, James (1994) *The Anti-Politics Machine: 'Development,' Depoliticization, and Bureaucratic Power in Lesotho* (Minneapolis: University of Minnesota Press).

Fukuyama, Francis (1992) *The End of History and the Last Man* (London: Hamish Hamilton).

Gills, Barry and George Philip (1996) 'Editorial: Towards Convergence in Development Policy? Challenging the "Washington Consensus" and Restoring the Historicity of Divergent Development Trajectories', *Third World Quarterly* 17, 4: 585–91.

Hulme, David and Michael Edwards, eds (1997) *NGOs, States and Donors: Too Close for Comfort?* (Basingstoke and London: Macmillan Press).

Huntington, Samuel P. (1965) 'Political Development and Political Decay', *World Politics* 27, 3 (April): 386–430.

Huntington, Samuel P. (1968) *Political Order in Changing Societies* (New Haven and London: Yale University Press).

International Monetary Fund (1995) *Annual Report 1995* (Washington, DC: International Monetary Fund).

International Monetary Fund (1996) *Annual Report 1996* (Washington, DC: International Monetary Fund).

Jubilee 2000 Coalition (1998) 'Bank Admits HIPC Conditions Wrong', *Debt Update* (March). Article posted on the International Political Economy Website (ipe@csf. colorado.edu), 14 March.

Kiely, Ray (1998) 'Neo Liberalism Revised? A Critical Account of World Bank Concepts of Good Government and Market Friendly Intervention', *Capital and Class* 64: 63–87.

Leftwich, Adrian (1994) 'Governance, the State and the Politics of Development', *Development and Change* 25.

Manzo, Kate (1991) 'Modernist Discourse and the Crisis of Development Theory', *Studies in Comparative International Development* 26, 2 (Summer): 3–36.

Mitchell, William C. (1969) 'The Shape of Political Theory to Come: From Political Sociology to Political Economy', in *Politics and the Social Sciences*, ed. Seymour Martin Lipset (New York: Oxford University Press).

Nelson, Paul J. (1995) *The World Bank and Non-Governmental Organizations: The Limits of Apolitical Development* (Basingstoke and London: Macmillan Press).

O'Brien, Donal Cruise (1972) 'Modernization, Order, and the Erosion of a Democratic Ideal: American Political Science 1960–70', *Journal of Development Studies* 8, 2: 351–78.

Packenham, Robert A. (1973) *Liberal America and the Third World: Political Development Ideas in Foreign Aid and Social Science* (Princeton, NJ: Princeton University Press).

Scott, Catherine V. (1996) *Gender and Development: Rethinking Modernization and Dependency Theory* (Boulder and London: Lynne Rienner Publishers).

Smith, Tony (1985) 'Requiem or New Agenda for Third World Studies?' *World Politics* 37, 4: 532–61.

Wade, Robert (1996) 'Japan, the World Bank, and the Art of Paradigm Maintenance: *The East Asian Miracle* in Political Perspective', *New Left Review* 217 (May/June): 3–36.

World Bank (1991) *World Development Report 1991: The Challenge of Development* (Oxford and New York: Oxford University Press).

World Bank (1994) *Governance: The World Bank's Experience* (Washington, DC: The World Bank).

World Bank (1995) *Index of Publications and Guide to Information Products and Services* (Washington, DC: The World Bank).

World Bank (1997) *World Development Report 1997: Rethinking the State* (Oxford and New York: Oxford University Press).

Part III

Political Capacities: The Arsenal of Democracy?

Chapter 8

Inaugurating the American Century: 'New World' Perspectives on the 'Old' in the Early Twentieth Century

Michael Heffernan

'Europe is the unfinished negative of which America is the proof.'
Mary McCarthy (1961)

Introduction

This volume examines the idea of 'the American century' from the vantage point of the current *fin-de-siècle* (Pfaff 1989). The phrase 'the American century' is often associated with Henry Luce, redoubtable publisher of *Life* magazine and author of a self-consciously prophetic article bearing that title written on the eve of the US entry into the Second World War (Luce 1941; see also McCormick 1989). In Luce's view, the USA was poised on the brink of an unprecedented era of global economic and cultural hegemony in 1941. US involvement in the conflagration engulfing the 'Old World' of Europe would necessarily propel Americans onto a global stage, he argued. America's economic primacy was such that the post-war world would inevitably reflect 'American' values of free enterprise, competitive commercialism and unfettered consumption. The global diffusion of an 'American way of life' was, Luce insisted, an inherently desirable evolution, a 'logical' aspiration for all people on the planet. The global projection of American culture was, for Luce, the culmination of the traditional American dream, a twentieth-century extension of the 'manifest destiny' which had compelled an earlier generation of restless Americans to spread out across the face of their new continent and create a single continental-scale power from the Atlantic to the Pacific, 'from sea to shining sea' (Meinig 1993, 1999).

There is a vast literature on the nature and implications of post-1945 American 'globalism', much of which is expertly reviewed in the other chapters in this volume.[1] The purpose of this chapter is to add an historical perspective; to explore the intellectual roots of America's global aspirations and the historical origins of the kind of global geopolitical imagination articulated by Luce in

1941 (McDougall 1997). The chapter focuses on the early, faltering and ultimately unsuccessful attempts by a small but influential group of East Coast
American intellectuals to formulate a new and distinctive American geopolitical vision of the world beyond their own continent. Particular attention will be
focused upon American perspectives on the crisis-stricken continent of Europe
during and immediately after the First World War. The 1914–18 war (into which
the USA was drawn in April 1917) was the first test for America's still hesitant
global ambitions, an early opportunity for expansionists to examine their
nation's potential as a global power and its capacity to influence the wider
geopolitical order. US contributions to the debate about the future political landscape of Europe after the First World War represented the first sustained attempt
by American intellectuals to devise a blueprint for a restructured Europe, to
develop and promote 'New World' solutions to the political problems of the
'Old World'.

'Tell Me What Is Right': The First World War and the House Inquiry

While the conversion of America's political elite to a more assertive global
geopolitics had to wait the mid-century crisis of 1939–45, the conditions which
might have spawned such an expansive vision were already in place at the last
fin-de-siècle. The great American historian of that age, Frederick Jackson Turner,
claimed that the cultural and political values of Americans, their characteristic
individualism and love of liberty, had been forged in the pioneering wilderness
of the unfolding American frontier. Many American commentators saw the
'closure' of the western frontier, symbolized by the intercontinental railway
connections between eastern and western seaboards, as the sublime resolution
of the nation-building project; as an opportunity for Americans peacefully to
develop the wide open spaces of the continent they now controlled. However,
a small but vocal minority in political, military and academic life interpreted
continental closure as merely a beginning, an invitation to seek out new frontiers beyond the self-imposed geopolitical limits of the Monroe Doctrine which
had restricted US interest to its own continent since 1823. 'Powerful and persuasive minds', as Williams (1982, 101–2) puts it, began to concern themselves
with the need 'to transform the idea . . . of continental empire into global
empire'. America's unique strength and high moral purpose stemmed directly,
it was claimed, from its constantly expanding frontier. America was an inherently expansive and ultimately global project. This view was poetically (*sic*)
expressed as early as 1867 by Secretary of State Frederick W. Seward in the
wake of the US purchase of Alaska:

> Our nation with united interests blest
> Not now content to poise, shall sway the rest;
> Abroad our empire shall no limits know,
> But like the sea in boundless circles flow.[2]

An America which stood still, Seward and his followers implied, which looked inwards rather than outwards, was doomed ultimately to decline.[3] This would be disastrous for the USA, to be sure, but also for the entire world which would be deprived of the unique advantages of an American way of life. John Locke had famously observed in the 1690s that 'In the beginning, all the world was America' (Locke 1967, 343; Arneil 1996). For American expansionists writing two centuries later, Locke's remark was in need of serious revision: 'In the end, all the world will be America.'

This was a minority view, to be sure, but its advocates drew sustenance from the astonishing dynamism of the American economy. Fuelled by new forms of industrial production and business innovations which culminated with the 'Fordist' revolution of the early 1900s, the American economy expanded at a breath-taking pace, quickly overspilling its continental limits. By 1913, American factories were producing a third of the world's industrial goods, more than twice what it had been in 1880 and a dominance which Britain had claimed a mere four decades earlier. The expansionist agenda, powerfully expressed by the American naval historian A. T. Mahan in his *The Influence of Seapower upon History* (1890), began to win increasing support in Washington in the last years of the nineteenth century. A critical turning-point came in 1898, when a previously isolationist Senate ratified the establishment of a protectorate in Cuba and a formal colony in the Philippines, the imperial spoils of the US victory in the preceding Spanish–American War (LaFeber 1993, 129–82). This paved the way for a steady overseas expansion of the American imperium into the Pacific, the Caribbean and Central America through the early years of the twentieth century. As Carville Earle (1999) reminds us, in 1860 the USA had not a single overseas territory; by 1920, 120,000 American troops controlled eight US protectorates and overseas colonies with a combined population which exceeded that of the western third of the USA itself.

The establishment of a continental-scale nation-state in the USA changed the ground rules which had previously governed the world geopolitical order. The new 'geopolitics'[4] which emerged during the 1890s and 1900s (as exemplified by the writings of Halford Mackinder, Rudolf Kjéllen and Friedrich Ratzel) was partly an attempt by European imperial theorists to grapple with the challenges and implications of a new, twentieth-century world order in which other continental-size states seemed destined to emerge, particularly in Europe and Asia. Such debates sustained, and were themselves intensified by, Germany's pitch for European hegemony from the early 1890s and by the resulting shifts in the European system of national alliances. This transformed a multipolar arrangement into a much less stable bipolar order in which Britain, France and Russia sought to impose an encircling alliance on Germany and its allies, Austria-Hungary and the Ottoman Empire (Heffernan 1998, 49–56).

These were the conditions which precipitated the firestorm of the First World War, the first global and industrial war. The isolationist lobby in the USA (which included several senior Republicans as well as powerful newspaper barons such as William Randolf Hearst) ensured American neutrality during the first 20 months of the war, though the tide of public opinion inexorably shifted towards the anti-German, 'pro-war' campaign in which American expansionists played a leading part. By 1916, the majority of the American public sided openly with the Allied cause and the bulk of American wartime exports to Europe were destined for either Britain or France. The sinking of the American liner *The Lusitania*, the interception of the so-called 'Zimmerman' telegraph outlining German geopolitical ambitions in central America, and the decision of the German High Command to wage unrestricted submarine warfare against all ships supplying enemy states together tipped the balance in favour of American mobilization. On 6 April 1917, six months after the re-election of President Woodrow Wilson, the USA declared war on Germany. Conscription was introduced the following month and the first American troops arrived in France in July. By the spring of 1918, when the German army launched its last-ditch attempt to break the deadlock on the Western Front, a quarter of a million American soldiers were arriving in Europe each month. By the Armistice on 11 November 1918, the American Expeditionary Force under General Pershing (which fought as a separate army independent of the joint Allied command) held over 20 per cent of the Allied front-line trenches. The two million American troops in France outnumbered even their British allies. Over 100,000 Americans died during the war and its immediate aftermath, the majority succumbing to the influenza pandemic rather than the fighting.

Woodrow Wilson, a man of strong Presbyterian principles, had an idealistic (some would say naive) faith in the power of reason and logic to solve even the most intractable problem. He earnestly believed that it was America's destiny to 'make the world safe for democracy'. Although he had taken his country into the war on the Allied side, Wilson insisted that the US would act as an honest, disinterested and objective arbitrator between rival European powers in the post-war negotiations. The war, tragic though it undoubtedly was, represented America's 'coming of age', he argued, an opportunity for the US to demonstrate to the 'Old World' its maturity and power (May 1960). Once American military prowess had helped to win the war, the nation's unique intellectual resources would then be mobilized to solve the problems of its 'parent' continent.

One of the most telling illustrations of Wilsonian idealism was 'The Inquiry' (known generally as the 'House Inquiry'), which Wilson established immediately after the declaration of war as a fact-finding, geopolitical 'think-tank' linked to (but independent of) the State Department. The Inquiry was charged with devising a distinctive American perspective on Europe's myriad geopolitical problems and was to become one of the most exhaustive and ambitious exercises in geographical and historical data collection ever attempted (Gelfand 1963; Walworth 1976). Its more familiar name derives form Colonel Edward Mandell House, one of Wilson's senior military advisers, who was given the task of overseeing the initial creation of the agency (George and George 1964). House's role was minimal, however, and he simply handed responsibility to a small team of academics who were instructed to recruit experts in different fields and prepare the ground for a future peace conference with the European powers, both allied and enemy. According to the wordy recollections of Sidney Edward Mezes, the Inquiry's Director (who was also House's brother-in-law), the rationale was simple (Mezes 1921, 6):

> The President felt that the United States was especially in need of such specialists at the [future Peace] Conference because of its traditional policy of isolation and the consequent lack, in its government departments, of a personnel thoroughly conversant, through intimate contact, with the inter-relations and internal composition of the European and Asiatic powers and their various dependencies.

The Inquiry's rather startling objective was the collection of an enormous corpus of historical, economic, environmental and ethnic data on the major European geopolitical problems. Ultimately, this was to become a catalogued, cross-referenced archive of stark, objective and incontrovertible fact, a mobile data repository which could be shipped across to Europe where it was generally assumed that a Peace Conference would eventually take place. Where the Europeans would rely on their familiar partisan arguments, the Americans would bring clear and indisputable facts to cut through the cant, hypocrisy and deception which had so frequently undermined European agreements of the past.[5]

Given these inductive and rigorously empirical objectives, the easiest way to organize the Inquiry's archive seemed to be according to geographical region. Partly for this reason, the decision was taken in November 1917 to move the Inquiry from its cramped offices in the New York Public Library to the elegant and spacious headquarters of the American Geographical Society (AGS) on Broadway, north of Central Park in Manhattan. The move was designed to underscore the Inquiry's scientific objectivity while also allowing access to the Society's unrivalled map collection, built up over the preceding 66 years and one of the finest in North America. The newly installed Director of the AGS was an ambitious 36-year geographer, Isaiah Bowman, subsequently President of Johns Hopkins University and an influential foreign policy guru through the inter-war years. Bowman had transformed the fortunes of the AGS since his arrival in the summer of 1915, increasing the membership from 1,000 to nearly 3,000 by the end of 1916. According to his friend and co-worker on the Inquiry, Charles Seymour (subsequently President of Yale University), Bowman became 'the presiding genius' behind the Inquiry 'by force of [his] initiative, personality and administrative skill'. Technically, however, Bowman was only the Inquiry's Secretary and 'Chief Territorial Specialist' under the directorship of first Walter Lippmann and then Mezes. Nevertheless it was Bowman who 'initiated and supervised the collection of documents . . . [and was] responsible for the selection of men to work on each aspect of the collective endeavour' (Seymour 1951, 2).

By the beginning of 1918, 150 academics were working for the various regional sections of the Inquiry, collecting data and writing briefing reports (table 8.1).[6] Weekly meetings of the regional subgroups produced lists of the maps, books, journals and pamphlets that were deemed necessary. These were either bought or borrowed from libraries across the USA and in Allied

Table 8.1 The House Inquiry

Director	**Dr S. E. Mezes**	College of the City of New York
Chief Territorial Specialist	**Dr Isaiah Bowman**	American Geographical Society
Regional Specialists:		
For the northwestern frontiers	**Dr Charles H. Haskins**	Harvard University
For Poland and Russia	**Dr R. H. Lord**	Harvard University
For Austria-Hungary	**Dr Charles Seymour**	Yale University
For Italian boundaries	**Dr W. E. Lunt**	Haverford College
For the Balkans	**Dr Clive Day**	Yale University
For Western Asia	**Dr W. L. Westermann**	University of Wisconsin
For the Far East	**Captain S. K. Hornbeck**	United States Army
For Colonial Problems	**Mr George L. Beer**	formerly of Columbia University
Economic Specialist	**Dr A. A. Young**	Cornell University
Librarian and Specialist in History	**Dr James T. Shotwell**	Columbia University
Specialist in Boundary Geography	**Major Douglas Johnson**	Columbia University
Chief Cartographer	**Professor Mark Jefferson**	State Normal College, Ypsilanti, Michigan

countries. H. A. Bumstead, the American military attaché in London, was given the difficult task of touring the capital's libraries and private collections in search of information. Money was no obstacle and Bumstead regularly offered bemused librarians substantial sums for the temporary loan of maps, journals or books. Confronted with an unhelpful response from the Secretary of the Royal Geographical Society, Bumstead requested permission for a team of stenographers from the American Embassy to visit the Society's library to transcribe selected articles which were unavailable in the US. In the early summer of 1918, Douglas Johnson, Professor of Physiology at Columbia and the Inquiry's boundary specialist, was despatched to London and Paris in search of more sensitive intelligence information. A 15-page letter from Johnson to Bowman dated 9 May 1918 details Johnson's meetings with Maréchal Joffre (the former Allied Commander in Chief) and Léon Bourgeois (head of the Service Géographique of the French army) in Paris followed by interviews with Sir William Tyrell (the Permanent Chief Secretary at the Foreign Office), Colonel Walter Coote Hedley (the Chief of the Geographical Section of the General Staff), Lords Rothermere and Northcliffe (the two dominant media barons) and A. J. Toynbee and R. W. Seton-Watson in London (D. W. Johnson 1918a). By early June, Johnson was able to send back to Mezes three large packages of 'detailed confidential information' which he had acquired from the French military authorities. 'I may say,' concluded Johnson, 'that they [the French military authorities] are anxious that the fact that these volumes have been transmitted to the Inquiry should not come to the attention of the military authorities of certain other countries' (D. W. Johnson 1918b).

Whatever the significance of this information, Johnson's ability to extract it from notoriously secretive military establishments was testimony both to his persistence and his public anti-German stance. As Chair of the American Rights League, he had campaigned in favour of American military support for the Allied cause through 1916 and early 1917 and had written several pro-war pamphlets, including an open letter in French about Germany's war guilt (directed to an unnamed German geologist) which was published in Paris in early 1917 (D. W. Johnson 1917a) and a 53-page diatribe which appeared in New York a few months later entitled *The Peril of Prussianism* (D. W. Johnson 1917b). The latter was the text of a lecture he delivered at the annual convention of the Iowa Bankers Association at Des Moines in June 1917.[7] It was accompanied by a remarkable series of seven maps showing the territorial expansion of Prussia/Germany which Johnson depicted as an amorphous black zone, spreading plague-like across the face of Europe beginning in the fifteenth century and ending with a nightmare future vision of a German-dominated Europe linked by the dreaded Berlin-to-Baghdad railway to a sprawling southern empire in the Middle East and the Persian Gulf (figure 8.1). America's sacred duty was to prevent this fearful prospect from coming to pass. The creation of a rival land empire, sustained by the industrial might of Germany, was self-evidently antithetical to American interests. The 'American ideal,' wrote Johnson, 'is based on the conception that the government is the servant of its citizens and exists for their benefit.' The 'Prussian ideal,' on the other hand, 'is based on the

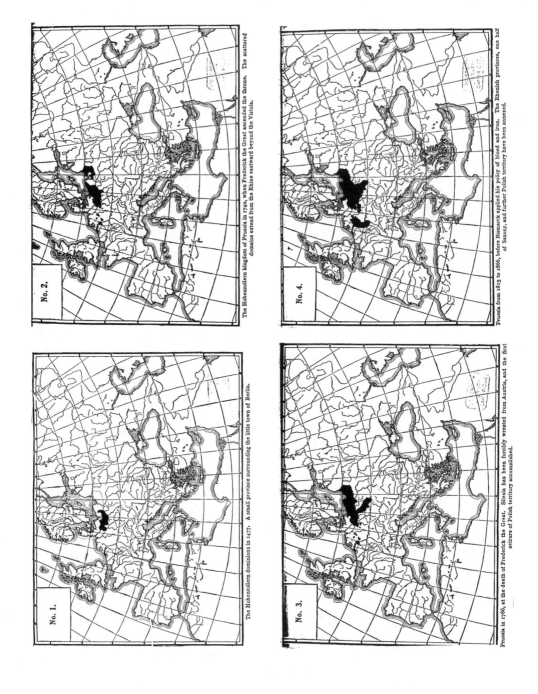

No. 1.

The Hohenzollern dominions in 1477. A small province surrounding the little town of Berlin.

No. 2.

The Hohenzollern kingdom of Prussia in 1740, when Frederick the Great ascended the throne. The scattered domains extend from the Rhine eastward beyond the Vistula.

No. 3.

Prussia in 1786, at the death of Frederick the Great. Silesia has been forcibly wrested from Austria, and the first seizure of Polish territory accomplished.

No. 4.

Prussia from 1815 to 1866, before Bismarck applied his policy of blood and iron. The Rhenish provinces, one half of Saxony, and further Polish territory have been annexed.

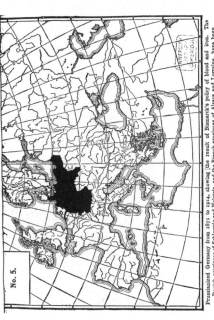

No. 5.

Prussianized Germany from 1871 to 1914, showing (the result of Bismarck's policy of blood and iron. The Danish provinces of Schleswig and Holstein, and the French provinces of Alsace and Lorraine, have been forcibly seized, and the smaller German states brought under Prussian domination.

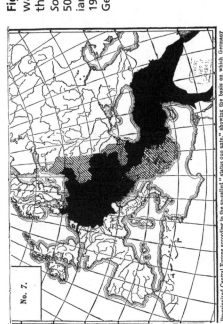

No. 7.

Prussianized Central Europe according to the so-called "status quo ante," showing the basis on which Germany is willing to end the war. Black areas show extent of absolute Prussian control; hatched areas, spheres of German influence where Prussian control would be dominant.

No. 6.

Prussianized Central Europe as it exists in the early part of 1917, showing the vast territory now completely under the domination of Prussian militarism.

Figure 8.1 Fear and loathing in a time of total war: Douglas Johnson's view of 'Prussianism' through the ages

Source: Johnson 1917b, opp. pp. 2, 10, 18, 26, 34, 42 and 50 respectively. These maps depict the extent of 'Prussian' influence in 1477, 1740, 1786, 1815–66, 1871–1914, 1917, and in some dread post-war world following a German victory in Europe.

conception that the individual is the servant of the government and exists only for the benefit of the government' (D. W. Johnson 1917b, 1, 3, 5). Two great land empires based on such fundamentally opposing ideals were *mutually antagonistic, and cannot long exist in the world side by side* . . . For Americans there is but one issue from this dilemma. The Prussian ideal must perish, the American ideal must live' (D. W. Johnson 1917b, 6, 49; his emphasis).

> America has decided that splendid isolation is no longer possible in a world rendered wondrous small by the swift steamship and express train, the telegraph and telephone, the cable and wireless telegraphy. We *are* our brother's keeper, and have entered on a policy of international cooperation, first to compel a just peace, and then to preserve it undisturbed from future assaults by autocratic militarism. We have pledged our faith and must fight to make the world safe for democracy.
>
> (D. W. Johnson 1917b, iv)

A few days after the November 1918 Armistice, the Inquiry's preparatory work was brought to an end and its impressive archive was squeezed into three army trucks and loaded onto the USS George Washington which steamed out of New York harbour in early December weighed down by its massive cargo. On board were the senior members of the Inquiry team together with Wilson himself, the first American President to visit Europe during his term of office. The American faith in their floating archive was summed up years later by Bowman:

> Where the experts of [other] nations came fully stocked with ideas, they did not have the mass of information assembled in flexible, workable form. Only the US delegation had such a resource, and we anticipated that this would give us a negotiating advantage even over the French, in whose capital city the fate of Europe and the Near East would be decided.
>
> (Bowman 1929)[8]

On 10 December, Wilson gathered a dozen of his advisers on a wind-swept deck in the mid-Atlantic, including Bowman who recorded the conversation for posterity in his notebook: 'Tell me what is right,' Wilson demanded, 'and I shall fight for it. Give me a guaranteed position' (Martin 1980, 87).

An Ideal World: The Geopolitics of Race and Language

The inductive empiricism of the Inquiry's activities reflected Wilson's claim that the USA was an open-minded, disinterested power, an honest broker between rival European nations. Maintaining this veneer of objectivity was extremely difficult, of course, for the USA had entered the war on the Allied side and senior members of the Inquiry had been in the vanguard of the anti-German, pro-war lobby. At the Peace Conference, however, every effort was made to reinforce the idea of America as a neutral, non-European power. The American delegation, headquartered in the elegant surroundings of the Hôtel Crillon on the Place de la Concorde, consistently sought independent and self-consciously American perspectives on the main geopolitical questions. The objective was to demonstrate America's firm grasp of European history, its mature and sophisticated awareness of geopolitical alternatives and, above all, its capacity to generate rational, 'scientific' solutions to Europe's geopolitical challenges. The American voice was to be everywhere reasoned, benign and distinctive.

The open, inductive nature of American thinking was presented as the inverse of the secretive, deductive and *a priori* reasoning which had hampered European international negotiations in the past. The heroic assumption was that perfectly rational solutions would logically emerge from the Inquiry's painstaking data collection and rigorous scientific method. The American faith in inductive reasoning largely explains why the Inquiry produced no single published document outlining the US position (though it would have been extremely difficult to produce such a statement in the short time available between the Armistice and the beginning of the Peace Conferences). That said, two themes dominated the American response to the European geopolitical dilemmas and strongly influenced the Inquiry's perspective on the most desirable post-war political geography – race and language. Both these topics, but particularly race, had been critical domestic considerations throughout the nineteenth century and were the source of intense and impassioned debate. The Civil War, which foreshadowed the industrial carnage of the First World War, was fundamentally a conflict about slavery and the rights of the black American population. The victory of the northern Union States appeared to reaffirm the progressive and democratic ideals which had inspired the foundation of the original American Republic. The subsequent demographic explosion during the late nineteenth century, fuelled almost entirely by successive waves of ethnically and linguistically diverse immigrants from Europe, further reinforced the comforting image of a liberal, assimilationist America, a racial 'melting pot'. At the same time, this unprecedented human influx had provoked widespread debate about the cultural, social and political implications and challenges of nation-building in such a multiracial and multilingual context. This remarkable recent history meant that by the First World War American intellectuals felt able confidently to assert their unique and progressive insights on the geopolitical significance of race and language for a reformulated Europe.

Race

The idea that racial criteria could be used 'scientifically' to define the 'natural' geopolitical limits for a reformulated Europe was immensely appealing to many American commentators. American racial theorists such as Nathaniel S. Shaler, Ellsworth Huntingdon, Madison Grant, W. Z. Ripley and Ellen Churchill Semple were of crucial importance here (Livingstone 1987; 1992, 216–59). Although it is difficult to generalize about such a diverse body of work, most American commentators tended to adopt an environmentalist and implicitly monogenetic approach to race. This posited a single, common humanity from which the immense variety of racial groups had emerged as a result of an endlessly variable external environment. As an environmentally determined attribute, race was as susceptible to change, modification and improvement as the environment in which it was fashioned. Race was a malleable, dynamic and developmental category. Improving the environment led to improvements in the race. This argument was a central tenet of the liberal, assimilationist American ideal which insisted that the millions of ethnically diverse immigrants that had flooded into the USA had, by their collective transformation of the American environment, improved and developed themselves as a result. The creation of a new, ordered environment on a new continent had spawned a new, hybrid race.

Such views were also common in Europe, of course, but they can be contrasted with a different and more quintessentially European perspective which saw racial characteristics as relatively fixed and unchanging, as essential or inherent attributes which were largely unaffected by external environmental conditions, at least in the short term. This view was connected to a polygenetic interpretation of humankind, the belief that different races were the modern descendants of discrete primeval subspecies of humanity. This was the view of racial difference advanced by the French 'father of racial ideology' Arthur de Gobineau and by Houston Stewart Chamberlain, the Germanized Englishman whose writings on 'Aryanism' so influenced Hitler and the Nazis. The implications of the polygenetic viewpoint were far more disturbing than those of the monogenetic perspective. For Gobineau, Chamberlain and their followers, there was a fixed and unchanging racial hierarchy. Certain races were simply incapable of development or improvement and, therefore, of nationhood. Such groups were destined always to remain under the suzerainty of other, superior races.

Both perspectives on race can be detected in nineteenth-century European imperial rhetoric about the non-European world. But in respect of the complex ethnic geography of Europe itself, the polygenetic viewpoint carried more obvious imperialist implications. The predominantly monogenetic racial theorizing of the American scholars on the Inquiry was therefore used to underpin an argument that the 'subject' races of Europe should be set free from the 'oppression' of tyrannical imperial regimes.

The practical problem, however, was how did one measure race? What mappable criteria could be used to establish the locations and limits of Europe's various racial groups? Ripley's work on the racial geography of Europe had posited the use of all manner of physiological and anthropological data, from skin pigmentation to phrenological indexes, to devise his simplified schema of three 'pure' European races, the Teutonic (or Nordic), the Alpine and the Mediterranean, which had intermingled to create the current complex ethnic mosaic (Ripley 1899). The enthusiasm about the geopolitical potential of racial mapping was such that some wartime commentators, both in the USA and in Europe, advocated a coordinated international survey of the physiological characteristics of Europe's population. This would be a 'scientific' prelude to the wholesale revision of the continent's borders to ensure the maximum fit between Europe's racial and political geography (H. J. Johnson 1919).

Language

Others were persuaded by non-racial indexes of national potential, notably language. One of the most intriguing documents to emerge from the Inquiry was Leon Dominian's book-length monograph, published in 1917 by the AGS, entitled *Frontiers of Language and Nationality in Europe*. Dominian, who was recruited onto the Inquiry by Bowman, insisted that the political geography of Europe should reflect the linguistic geography of the continent, which he summarized in a detailed map (figure 8.2). Language was the best measure of national cohesion because 'words express thoughts and ideals', claimed Dominian. 'The advance of civilization,' he continued,

> has been marked by the progress of nationality, while nationality has been consolidated by identity of speech . . . Men alone cannot constitute nationality. A nation is the joint product of men and ideas. A heritage of ideas and traditions held in common and accumulated during the centuries becomes, in time, the creation of the land to which it is confined. Language, the medium in which is expressed successful achievement or hardship in common, acquires therefore cementing qualities. It is the bridge between the past and the present.
>
> (Dominian 1917, 4)

Dominian argued that there were about 24 language regions in Europe which had sufficient geopolitical cohesion to justify national status. The thorny problems of bi- or even multilingual populations, concentrated as they often were in the most geopolitically disputed territories, were passed over in silence. Although he differed from the racial cartographers, Dominian's analysis exhibited the same belief in a geopolitical panacea, a touchstone which could ensure European peace and stability and which could be uncovered by the application of scientific principles based on rational, objective survey and observation.

An American Europe? The Paradox of National Self-determination

The quest for a modern 'scientific' approach to the geopolitical problems of Europe produced a general consensus on the Inquiry that a reshaped Europe must acknowledge and enshrine the right to national self-determination of a far larger number of European groups, whether defined by race, language or some other criterion. The American vision of a rejuvenated Europe was neatly summarized by the geographer Albert Perry Brigham in 1919: 'twenty-five human groups [in Europe] . . . show such unity of purpose and ideal, such community of interest, of history, and of hopes, and each in such reasonable numbers, that they have embarked or deserve to embark on a career of nationality' (Brigham 1919, 218).

The idea of national self-determination was, of course, popularly associated with Wilsonian idealism and reflected long-established American concerns with race and language. Yet there is a paradox at the heart of the American advocacy of national self-determination for the minority European peoples. This

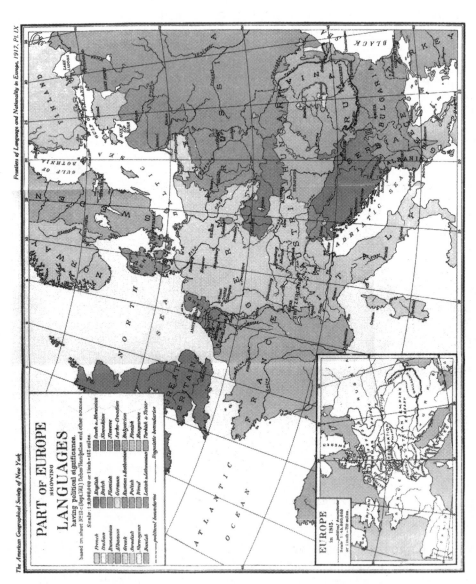

Figure 8.2 Language and national credentials: Leon Dominian's linguistic geography of Europe

Source: Dominian, 1917, opp. 334.

was advanced, to be sure, as an anti-imperial policy and was therefore consistent with America's traditional posture as an anti-imperial nation forged out of a struggle for independence from a European colonial power. But, the subsequent development of the USA had firmly rejected the idea of separate national identities within the new nation. The different immigrant races that made up the American population were to be fused into a new, hybrid people, precisely the opposite of what America seemed to be proposing in Europe. Although cloaked in the language of universal rights, America's geopolitical idealism in respect of Europe was deeply contradictory. If the American experience was genuinely to influence the creation of a new Europe, a more consistent policy (which was not to emerge in the USA until after 1945) would have been to argue for a greater measure of European economic and political integration, an inchoate United States of Europe based on the ideal of the USA.

National self-determination was, in any case, an extremely contentious issue, as it challenged the geopolitical integrity not only of sprawling multi-ethnic 'enemy' states such as Austria-Hungary and the Ottoman Empire but also Allied states, notably Britain, where the struggle for national independence in Ireland reached its bloody climax following the Easter Rising of 1916. Wilson was well aware of this and consequently shrank from using the term 'self-determination' in his 14-point plan, issued on 8 January 1918 to honour his election pledge 'to make the world safe for democracy' and to initiate what we now call 'the peace process'. The vaguer notion of 'autonomous development' was invoked in points 10 and 12 with respect to 'the peoples of Austria-Hungary' and the 'nationalities' of the Ottoman Empire respectively. Even the explicit call in point 13 for 'an independent Polish State . . . [to] include the territories inhabited by indisputably Polish populations' with its own 'free and secure access to the sea' made no explicit reference to self-determination and contained no geographical details (Henig 1995, 75–6).

Despite the elaborate preparations of the Inquiry and other advocates of national self-determination, no major European political leaders believed this idea could be applied as a universal guiding principle at the Peace Conference. The outcome of such a policy – a 'Balkanized', fragmented Europe of racially defined micro-states – was geopolitically impractical and inherently unstable, a recipe for permanent ethno-nationalist territorial squabbling. Self-determination was imposed selectively by the 'victorious' Allied powers, either to undermine the territorial integrity of the Central Powers and Bolshevik Russia or to create 'buffer-states', dependent on western European support, which were to prevent future German expansionism to the east (Poland) and south (Czechoslovakia and Yugoslavia) or Bolshevik expansion to the west (Poland, Finland and the Baltic states). These 'buffer-states' needed to be strong enough to perform their appointed task and, in the case of Czechoslovakia and Yugoslavia, large multi-ethnic arrangements were promoted rather than small, racially cohesive states. The map of Europe fashioned in 1919 was, therefore, the outcome of two logically inconsistent strategies deployed selectively by the Allies to achieve highly conventional geopolitical objectives. Germany and Bolshevik Russia were penalized territorially while Austria-Hungary and the Ottoman Empire were entirely dismembered in the name both of a traditional, punitive geopolitics and of national self-determination. The latter justification was self-serving and duplicitous for it was applied only in relation to the former Central Powers.

The American advocacy of national self-determination was also doomed by the collapse of Wilson's support at home. By the time he left for the Peace Conference, Wilson had probably narrowed his sights to a single objective: the establishment of a League of Nations. Although this was finally agreed, Wilson was unable to convince the necessary two-thirds majority of the US Senate, which had retreated into a more familiar isolationist stance, to ratify the Treaty of Versailles. The new League, which Wilson and the Inquiry had done so much to champion, was therefore left under the dominating influence of Britain and France, old imperial powers whose attitude to the internationalism which the League supposedly represented was, at best, luke-warm.

The earnest hopes of the various members of the House Inquiry were quickly dashed by their experience of European *realpolitik* in 1919. A new and much harsher realism can be detected almost immediately in the post-war proclamations of American expansionists. Soon after the appearance of his famous survey of *The New World* (Bowman 1921), Isaiah Bowman wrote to the leading French geographer Albert Demangeon (whom he had met in Paris) on 5 January 1922, lamenting the ignorance and knee-jerk isolationism of his compatriots:

> The majority of our people have never seen the ocean. It is remote and unimportant to them . . . how much more remote are the islands that lie far distant from our shores? This is a fundamental truth that no political administration can ignore. It will take a great many conferences, a great many books, and many years of newspaper writing to educate Americans to the importance of the outside world . . . We have another handicap to overcome and that is the feeling that democracy can do no wrong. In our haste to condemn the tyranny of kings we shall, I fear, forget that democracy too has its tyrannies.
>
> (Bowman 1922a)

A month later, on 9 February, Bowman wrote to Hugo W. Koehler, in the midst of Civil War in Ireland following the seccession of the 26 'southern' counties from British rule. Here he described

> the great wave of disappointment that swept the world when it was discovered, in so many different places, that the freedom that people gained from their so-called oppressors obliged them to run their own public affairs. Not only were they inexperienced in doing so; they didn't want to take the trouble . . . or they were simply too ignorant . . . Now Ireland is 'free', is it really Ireland that is free or is it England?
>
> (Bowman 1922b)

Conclusion

The implications of Bowman's comments are clear. America's early skirmish with global, and more particularly European, power politics had demonstrated an idealism bordering on naiveté mixed with a conceptual confusion, particularly about the geopolitical role of race and ethnicity. The US had also defended abstract political ideals, most notably democracy, and had largely ignored the harsh geopolitical realities of European and global affairs.[9] The failure of the American pitch for a planetary political role in 1919 reinforced in many quarters an older isolationism between the wars, but it did not entirely undermine an expansionist American agenda. The soaring ambition which had characterized America's early brush with geopolitical 'globalism' in the opening years of the twentieth century was perhaps better expressed after 1920 by the USA's steadily growing economic and cultural hegemony down to 1945. The more proprietorial form of American imperialism, hinted at in Bowman's quotations above, was by no means moribund, however, and was to have an enduring and ultimately dominant influence on the American geopolitical imagination (Ambrose 1990; Gaddis 1987; Harper 1994; Loth 1988; McCauley 1995). As Akira Iriye (1993) reminds us, the 'globalizing of America' really began with the American experience of the First World War.

Notes

1   Ambrose (1990) and Cohen (1993) are outstanding and very readable starting points.
2   Quoted in LeFeber (1993, 15).
3   Interestingly, the rhetorical preoccupation of late nineteenth-century American imperialism with the frontier, itself a reworking of classical Roman antecedents, began to influence European imperial theorists during the 1890s and 1900s. The appropriation of the Jacksonian 'frontier thesis' to the British imperial experience is perfectly illustrated by the writings of the English Tory grandee, George Nathaniel Curzon, sometime Foreign Secretary and Vice-Roy of India. See Curzon (1907) and, for related commentaries, Bassin (1993) and Kearns (1984).
4   The term 'geopolitics' was first coined by the Swedish political scientist Rudolf Kjéllen in 1899. See Parker (1998); Atkinson and Dodds (1999); Raffestin et al. 1995.
5   For a lively general discussion of the faith which has been placed in information and the 'archive' in the maintenance of state power and imperial authority, see Richards (1993).
6   Geographers were especially prominent. Apart from Bowman, Douglas Johnson and Mark Jefferson, the Inquiry also drew on the expertise of W. M. Davis, Wallace Attwood, Nevin Fenneman, Lawrence Martin and Ellen Churchill Semple, the last named being the only woman to be so employed. According to Seymour (1951, 3), the Inquiry gave a major boost to American geography: 'the sense of the relations of geography to all fields pertinent to peace spread contagiously among economists, historians and lawyers'.
7   A German version was also published 'for circulation among the large German-born population' of the USA.
8   Bowman's biographer, Geoffrey Martin, agrees with this assessment: 'The American delegation,' Martin claims, 'was the best equipped, able to be objective to a degree not possible in some of the other twenty-six delegations as the United States had not been close to the theater of war' (Martin 1980, 89).
9   The 'gap' between 'democratic ideals' and 'geographical realties' was also the theme of Mackinder's (1919) analysis of the post-1918 world.

References

Ambrose, S. E. (1990), *Rise to Globalism: American Foreign Policy since 1938* (Harmondsworth).

Arneil, B. (1996), *Locke on America* (Oxford).

Atkinson, D. and Dodds, K. (eds) (1999), *Geopolitical Traditions? Critical Histories of a Century of Geopolitical Thought* (London).

Bassin, M. (1993), 'Turner, Solov'ev, and the "Frontier Thesis": The Nationalist Significance of Open Spaces', *Journal of Modern History* 65: 473–511.

Bowman, I. (1921), *The New World: Problems of Political Geography* (New York).

Bowman, I. (1922a), Letter to Albert Demangeon, 5 Jan. (American Geographical Society, Isaiah Bowman papers).

Bowman, I. (1922b), Letter to Hugo W. Koehler, 9 Feb. (American Geographical Society, Isaiah Bowman papers).

Bowman, I. (1929), Letter to Frank Debenham, 12 July (American Geographical Society, Isaiah Bowman papers).

Brigham, A. P. (1919), 'Principles in the Determination of Boundaries', *Geographical Review* 7, 4: 210–19.

Cohen, W. I. (1993), *The Cambridge History of American Foreign Relations. Volume IV: America in the Age of Soviet Power, 1945–1991* (Cambridge).

Curzon, G. N. (1907), *Frontiers* (Oxford).

Dominian, L. (1917), *Frontiers of Language and Nationality in Europe* (New York).

Earle, C. (1999), *Space, Time, and the American Way* (Baton Rouge).

Gaddis, J. L. (1987), *The Long Peace: Inquiries into the History of the Cold War* (Oxford).

Gelfand, L. E. (1963), *The Inquiry: American Preparations for Peace 1917–1919* (New Haven).

George, A. L. and George J. L. (1964 [1956]), *Woodrow Wilson and Colonel House: A Personality Study* (New York).

Gruber, C. S. (1976), *Mars and Minerva: World War I and the Uses of Higher Learning in America* (Baton Rouge).

Harper, J. L. (1994), *American Visions of Europe: Franklin D. Roosevelt, George F. Kennan, and Dean G. Acheson* (Cambridge).

Heffernan, M. (1998), *The Meaning of Europe: Geography and Geopolitics* (London).

Henig, R. (1995), *Versailles and After, 1919–1933* (London).

Iriye, A. (1993), *The Cambridge History of American Foreign Relations. Volume III: The Globalizing of America, 1913–1945* (Cambridge).

Johnson, D. W. (1917a), *Lettre d'un Américain à un Allemand sur la guerre et les responsabilités de l'Allemagne* (Paris).

Johnson, D. W. (1917b), *The Peril of Prussianism* (New York).

Johnson, D. W. (1918a), Letter to Isaiah Bowman, 9 May (Archives of the American Geographical Society, Douglas Johnson papers).

Johnson, D. W. (1918b), Letter to Sidney Mezes, 3 June (Archives of the American Geographical Society, Douglas Johnson papers).

Johnson, H. J. (1919), 'The Anthropologist's Approach: Race, Language and Nationality in Europe', *Sociological Review* 11: 37–46.

Kearns, G. (1984), 'Closed Space and Political Practice: Frederick Jackson Turner and Halford Mackinder', *Environment and Planning D: Society and Space* 21: 23–34.

LaFeber, W. (1993), *The Cambridge History of American Foreign Relations. Volume II: The American Search for Opportunity, 1865–1913* (Cambridge).

Livingstone, D. N. (1987), *Nathaniel Southgate Shaler and the Culture of American Science* (Tuscaloosa).

Livingstone, D. N. (1992), *The Geographical Tradition: Episodes in the History of a Contested Enterprise* (Oxford).

Locke, J. (1967 [1698]), *Two Treatises on Government* (Cambridge).

Loth, W. (1988), *The Division of the World 1941–55* (London).

Luce, H. (1941), *The American Century* (New York).

Mackinder, H. J. (1919), *Democratic Ideals and Realities: A Study in the Politics of Reconstruction* (London).

Martin, G. J. (1980), *The Life and Thought of Isaiah Bowman* (Hamden, CT).

May, H. (1960), *The End of American Innocence: A Study of the First Years of Our Own Time 1912–1917* (New York).

McCauley, A. (1995), *The Origins of the Cold War 1941–1949* (London).

McCarthy, M. (1961), 'America the Beautiful', in *On the Contrary* (New York).

McCormick, T. L. (1989), *America's Half-Century* (Baltimore).

McDougall, W. A. (1997), *Promised Land, Crusader State: The American Encounter with the World since 1776* (Boston).

Meinig, D. W. (1993), *The Shaping of America: A Geographical Perspective on 500 Years of History. Volume II: Continental America, 1800–1867* (New Haven).

Meinig, D. W. (1999), *The Shaping of America: A Geographical Perspective on 500 Years of History. Volume III: Transcontinental America, 1850–1915* (New Haven).

Mezes, S. E. (1921), Preparations for peace, in E. M. House and C. Seymour (eds), *What Really Happened in Paris: The Story of the Peace Conference, 1918–1919, by American Delegates* (New York).

Parker, G. (1998), *Geopolitics Past, Present and Future* (London).

Pfaff, W. (1989), *Barbarian Sentiments: How the American Century Ends* (New York)

Raffestin, C., Lopreno, D. and Pasteur, Y. (1995), *Géopolitique et histoire* (Lausanne).

Richards, T. (1993), *The Imperial Archive: Knowledge and the Fantasy of Empire* (London).

Ripley, W. Z. (1899), *The Races of Europe: A Sociological Survey* (New York).

Seymour, C. (1951), *Geography, Justice and Politics at the Paris Peace Conference* (New York).

Walworth, A. (1976), *America's Moment: 1918 – American Diplomacy at the End of World War I* (New York).

Williams, W. A. (1982), *Empire as a Way of Life: An Essay on the Causes and Character of America's Present Predicament, Along With a Few Thoughts About an Alternative* (Oxford).

Chapter 9

European Integration and American Power: Reflex, Resistance and Reconfiguration

Michael Smith

Introduction

From the earliest days of the European integration project, one of its central preoccupations has been with the impact of US power and the generation of effective responses to that power. This chapter aims to explore the complex historical evolution of relations between the European Communities and now the European Union, and the United States, with a particular emphasis on the reception of US power in the 'new Europe'. In doing so, it takes as its starting point the inescapable economic, political and security dominance of the United States in the late 1940s, and traces the ways in which that dominance has changed and been challenged during successive decades.

The chapter takes as its central theme the interplay of three components in the reception of US power by the European integration project: reflex, resistance and reconfiguration. In the earliest days of European integration, it is possible to see the project with some accuracy as a reflex not only of US power but also of the development of the Cold War. As both the integration project and the US position in the world arena developed during the 1960s, 1970s and 1980s, it is notable that this generated among Europeans a propensity to resistance, which had important implications for the development of European integration. From the mid-1970s onwards, but particularly from the early 1980s, we are presented with a set of radical shifts both in the world arena and in transatlantic relations which can be seen as leading to a reconfiguration of essential elements in the relationship between US power and European integration. The argument of the chapter is that whilst deposits of reflex and resistance are still undeniably present in EU–US relations, there is an important sense in which reconfiguration has now shifted the foundations of the relationship.

As noted above, the structure of the chapter is essentially 'historical': moving from the initiation of the European integration project in the 1940s through a series of phases. It explores the Gaullist resistance of the 1960s, the growth of uncertainty and risk in the 1970s and its apotheosis in European responses to Reaganism during the 1980s, before focusing on the still-incomplete reconfiguration of the late 1990s. It concludes on what might seem an unexpectedly positive note: that as far as European integration is concerned, the 'American (half)century' has provided an essentially positive stimulus to redefinition and the emergence of a partially formed European identity.

The 1940s and 1950s: European Integration as American Project

It is almost trite to point to the ways in which at the end of the Second World War American power not only predominated on the world scale but also had come to replace that of the formerly leading European countries. It is rather less conventional to point to the scope of the power as opposed to its scale; it encompassed not only the central military and economic dimensions, but also – and increasingly – ideological and cultural dimensions of a novel type. This power, that is to say, was not merely hegemonic, but also pervasive. It encouraged a strong promotion of essentially modernist beliefs in societal organization, ranging through the mechanisms of government, the economy and communications from the national to the local level. Not only this, but it underpinned powerful assumptions about the relationships between democracy, capitalism and peace or progress, which were of a different type from those that had characterized continental Europe during the first half of the twentieth century. In this way, the context for the initiation of the European integration project was fundamentally shaped by the pervasiveness of US power. From the late 1940s on, it was also shaped by the perceived challenge to that power constituted by the Soviet Union and the emergent Soviet bloc (DePorte 1986).

If these foundations are taken as read, then the integration project, beginning with the European Coal and Steel Community (ECSC) established in 1951, is rightly seen as an expression of American power, and as a means of transmitting that power more effectively into the European political economy (Hogan 1987; Hoffmann and Meier 1984). Alongside the more overtly security organizations such as the North Atlantic Treaty Organization (NATO), the embryonic European Communities expressed the key elements of US power: they were based on presumptions of liberal democracy and open markets, they were essentially modernist in their focus on the creation of institutions for a market-centred society, and they reflected the need to stabilize and reconstruct the (west) European economies as part of the Cold War international system. It is often argued that the well-springs of European economic integration are to be found in security issues, particularly the desire to prevent a recurrence of war between France and Germany; but this explanation must be seen firmly in the context of the overarching requirements of US foreign policy and the Cold War system. The integration project in this perspective is essentially an American project – one of the ways in which 'hybrid Americas' based on specific forms of industrial organization and cultural convergence were extended during the 1950s, against the background of an increasingly global ideological and security competition (Ellwood 1995; van der Pijl 1984).

Was European integration just a reflex of US power, though? Even in the 1950s, when such an explanation might be at its strongest, there was significant evidence that other forces were at work. To be sure, the reception of US power both in its economic form (such as the Marshall Plan) and in its more explicitly military dimensions was fundamental to the development of the integration project; it formed the backcloth against which the European Defence Community was proposed and failed (1951–4) and against which both the European Economic Community and the European Atomic Energy Community were established after the Treaties of Rome (1957) (Heller and Gillingham 1992; Riste 1985). More than this, it also shaped the international political economy to which European institutions were inexorably subject, through the Bretton Woods system for regulation of world trade and payments (Calleo and Rowland 1973). But to focus on these dominating forces is to miss other signs. From the outset, there were present in European integration the seeds of resistance: as Alan Milward has argued, one effect of the project was to strengthen both the national and the collective capacities to shape the international milieu (Milward 1992). For France and Germany, the project was a road not only to Fordist forms of industrial development but also to the enhancement of national status and legitimacy; for smaller member states, the Communities provided a channel for resistance to the growing power of the global markets, as well as an alternative political focus in Cold War Europe. Whilst during the 1950s, these inclinations did not lead to any significant challenge to US power via the Communities, or to any marked reconfiguration of the international milieu, it is inaccurate to present the picture as one of unmitigated US dominance.

The 1960s: Integration as a Road to Resistance

The 1960s is often portrayed as the period in which European resistance to American power became both overt and effective. Its most explicit expression, that of Gaullism in France, had not been absent in the 1950s; there was in French political and cultural life a deep-seated feeling of anti-Americanism, which was then played upon by General de Gaulle in his rise to power during the later parts of the decade (Harrison 1981). As already noted, both the French and the Germans (but in different ways) saw European integration as the key to recovery of national status; for the Gaullists, this led to confrontation with US power both in the context of NATO and in the world political economy, whilst for the Germans it led to a complex balancing act between Atlanticism and Europeanism (Calleo 1965). During the early 1960s, the French first demanded parity of esteem and institutional status in NATO, and later distanced themselves from the Organization, seeing it as part of an Anglo-Saxon 'conspiracy'; a corollary of this position was continued resistance to British entry into the integration project. At the same time, they mounted a challenge to the dominance of the dollar in the international monetary system, and campaigned for stronger measures to defend Europeans against US power in international trade and forms of cultural transmission. Not all of this campaign was conducted through the European integration project: in fact, the French ended the 1960s at odds with the European institutions as well as with the Americans and the Bretton Woods system. In formal institutional terms, European integration expressed the French position rather little, except in terms of the EEC's Common Agricultural Policy (CAP), which was seen as an affront by many American agricultural interests. Attempts to construct a common European foreign policy, or to enhance monetary collaboration in the EEC, seemed stillborn not least because they were tainted by association with the Gaullist position (Serfaty 1968; Morse 1973).

Although the French were the most prominent 'resisters' of the 1960s, this did not mean that there was no inclination to resist US power among other European countries. It was during the 1960s that European integration began to exercise influence through the completion of the customs union and the erection of the Common Commercial Policy. As a result, not only did the EEC become a force to be reckoned with in international trade negotiations, it also became the focus of concerns with what in the 1990s became known as 'market access'. Quite simply, the development and dynamism of the Common Market acted as a magnet to investors and traders, and as a result, the institutional leverage of the Community was increased. The Americans found in the Kennedy Round of trade negotiations between 1963 and 1966 that the Europeans were able to dig their heels in and threaten to undermine the negotiations as a whole; they also found that for every US farmer outraged by the CAP, there was a multinational investor eager to penetrate the new European market (Calleo and Rowland 1973; Diebold 1972). Although the French and others saw the activities of US multinationals as a threat to national or collective economic management, it can also be argued that these very activities added to the legitimacy of the EEC as an institutional manager, and that in the longer term they increased the Europeans' powers of resistance.

What was clear from the 1960s, though, was that whilst the EEC had established itself as an economic force, the Europeans were incapable of collective action or commitment in the sphere of 'hard security'. Here, NATO seemed to hold unchallenged sway, with the British and the Germans among others only too eager to consolidate the US commitment it enshrined. The feeble stirrings of what came to be called European Political Cooperation (EPC) were only beginning to be felt by the end of the decade, and had no effect on the reigning preoccupations in Cold War terms: Vietnam, nuclear weapons and the US–Soviet confrontation (Grosser 1980). Whilst resistance might be possible in certain economic spheres, submission continued to be the order of the day in security matters, reflecting the pervasive penetration of Cold War values into European governing elites.

The 1970s: Beyond Resistance?

Whilst the 1960s might for some have given resistance a bad name, because of the apparent failures of Gaullism, it became apparent in the 1970s that a number of potent forces were driving the Europeans not only to contemplate resistance to US power but also more than ever to see US power as the problem rather than a potential solution in a number of international milieux. By the end of the 1960s, two crucial axes of international relations were pervaded not so much by a sense of the stability generated by US power as by a sense of the risks and uncertainties produced by an increasingly erratic America. The first of these axes was that of 'hard security'. As already noted, the European integration project had flourished at least in part because of a reluctance to get involved in such matters: the EEC's status as a 'civilian power', bereft of the military apparatus of statehood, gave it a comfortable niche in a turbulent world (Duchene 1976). But the US involvement in Vietnam, to which many European leaderships were overtly or covertly opposed, began to give the lie to this easy fiction. At the same time, and partly for the same reasons, Europeans began to see the risks inherent in US policies for the international monetary and trading systems. Most spectacularly, the declaration by President Nixon in 1971 of a New Economic Policy which entailed an effective devaluation of the dollar and various restrictive trade measures aimed at the Europeans, created an awareness that the US was a 'predatory hegemon' in the international economy, rather than the benevolent sponsor of multilateral economic institutions (Calleo 1982; Vernon 1973).

Significantly, these intimations of American mortality came at the same time – in the early 1970s – as the EEC was expanding and becoming aware of the politicization of its previously comfortable 'civilian' existence. The departure of President de Gaulle cleared the way for the entry of Britain, Denmark and Ireland into the Communities; the British in particular were supportive of increased diplomatic coordination between Community members (always with the proviso that NATO dealt with 'hard security'). Alongside this move to expand went attempts to commit Community members to Political Union and Economic and Monetary Union by the 1980s. It would be wrong to overplay the part played in any of these developments by the enhanced perception of risks emanating from US policies; many of the projects also ran at least temporarily into the sand because of the extreme turbulence of the world political economy during the 1970s. But it is at least arguable that the 1970s saw a sea-change in European reception of US power, mediated by the development of the integration project. Where on balance this power had previously been seen as reassuring, now there were at least doubts about the collateral risks entailed in easy acceptance of US predominance (Grosser 1980; Serfaty 1979). Equally, whereas previously the capacities of the Communities had been seen as essentially economic and politically self-limiting, there was now a new willingness at least to discuss the possibility of a European identity on the world stage. Indeed, the tensions created by US policies in the Middle East during and after the 1973 October War can be seen as a vital catalyst to perceptions of a new identity of interests among members of the Communities (Allen and Pijpers 1982). The informal diplomatic coordination of European Political Cooperation became progressively more explicit during the 1970s, not least as the result of US disinterest in the Conference on Security and Cooperation in Europe between 1970 and 1975; this Conference produced the Helsinki Declaration, which was to become a potent vehicle for discussion of 'soft security' issues and thus a focus around which the development of European interests could centre.

The 'New Cold War': Resistance as a Way of Life

From the mid-1970s onwards, it could be argued, the direction of European resistance to US power was determined; but it took the influence of Reaganism and the 'New Cold War' during the late 1970s and early 1980s to give it a decisive twist. The new phase did not begin with the election of President Reagan in 1980, but his presidency did denote an intensification both of the security tensions and of economic confrontation between a resurgent United States and the European integration project. As a result, awareness of difference and of a nascent European identity in security as well as economic and diplomatic issues was given a significant new impetus (Allen and Smith 1989; Freedman 1983).

The preconditions for the new awareness of European identity were established as early as the mid-1970s. The partial breakdown of the Bretton Woods system, and the end-game in Vietnam, had demonstrated not only the risks entailed in US predominance but also the limitations of American leadership. Although it was possible to argue that (for example) the oil price crisis of 1973–5 had re-established US hegemony, in fact it had demonstrated the self-serving nature of US policies as well as the limits to its leverage in areas where economics and security were closely allied (Chace and Ravenal 1976; Calleo 1982). The breakdown of the international monetary order had illustrated not only the problems of European monetary coordination, but also the extent to which US hegemony had been eroded. Although it was very difficult to see at the time, the context was one in which at least a limited reconfiguration of institutions and practices in the international political economy might be undertaken, and in which the changing security agenda had moved at least a little way away from the areas of greatest US strength.

Did this mean that the Europeans, through integration, had developed a comparative advantage in the types of international issues that were coming to predominate? This certainly did not seem to be the case as US–Soviet tensions grew during the early 1980s. Rather, the conflicts in Afghanistan, the Persian Gulf and the Horn of Africa seemed to have globalized the Cold War, thus neutralizing the EEC's regional and civilian assets. The re-nuclearization (or rather, the reorientated nuclearization) of western Europe itself, through the deployment of Cruise and Pershing missiles during the early 1980s, seemed to demonstrate that even where there was significant domestic resistance to US power, the Americans could call in their support from European governments (Joffe 1987; Kahler 1987). This was a hard-won and some might say pyrrhic victory, but as many Reaganites would argue, the prerequisite of Soviet defeat and the collapse of the Soviet economy. Where was the EEC in all of this turmoil? One answer is that on matters of 'hard security' the Communities were still the prisoners of a structure in which not only was NATO the central focus but several key EEC member states were enthusiastic co-optees to US policies. But there are more subtle and perhaps more significant answers. The increasing use of economic and technological sanctions as forms of punishment and coercion might have reinforced US power in the short term, but in the longer term it moved matters onto ground where the Communities had at least some standing and institutional clout (Allen and Smith 1989). At the same time, the persistence of east–west contacts through the Helsinki follow-up process, in which EPC continued to play a major role for Community members, brought together issues of 'soft security' at the European level, in ways which reinforced the sense of difference from US interests. Even in 'hard security' areas, the attempts by the French to reinvigorate the Western European Union (WEU) as a specifically European defence forum met with a positive if muted response from Britain and Germany (Treverton 1985).

It is possible to argue on this basis that gradually the Community dimension was being brought towards the centre of the international stage, and that the confrontational aspects of US foreign policy during the early 1980s acted as a catalyst to the enhancement of awareness of a European identity; also, that such awareness was able for some purposes to build on the institutional assets of European integration, which as in previous times provided a basis for strategies of resistance. Certainly, it is the case that the Communities provided a kind of refuge for European governments subject to attack by the Americans on grounds of their cowardice in the face of the enemy. This is not to argue that wholesale defection from the Atlantic alliance had taken place or was likely to take place; rather, it is to argue that increasingly the European integration project was seen in transatlantic terms as politicized and thus as capable of being used as the basis for differentiation from US policies that were seen as risky or counter-productive. Importantly, this tendency was accompanied by developments in the more 'civilian' parts of the Community's activities, with the inauguration of the Single Market Programme during the mid-1980s as an explicit counter to the competitive challenge of a confrontational USA and a dynamic Japan (Smith 1992). The issue was, to what extent might the habit of resistance in political and diplomatic affairs come together with the reconfiguration of the European integration project in its more conventional economic sense?

The 1980s and 1990s: Towards Reconfiguration?

If one accepts the logic of the argument so far – that the inclination to resist US power was heightened during the 1970s and 1980s by heightened perceptions of risk, and that this combined with a series of international conjunctures to create 'space' for self-assertion on the part of the EC and its members – then it is appropriate to ask whether this made of the EC a 'power' in a sense approaching that of the United States. Clearly, there is evidence that economically, the EC by the 1980s had become a 'power' on the scale of the United States; but this power was limited partly by institutional constraints and partly by the influence of the EC member states, as well as by the convenience of 'civilian power' status (Hill 1991). Above all, the limitations to a possible reconfiguration both of the EC and of the European-American system could be found in the persistence of the Cold War and of the primacy accorded to Superpower relations and security issues. Whatever else the Communities and their members might aspire to, they would not aspire to a form of separatism that would remove the security overlay of American power and – through NATO – of transatlantic institutions. There was a self-serving logic to this, sometimes decried by the US as a form of free-riding on the US nuclear and broader security guarantee (Peterson 1996).

But by the end of the 1980s, the ostensible need for an American security guarantee, and the overlay of Superpower relations, had weakened if it had not been eliminated. The disappearance of the Soviet bloc and then of the Soviet Union between 1989 and 1991 radically changed the status of transatlantic security relations, and of the EC within them. Not only this, but it chimed with a much longer-standing trend in the transformation of security issues more broadly, to which attention was drawn above. As a result, the Community appeared to many not as a reflex of US policies, or as a 'defence mechanism' against US excesses, but increasingly as the salient organizational form for the 'new Europe', both economically and in terms of security (Allen and Smith 1991–2; Smith and Woolcock 1993). The success of the Single Market Programme, the political and economic stability of the EC and its members, and the apparent marginalization of the United States in many areas of European politics meant that there were strong arguments for a radical reconfiguration of the institutions and practices of transatlantic relations. If US power was to be received into Europe, it would be so according to many on new terms; terms increasingly set by an expanding and ever more powerful EC (Peterson 1996).

It has to be said that this perception was not only current in the Community. American policy-makers, among them Secretary of State James Baker, expected in the first flush of the post-Cold War era that there would be a need in a sense to 'contain' the EC as it emerged to manage key areas of the new Europe. In December 1989, speaking in Berlin, Baker called for a new and more active EC engagement in the management of the new Europe, complementing but not replacing NATO and the CSCE. Implicit in his prescription was the need to retain those institutions in which the US had a commanding (or at least an established) presence. US power, it appeared, was to be retained despite the hopes of some that it would retreat, and despite the very real pressures from the home front of a new isolationism. One of the key elements in this strategy, as noted above, was to be the consolidation of existing transatlantic institutions and the creation of new EC–US mechanisms for consultation and coordination (Smith 1996). At the same time as this process was set in train, the European Community itself was undergoing a wide-ranging process of institutional reform, which created among other things a Common Foreign and Security Policy (CFSP) and the foundations for Economic and Monetary Union (EMU) (Rummel 1992). The roots of the first can be found to a large extent in the uncertainties created by US power and its reception in the 1980s, leading to pressures for steps going beyond the diplomatic coordination of EPC; the origins of the second lay in important ways in the 'Nixon shock' of 1971, and in sustained European efforts to insulate themselves from the irresponsibilities of US economic policy management and the resulting instability of the dollar during the 1970s and 1980s. It appeared that these developments might give the EC (to become the European Union in November 1993) enhanced credentials in precisely those areas where previously it had been most dependent on the exercise of US power: 'hard security' and the monetary dimension (Hill 1993).

The first fruits of the EC–US institutional reform process came in November 1990, with the agreement of the Transatlantic Declaration – a document which fell short of US calls for a formal treaty, but which established or consolidated a wide-ranging infrastructure of transatlantic networks (Schwok 1991). The growth of these networks in turn led by the middle of the 1990s the calls for a new institutional agreement, covering large areas of the 'new agenda' in security as well as economic affairs. As a result, in December 1995, the EU and the US agreed on a New Transatlantic Agenda of principles and aims, accompanied by a Transatlantic Action Plan listing as many as one hundred and fifty areas for 'joint action'. Many of these fell into precisely the areas where the increased politicization and political weight of the EU was likely to have significant impact, such as environmental issues, the regulation of dual-use technologies, the management of transnational criminal problems, and so on. A number of others seemed to betoken a shift in the balance of multilateral economic management, with the EU taking a role alongside the United States in the running of the reformed Bretton Woods institutions such as the newly-formed World Trade Organization (Smith 1998; 1999).

There is considerable *prima facie* evidence in this process for an institutional reconfiguration of the links between the US and the European integration project. A revitalized integration process, located in a transformed Europe, could work on different terms with a United States whose leverage and legitimacy had been eroded by a combination of its own policies and changes in the global milieu (Gompert and Larrabee 1997). But as always, there are important variations and limitations to be noted. The most dramatic of these, not surprisingly, occurred in the area of 'hard security', and the most harrowing were to be seen in the treatment of conflicts in the former Yugoslavia. Both Europeans and Americans felt at the onset of these conflicts that they were an opportunity if not a test case for European (meaning European Community/Union) initiative. It became clear during the mid-1990s that whilst the EC/EU could aspire to some diplomatic influence, and to coercive action through economic and diplomatic sanctions, its members were either unwilling or unable collectively to grasp the nettle of military involvement. Alongside the unwillingness of the Americans themselves, this had the effect of extending and intensifying the conflict; in the end, it appeared that as in the Gulf War of 1991 against Iraq, only the US could build, sustain and implement a coalition centred on military action. When it came to peace-building on the basis of a military settlement, the EU was not insignificant; but it could be argued that its influence was essentially secondary still, and in many ways still a reflex of US policies (Buchan 1993; Zucconi 1996).

A cynic might also argue that such reflexive policies were to be observed even in the handling of economic aspects of security. The EU was undeniably central to the stabilization of the 'new Europe', both through the promise of membership for ex-members of the Soviet bloc and through the provision of economic and technical assistance. Despite this centrality, many of the EU's actions demonstrated conservatism and an inward-looking logic, for example in dealing with the countries of the former Soviet Union; some drew a sharp contrast between such conservatism and the relatively rapid 'reinvention' and enlargement of NATO (Allen 1997). Political and economic crisis in Russia during 1997/8 again hit the EU on a weak spot: those who were heavily committed, such as the Germans, had a major incentive to give support, whilst those with more marginal involvement were content with an apparently more comfortable distancing from the high politics of the crisis. Further afield, despite the predictions of some US commentators that EMU would make the EU into a monetary superpower, the Europeans were cautious and often inactive in the light of Asian financial turbulence; the head of the European Central Bank was heard to opine that Europe would remain an oasis of calm whilst global melt-down went on around it.

More positive evidence for an effective reconfiguration of power relations in the world political economy, and for the role of the EU, could be found in trade relations. The Uruguay Round of trade negotiations, which ended in 1993, demonstrated not only that agreement between Americans and the EC was indispensable to the liberalization of world trade, but also that the EC, acting as the agent of the EU, could operate in a strategic fashion to shape the emerging global trading system (Woolcock and Hodges 1996; Paemen and Bensch 1995; Smith 1998). In a variety of emerging sectors, including telecommunications, information technology and financial services, the EC could act at least as the partner of the US, and often as the builder of coalitions that could overcome US reluctance to open its markets. Importantly, a number of areas in which the EC could exercise a new leverage concerned not merely trade but also cultural and political symbols that had previously been appropriated by the Americans; thus, the trade in audiovisual products was a focus of EC resistance (led by the French), and the whole area of competition policy and merger control saw a new EC influence.

The reconfiguration of power and its projection in EU–US relations remained incomplete, however. It is apparent from the argument above that whilst the EU, and the EC as its agent, could operate to shape important areas of the global political economy, by the end of the twentieth century there were still important areas in which the EU's role remained a reflex of US power, alongside areas in which it could be used as a vehicle for resistance but not as the basis for active projection of 'European' power or values. The result was a tantalizing balance between a sense of European weight and leverage (both inside and outside the Union) and a guilty feeling of continued reliance on US power, which would not simply be removed by the inauguration of EMU or by the elaboration of a 'European defence identity'.

Conclusion

This chapter has taken the view that the European integration can be seen in three ways as a phenomenon of the 'American century'. First, it can be seen as a reflex of US power and as part of global US power projection; as such, the integration project has served a vital function in the maintenance of US predominance. Second, the project can be seen as a focus of resistance to US power and its projection; as such, the project has increasingly demonstrated the capacity to create a 'Europe that can say no', but has suffered also from self-limiting forces and a form of conservative bias in the face of rapid change. Finally, the project has served as the generator of substantial but not complete reconfiguration of the power relations between the EU and the US, expressed in terms not only of institutions but also of ideas and practices. In such ways, it has gone beyond reflex and resistance, and has been able to establish itself as an alternative but not exclusive form of international agent. Its failings are many and manifest, but ironically the US has played a major role in creating the momentum which has carried it thus far, to a point at which it is no longer ludicrous to talk of 'Europe' as an international form and focus for action.

References

Allen, D. (1997) 'EPC/CFSP, the Soviet Union, and the Former Soviet Republics: Do the Twelve have a Coherent Policy?' In Regelsberger, E., de Schoutheete de Tervarent, P. and Wessels, W. (eds), *Foreign Policy of the European Union: from EPC to CFSP and Beyond*. Boulder, CO and London: Lynne Riener: 219–35.

Allen, D. and Pijpers, A. (eds) (1982) *European Foreign Policy and the Arab–Israeli Dispute*. The Hague: Nijhoff.

Allen, D. and Smith, M. (1989) 'Western Europe in the Atlantic System of the 1980s: Towards a New Identity?' In Gill, S. (ed.), *Atlantic Relations: The Reagan Era and Beyond*. Brighton: Wheatsheaf: 88–110.

Allen, D. and Smith, M. (1991–2) 'The European Community in the New Europe: Bearing the Burden of Change'. *International Journal*, XLVII (1), Winter: 1–28.

Buchan, D. (1993) *Europe: the Strange Superpower*. Aldershot: Avebury.

Calleo, D. P. (1965) *Europe's Future: The Grand Alternatives*. New York: Horizon Press.

Calleo, D. P. and Rowland, B. C. (1973) *America and the World Political Economy: Atlantic Dreams and National Realities*. Bloomington and London: Indiana State University Press.

Chace, J. and Ravenal, E. (eds) (1976) *Atlantis Lost: the United States and Europe after the Cold War*. New York: New York University Press.

DePorte, A. W. (1986) *Europe between the Superpowers*. 2nd edn. New Haven: Yale University Press.

Diebold, W. (1972) *The United States and the Industrial World: American Foreign Economic Policy in the 1970s*. New York: Praeger.

Duchene, F. (1976) 'The United States and European Community'. In Rosecrance, R. (ed.), *America as an Ordinary Country*. Ithaca: Cornell University Press: 87–109.

Ellwood, D. (1995) *America and the Reconstruction of Western Europe, 1945–1955*. London: Longman.

Freedman, L. (ed.) (1983) *The Troubled Partnership: Atlantic Relations in the 1980s*. London: Heinemann.

Gompert, D. C. and Larrabee, F. S. (eds) (1997) *America and Europe: A Partnership for a New Era*. Cambridge: Cambridge University Press.

Grosser, A. (1980) *The Western Alliance: The United States and Western Europe Since 1945*. London: Macmillan.

Harrison, M. (1981) *The Reluctant Ally: France and European Security*. Baltimore and London: Johns Hopkins University Press.

Heller, F. H. and Gillingham, J. R. (1992) *NATO: The Founding of the Atlantic Alliance and the Integration of Europe*. New York: St. Martin's Press.

Hill, C. (1991) 'European Foreign Policy: Power Bloc, Civilian Model – or Flop?' In Rummel, R. (ed.), *The Evolution of an International Actor: Western Europe's new Assertiveness*. Boulder, CO: Westview Press: 31–55.

Hill, C. (1993) 'The Capabilities–Expectations Gap, or Conceptualising Europe's International Role'. *Journal of Common Market Studies*, 31(3): 305–28.

Hoffmann, S. and Meier, C. (eds) (1984) *The Marshall Plan: A Retrospective*. Boulder, CO: Westview Press.

Hogan, M. J. (1987) *The Marshall Plan: America, Britain, and the Reconstruction of Western Europe, 1947–1952*. Cambridge: Cambridge University Press.

Joffe, J. (1987) *The Limited Partnership: Europe, the United States and the Burden of Alliance*. Cambridge, MA: Ballinger.

Kahler, M. (1987) 'The United States and Western Europe: The Diplomatic Consequences of Mr Reagan'. In Oye, K., Rothchild, R. and Lieber, R. (eds), *Eagle Resurgent? The Reagan Era in American Foreign Policy*. Boston: Little Brown: 297–333.

Milward, A. (1992) *The European Rescue of the Nation-State*. London: Routledge.

Morse, E. L. (1973) *Foreign Policy and Interdependence in Gaullist France*. Princeton, NJ: Princeton University Press.

Paemen, H. and Bensch, A. (1995) *GATT to WTO: The European Union in the Uruguay Round*. Leuven: Leuven University Press.

Peterson, J. (1996) *Europe and America in the 90s: Prospects for Partnership*. 2nd edn. London: Routledge.

Riste, O. W. (ed.) (1985) *Western Security: The Formative Years*. Oslo: Universtitatsforlaget.

Rummel, R. (ed.) (1992) *Toward Political Union: Planning a Common Foreign and Security Policy in the European Community*. Boulder, CO: Westview Press.

Schwok, R. (1991) *EC–US Relations in the Post-Cold War Era: Conflict or Partnership?* Boulder, CO: Westview Press.

Serfaty, S. (1968) *France, de Gaulle and Europe*. Baltimore: Johns Hopkins University Press.

Serfaty, S. (1979) *Fading Partnership: America and Europe after Thirty Years*. New York: Praeger.

Smith, M. (1992) ' "The Devil you Know": The United States and a Changing European Community'. *International Affairs*, 68(1): 103–20.

Smith, M. (1996) 'The United States and Western Europe: Empire, Alliance and Interdependence'. In McGrew, A. (ed.), *The United States in the Twentieth Century: Empire*. London: Hutchinson, for the Open University Press: 97–136.

Smith, M. (1998) 'Competitive Co-operation and EU–US Relations: Can the EU be a Strategic Partner for the United States in the World Political Economy?' *Journal of European Public Policy*, 5(4): 561–77.

Smith, M. (1999) 'The United States, the European Union and the New Transatlantic Marketplace: Public Strategy and Private Interests'. In Philippart, E. and Winand, P. (eds), *Decision-Making in United States–European Union Relations*. Manchester: Manchester University Press.

Smith, M. and Woolcock, S. (1993) *The United States and the European Community in a Transformed World*. London: Routledge, for the Royal Institute of International Affairs.

Treverton, G. (1985) *Making the Alliance Work: The United States and Western Europe*. London: Macmillan.

van der Pijl, K. (1984) *The Making of an Atlantic Ruling Class*. London: Verso.

Vernon, R. (1973) 'Rogue Elephant in the Forest: An Appraisal of Transatlantic Relations'. *Foreign Affairs*, 51(3): 573–87.

Woolcock, S. and Hodges, M. (1996) 'EU Policy in the Uruguay Round'. In Wallace, H. and Wallace, W. (eds), *Policy-Making in the European Union*. 3rd edn. Oxford: Oxford University Press: 301–24.

Zucconi, M. (1996) 'The European Union in the Former Yugoslavia'. In Chayes, A. and Chayes, A. H. (eds), *Preventing Conflict in the Post-Communist World: Mobilizing International and Regional Organizations*. Washington, DC: Brookings Institution: 237–78.

Chapter 10

The United States, the 'Triumph of Democracy' and the 'End of History'

R. J. Johnston

Liberal democracy and capitalism remain the essential, indeed the only, framework for the political and economic organization of modern societies.

(Fukuyama, 1995, 353)

The end of the twentieth century has been marked by United States pre-eminence as a world economic, political and military power and by the largest number of countries ever practising a democratic form of government. It is tempting to correlate the two, to argue that the widespread transition to democracy is a consequence of American hegemony and persuasion, that the country which has long had a democratic form of government has actively and successfully promoted what Fukuyama (1992) called 'the triumph of democracy' throughout the world that it now dominates.

American economic hegemony was largely confined to the second half of the twentieth century whose first half was dominated by the long decline of British power. Although the Americans were well placed to succeed after the First World War, their Congress declined to do so, leaving a 'global power vacuum' which they only filled after 1945 (Taylor, 1993). Their position was challenged for the next 45 years by the Soviet Union and its alternative view of how to organize economic and political life. The Soviet system was established at the end of the First World War, but only expanded after the Second when its alliance with the victorious 'western' nations allowed extension of its sphere of influence. It ended with the 'collapse of Communism' in the late 1980s and the adoption of democratic forms of government in most of the countries formerly identified as Communist, at the same time as many non-Communist authoritarian governments elsewhere (notably throughout Latin America) were yielding to popular pressure for democracy to be restored.

It is thus extremely tempting to equate this major transition with American world power, especially as successive United States administrations since 1945 have been much involved in rhetorical pro-democracy arguments and policies. This chapter evaluates that equation. In doing so, it focuses only on the form of government employed in a country and leaves the wider issue of democratization, which involves the development of autonomous, self-reflexive and self-determining human beings (Pateman, 1996; Slater, 1998), to be discussed elsewhere. Democracy is traditionally associated with three sets of freedoms, according to Dahl (1978):

1. In the formulation of preferences, involving the freedoms to form and join organizations, of expression, of information, to compete for votes and to stand for public office;
2. In signifying preferences, through free and fair elections; and
3. In the equal weighting of preferences, and their relationship to policy-making.

These three, and especially the second, are often termed the principles of liberal democracy, and dominated American attempts to promote democracy during the second half of the twentieth century. Despite wider concerns over human rights – especially during the Carter Presidency (1977–80) – there was less attention to the much wider range of issues involved in democratization: the American goal, it seems, was to ensure that countries were run by popularly elected governments. Whether this was because of a deep belief in the merits of democracy, or merely because it best suited American interests, is central to the evaluation that follows, which looks in turn at how the United States took up promoting the democratic cause, at the absence of any clear logic underpinning its linkage with economic development, and at the success of its policies.

Creating United States World Hegemony and the Preconditions for Democracy

In 1947, President Truman told Congress that its country's role was to promote its conceptions of democracy and freedom:

> At the present moment in world history nearly every nation must choose between alternative ways of life. The choice too often is not a free one. Our way of life is based upon the will of the majority, and is distinguished by free institutions, representative government, freedom of speech and religion, and freedom from political oppression. The second way of life is based upon the will of a minority forcibly imposed upon the majority. It relies upon terror and oppression ... I believe it must be the policy of the United States to support free peoples who are resisting attempted subjugation by armed minorities or by outside pressures ... The free peoples of the world look to us for support in maintaining their freedoms. If we falter in our leadership, we may endanger the peace of the world – and we shall surely endanger the welfare of this Nation. Great responsibilities have been placed upon us by the swift movement of events.

Such sentiments have been reflected in American foreign policy ever since, in its creation of regional security pacts (such as NATO, SEATO and ANZUS), for example, plus its formulation and application of domino theory to justify involvement in several parts of south-eastern Asia in the 1950s to 1970s to stem the advancing tide of Communism.

Resisting the tyranny of the minority involved promoting liberal democracy, as set out in what became known as Truman's Point Four, part of his January 1949 Presidential inaugural address. In words attributed to his adviser, Clark Clifford (McDougall, 1997, 181), he said that

> we must embark on a bold new program for making the benefits of our scientific advances and industrial progress available for the improvement and growth of underdeveloped areas . . . The old imperialism – exploitation for foreign profit – has no place in our plans. What we envisage is a program of development based on the concepts of democratic fair-dealing . . . Democracy alone can supply the vitalizing force to stir the peoples of the world into triumphant action, not only against their human oppressors, but also against their ancient enemies – hunger, misery and despair.

The United States was not seeking to establish a new, closed trading area over which it would exercise economic and political hegemony, as in the predecessor British Empire (indeed, it encouraged the break-up of that Empire and from 1944 blocked attempts to extend the British sphere of influence: Charmley, 1993). It was promoting a world of free states, where liberal democracy would prevail and economic prosperity advance rapidly – on the assumption that liberal democracy was a vital ingredient in the advance of economic welfare.

Less than a decade after Truman's ringing pronouncement, however, Presidential advisers such as Henry Kissinger and Walt Rostow were sceptical either that democracy was a necessary precursor to economic development, and so should be actively promoted, or even that it should necessarily follow a successful programme of economic development, whatever the existing form of government. Kissinger (1961, 290–1), for example, contended that 'In all the traditional democratic societies, the essentials of the governmental system antedated the industrial revolution', and his argument, accepted by Eisenhower, was that the main goal of foreign aid programmes should be to bolster resistance to Communism, not necessarily to advance liberal democracy.[1]

Kissinger's case was enhanced by Rostow's (1990) for the role of foreign investment as the basis for 'take-off' into 'self-sustaining growth', a cause accepted and adopted by the Kennedy administration and its successors. It was put into operation in Latin America through the Alliance for Progress aid programme, inaugurated in Chile in 1961, whose first goal was 'to improve and strengthen democratic institutions through application of the principle of self-determination by the people' (Smith, 1994, 217). The programme involved $20 billion of US investment over ten years, because of its conviction that:

> free men working through the institutions of representative democracy can best satisfy man's aspirations, including those for work, home and land, health and

schools. No system can guarantee true progress unless it affirms the dignity of the individual which is the foundation of our civilization.

Whatever its achievements in other fields, such as land reform, the Alliance failed to deliver its democratic promises – there were nine military coups against civilian governments in its first five years (Smith, 1994, 223). In part, Smith argues, this was because the programme had no clear conception of the link between economic development and democratization: it simply assumed that the former would lead to the latter, but socio-economic changes did not appear to breed democracy, at least immediately (it could be argued that the economic changes initiated in the 1960s laid the foundations for democratization in the 1990s). And when it did, many of the democratically elected governments were left-leaning and relatively incompetent; American security and economic interests in such circumstances were better met by military dictatorships.

The policy in the 1960s became what McDougall (1997) calls 'global meliorism', exemplified by Kennedy's 1951 argument that in Indochina the goal should be 'To check the southern drive of Communism . . . but not only through reliance on force of arms. The task is rather to build strong native anti-Communist sentiment' (quoted in Rostow, 1985, 61–3). This would involve showing that 'economic growth and political democracy can develop hand in hand' (quoted in Rostow, 1972, 185): the Vietnamese were to be offered 'a revolution – a political, economic and social revolution far superior to anything the Communists can offer', and India was to provide an answer to the rhetorical question 'Shall these new powerful states emerge to maturity from a totalitarian setting or from a democratic setting, built on human values shared with the West?' Kennedy's advisers were uncertain whether economic growth led to democracy or, alternatively, whether stable representative government must be present before the Rostovian take-off, however: McDougall (1997, 186) expresses their question – 'Must an alien society be modernized to prepare the ground for democracy, or would the planting of popular government suffice to flower social development?' But events were pressing, not least in Vietnam, and politicians could not wait; American policy had to be made and implemented as part of the larger goal of resisting the spread of Communism.

Democracy, Education and Economic Development

Can democracy simply be 'created in' in or 'imposed upon' a country, or are there preconditions which are at least highly desirable if not necessary for it to flourish? The previous discussion suggests that leading American political advisers were uncertain whether economic development precedes or follows the adoption of a democratic form of government. It also indicates that their main concern was primarily political, with ensuring that Communism was contained, and then economic – with creating a large market system that the United States could dominate. Democracy was of interest to the extent that it helped to deliver a secure world in which capitalism flourished and the United States predominated.

There is a massive literature, stemming from a seminal paper by Lipset (1959; see also Lipset, 1994), demonstrating a strong correlation between economic development and democratic forms of government – the more-developed countries are more likely to be democracies, and to have the best-educated populations – but little detailed evidence showing a causal link in either direction (see, for example, Londregan and Poole, 1996; Vanhanen, 1997). The residuals from the relationship are many and varied, leading to searches for reasons why transitions to democracy have either not been made or been difficult to sustain (see Haggard and Kaufman, 1995): in Latin America, for example, Rostow (1971) argued that the region's political culture (including a relative lack of voluntary cooperation and the dominance of the church and landlords over many aspects of life and livelihood) was a barrier to a permanent shift.

In reviewing this large literature, Fukuyama (1992; see also Allison and Beschel, 1991) identified no clear link flowing from economic development and education through to liberal democracy: he concluded that although the choice of democracy is undoubtedly favoured by economic development and mass education, the selection of a particular form of government is autonomous and resides in a separate desire – for dignity and what he terms *thymos* – 'the desire to be valued and recognised' (p. 163). The transition to democracy can be thwarted in some circumstances, with several cultural factors potentially assisting resistance to it:

1. National disunity within a country, because each of its constituent groups has such a highly developed sense of its own national identity that they fail to recognize the rights of the other groups;
2. Some religions, especially those which do not recognize freedom of conscience, are intolerant of different and of egalitarian claims;
3. Some cultures have class structures in which the main groups are mutually hostile and self-regarding; and
4. Some cultures are more open than others to freedom of association and a civil society which encourages voluntary organizations outwith the state superstructure.

National unity, permissive religions, open class structures and grass-roots 'self-government' are all crucial conditions within which liberal democracy is likely to flourish, therefore, leading Fukuyama to conclude that (p. 219):

> The strongest contemporary liberal democracies – for example, those of Britain or the United States – were ones in which liberalism preceded democracy, or in which freedom preceded equality. That is, liberal rights of free speech, free association and political participation in government were practised among a small elite – largely male, white, and landed – before they spread to other parts of the population. The habits of democratic contestation and compromise, where the rights

of losers are carefully protected, were more readily learned first by a small, elite group with similar social backgrounds and inclinations, than by a large and heterogeneous society full of, say, long-standing tribal or ethnic hatreds. This kind of sequencing allowed liberal democratic practice to become ingrained and associated with the oldest national traditions. The identification of liberal democracy with patriotism strengthens its thymotic appeal for newly enfranchised groups, and binds them to democratic institutions more firmly than had they participated from the start.

Such cultural conditions are not sufficient to bring liberal democracy about, however: this can only occur if the country is wisely led by people committed to democratic values (which may involve neutralizing the role of the armed forces, deeply involved in many of the twentieth-century authoritarian regimes). Nor are they necessary: stable democratic institutions have occasionally emerged in unlikely cultural milieux, such as Japan some decades ago and Soviet Russia more recently.

The United States and the Global Transition to Democracy

So has the near-world-wide transition to democracy by the end of the twentieth century been a major American contribution to the starting conditions for the next century and millennium, as a product of deliberate strategy? Has it been just a by-product of other American imperatives? Or is it merely a spurious correlation? A brief chapter can only sketch an answer, whose detail if not overall thrust must necessarily be provisional. The general argument advanced here is that creation of stable democracies requires much more than pressure from a hegemonic power, however much this is accompanied by aid packages. Furthermore, if stable democracies are hard to create and sustain, it may not be in a hegemonic power's interests, whether economic or political, to promote them, in which case the correlation identified at the outset of this chapter may indeed be spurious.

The United States continues to promote and press for democracy, however. Its apparent success is testified by data on the number of countries classified as democracies by the mid-1990s, at the end of a period termed the 'Third Wave of democratization' by Huntington (1991, 3), initiated, he claims, by the 1974 military coup against General Salazar's authoritarian regime in Portugal. Vanhanen (1997) lists 172 by the mid 1990s; O'Loughlin et al. (1998), claim that 56 per cent of all polities are now democracies and list 61 countries in Latin America, Africa and Eastern Europe which became 'more democratic' between 1978 and 1994.[2] Huntington's data for all countries with populations of more than one million only show that nearly half of them were democracies in 1990, however (figure 10.1), stressing that much of the recent change has been in the world's smaller states (IDEA, 1997, lists 211 separate independent states and related territories with democratic electoral systems).[3] American policy from the establishment of Point Four on appears to have succeeded.

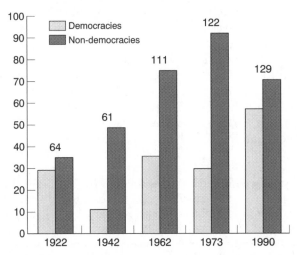

Figure 10.1 The number of democracies among countries with populations exceeding one million
Source: Huntington (1991, 26).

Huntington (1991, 45–6) identified five factors that contributed to the 'Third Wave':

1. The growing legitimacy problems of many authoritarian regimes;
2. Unprecedented global economic growth in the 1960s, which 'raised living standards, increased education, and greatly expanded the urban middle class';
3. Major shifts in the doctrine and attitude of the Roman Catholic church;
4. Changing policies of external actors, including the new attitude of the European Community towards its expansion, Gorbachev's policy shifts towards the 'Soviet empire' in the 1980s, and 'the major shift in US policies beginning in 1974 toward the promotion of human rights and democracy in other countries'; and
5. 'Snowballing' or 'demonstration effects' (see O'Loughlin et al., 1998).

Regarding the fourth, he noted that 15 of the 29 democratic regimes identified by Dahl (1978) in 1970 had been instituted during periods of foreign rule: their democracy was an 'ex-colonial inheritance'.[4] Since then 'the major sources of power and influence in the world – the Vatican, the European Community, the United States, and the Soviet Union – were actively promoting liberalization and democratization' (Huntington, 1991, 86), with American influence strongest in Latin America and Asia. He traces the origin of this shift by the American government (following the *realpolitik* era of Nixon and Kissinger, when promoting democracy was not a high priority) to the 1973–4 hearings and report of a Congressional Subcommittee on International Organizations and Movements which recommended that promotion of human rights become a major American foreign policy goal. This was accepted by Congress, which added human rights amendments to the Foreign Assistance Act, the Mutual Assistance Act, the Trade Reform Act and the International Financial Institutions Act, preventing assistance being given to countries guilty of gross violations of human rights 'unless the President found that there were compelling reason to do so' (p. 92). Later legislation extended the link between US aid and democratization: the objective in the 1989 Support for East European Democracy Act is that (Allison and Beschel, 1992; on 'conditioned aid' more generally, see Nelson, 1992 and Hook, 1998):

> The President should ensure that the assistance provided to East European countries pursuant to this Act is designed – (1) to contribute to the development of democratic institutions and pluralism characterized by: (a) the establishment of fully democratic and representative political systems based on free and fair elections, (b) effective recognition of fundamental liberties and individual freedoms, including freedom of speech, religion, and association, (c) termination of all laws and regulations which impede the operation of a free press and the formation of political parties, (d) creation of an independent judiciary, and (e) establishment of non-partisan military, security, and police forces.

A President who fully supported these goals emerged in 1976 with the election of Jimmy Carter, whose commitment to human rights was a response to the moral issues of Vietnam and Watergate: he believed not only in the need for changed attitudes and behaviour within the United States but also that reform of its foreign policies would give 'the country back its sense of purpose while correcting the manifest abuses of unjust and repressive regimes abroad, which so often gave the communist appeal its strength' (Smith, 1994, 240). Carter's pro-human rights policy was firmly embedded in his belief that the expansion of democracy overseas would aid American interests: 'authoritarian governments were poor custodians of American security interests' (Smith, 1994, 260). He was followed by Reagan, who was less concerned with promoting human rights under authoritarian regimes than with countering Communism, and he provided support for a number of non-democratic causes (such as Renamo, Unita and the Mujahadin). By the end of his first term (1984), however, he had established a National Endowment for Democracy (NED).

Reagan became convinced of the value of democracy by the success of bodies such as Solidarity in Poland: it was a means to two ends – the defeat of authoritarianism, especially Communist authoritarianism, which threatened American security; and the promotion of free-market economics, within which American businesses could thrive. Thus his administrations' policies in El Salvador, Guatemala and Honduras, for example, combined military assistance with the promotion of democracy to counter the threat of Communist expansion backed by Cuba and the Soviet Union.[5] Within Latin America generally, however, Carothers (1991, 116) has argued that 'The Reagan administration's claim that the United States played an important role in the resurgence of democracy . . . in the 1980s is largely unsubstantiated'[6]: even the later pro-democracy policies under NED 'had little real substance' (p. 115), and the concept of democracy promoted was minimal – as long as the government came to power through 'reasonably free and fair elections' (p. 117) it would be supported, irrespective of how independent that government was of major interest groups (including the military) and how much political participation there was in the country other than in the electoral process.

Huntington identifies six mechanisms by which the United States promoted democracy during this Third Wave:

1. Pro-democracy pronouncements by Presidents and other leaders, both generally and in specific circumstances and places, backed up by advocacy in media such as Voice of America and Radio Free Europe;
2. Economic pressures and sanction – in the 1990s, for example, the USAID Democracy Initiative allocated aid to countries according to their progress towards democratic targets (Shin, 1994, 165);[7]
3. Diplomatic action by 'a new breed of "freedom-pusher" US ambassadors';
4. Material support for 'democratic forces' such as Solidarity;
5. Military action (or its threat) in a small number of cases, such as the Dominican Republic (1978), Grenada (1983) – on which see also Whitehead (1991) – the Philippines (the 1989 election of Aquino), and Haiti (1995);
6. Multilateral diplomacy, as with debates on the Helsinki Accords.

But did they help? Both Huntington and Smith (1994) contend that American action in putting human rights and democratization high on the international political agenda stimulated discussion and media coverage and 'strengthened the overall global intellectual environment favourable to democracy' (Huntington, 1991, 95). In many cases, notably in Latin America, there was much evidence of American interference in the domestic politics of authoritarian regimes designed to aid pro-democratization forces, alongside what Huntington (1991, 95) terms the 'democratic version of the Brezhnev doctrine: within its area of influence it would not permit democratic governments to be overthrown'. Huntington argued that US support was critical to democratization in 10 countries during the third wave (nine of them in Latin America: the other was the Philippines) and a contributing factor in six others (only one – Bolivia – in Latin America; the others were Chile, Korea, Poland, Portugal and Taiwan): 'As with the Catholic Church, the absence of the United States from the process would have meant fewer and later transitions to democracy' (p. 98).

But why was the United States so active during this Third Wave, and was it really that effective? Was it simply a belief in the strength of a particular form of government that led President Bush, for example, to 'define our new mission to be the promotion and consolidation of democracy' (statement by Secretary of State James Baker, quoted in Huntington, 1991, 284), or was it a valid means to a national end? From 1945 on, the United States was especially concerned with the 'defeat' of Communism and its threat to the American sphere of influence, in which democracy may or may not have been the best counter: 'The end of the Cold War and of the ideological competition with the Soviet Union could remove one rationale for propping up anti-communist dictators' (Huntington, 1991, 284), not only in countries surrounding the Soviet bloc, where fear of the domino effect underlay the support for non-democratic regimes in South Korea, South Vietnam and elsewhere, but also in Latin America. There 'left-wing' governments from Jamaica to Chile, let alone Cuba, were treated with great suspicion as potential bases for communist advance towards the USA, and the cause of democracy was considered secondary to anti-Communist resistance (as illustrated by American involvement in the overthrow of Salvador Allende's popularly elected left-wing government in Chile by a military dictatorship in 1973). But the costs of the Cold War were an increasing drain on the American economy (as well as the Russian), and the Polish and Czech examples stimulated the Americans to promote the cause of democracy in the Soviet empire as a potent means of undermining its legitimacy: its success, exemplified by the fall of the Berlin Wall in 1989 and Yeltsin's ousting of Gorbachev in 1991, was unexpected in its suddenness and extent, but extremely welcome.

Democracy was a means to an end in another respect. America's economic and political strength has been built on an open world economy and increasingly liberal trade regimes: the United States has never sought to establish an Empire like its predecessor global hegemons. In recent decades it has led in building an international economic order that has fostered a massive expansion in world trade. American governments have generally believed that liberal trading arrangements, from which they will benefit, are more likely to be sustained in a system of democratic than authoritarian states, because of the wider distribution of prosperity and the larger markets that follow. But market stability is even more important, and if authoritarian regimes can deliver this better than democratic governments, American administrations are unlikely to promote the democratic cause. This is illustrated by the role of the World Bank in recent decades, of which the USA is by far the most powerful member; according to Caufield (1996, 209), it favours strong governments whatever the type of regime:

> The reason for the Bank's enthusiasm for dictators is simple: Autocratic governments are more capable of instituting and seeing through the unpopular reforms the Bank often prescribes for its clients than are democratically elected governments, which rely on the support of the public . . . [hence its] long history of demonstrating markedly more enthusiasm for dictatorships than for democracies.

To some extent, the Bank acts as a surrogate for the United States, by sustaining non-democratic regimes that might create the conditions for economic development which would benefit American trading interests. According to Cobbs (1991, 290) 'Businessmen on the whole do not "care" whether Latin America is democratic, dictatorial, or some place in between − or if they do care it does not usually affect their investment decisions': they are neutral on political issues there because this is 'a useful device for avoiding confrontation with host governments'.

The answers to the questions posed at the start of this section thus almost certainly include 'no' to the third: the relationship between American hegemony and democratic transition is not a spurious one. Regarding the other two, there is no doubt that the promotion of democracy has been part of American strategy, at least since the mid-1970s, but not simply because of an altruistic commitment to human rights, though this has been part of the foundation to American policy. The crucial issues have been the end of the Cold War and its debilitating arms race, involving the collapse of Communist regimes (notably in the Soviet Union) and the promotion of world trade through global economic development, from which American interests should benefit enormously. To these ends, the promotion of democratic regimes has been undertaken pragmatically, where it seems the best route to the goal. The defeat of communism illustrated this pragmatism: within the 'Soviet empire' support for nascent democratic movements seemed the best way of promoting the desired collapse in the 1980s; outside that bloc, authoritarian regimes were often perceived as the best bulwark to Communist advance, and democratic challenges to their legitimacy (usually though not always on economic as well as human rights grounds) were much more likely to be tolerated later than earlier.

The limits to American influence were to some extent geographic. In assessing the prospects for further democratization, Huntington (1991, 285) notes that by 1990:

> The countries in Latin America, the Caribbean, Europe, and East Asia, which were most susceptible to American influence, had, with a few exceptions, become democratic . . . The undemocratic countries in Africa, the Middle East, and mainland Asia were less susceptible to American influence.

(Which does not mean that the Americans did not try: they provided massive support to Iraq during its long conflict with Iran in the 1980s, only then to lead the fight against the 1991 Iraqi invasion of Kuwait!) Where the American model was influential it was not only because it 'stood for freedom', he claims, but also because 'it conveyed an image of strength and success'. Does it still convey that (p. 287)?

> At the end of the 1980s many argued that 'American decline' was the true reality. Others argued to the contrary. Virtually no one, however, denied that the United States faced major problems: crime, drugs, trade deficits, budget deficits, low savings and investment, decreasing productivity growth, inferior public education, decaying inner cities. People about the world could come to see the United States as a declining power, characterized by political stagnation, economic inefficiency, and social chaos. If this happened, the perceived failures of the United States would inevitably be seen as the failures of democracy. The world-wide appeal of democracy would be significantly diminished.

The third wave of democratization may then be followed by a third reverse wave (in the same way that Huntington identified a reverse wave following each of the previous two):[8] indeed, Huntington has detected the start of such a trend, with anti-democratic regimes being established in Nigeria, Sudan and Suriname (which had become democratic during the third wave), joining Burma, Fiji, Ghana, Guyana, Indonesia and Lebanon (all democratic for some time during the second wave). Could the United States prevent this? Would it want to if it did not threaten its national security and trading interests? Events in mid-1998 show that the United States is prepared to act in the former case, when it imposed economic sanctions on India and Pakistan following their nuclear test programmes,[9] but . . .

Democracy . . . What Kind of Democracy?

There is only limited evidence that the United States has had a major direct influence on the transition to democracy in a large number of countries during recent decades, therefore. Moreover, where it did achieve its goal it was relatively unconcerned over the details of the type of democratic system established. A large number of different electoral systems meets the definition of democracy as involving 'free, fair and honest periodic elections . . . in which virtually all of the adult population is eligible to vote'. The American system for electing the two houses of its legislature was inherited from British colonial rule: Representatives and Senators are elected by the first-past-the-post system (fptp) in single-member constituencies.[10] On the remainder of the American continent, however, only two other countries use that system – Belize and Canada. All but three others have proportional representation systems (IDEA, 1997). Similarly, in the former Soviet Union only one republic – Kazakhstan – has adopted fptp, with most of the others introducing either the French two-round variant of fptp or what IDEA terms the semi-proportional system of parallel fptp. Estonia and Latvia, along with all of the Eastern European ex-Communist countries except Albania, have adopted proportional systems.

This widespread adoption of electoral systems other than the American implies that countries were aiming for a different form of democracy, with (potentially) a wide range of parties elected through proportional representation leading to coalition governments and compromises among different sections within society. This contrasts with the USA – totally dominated by two catch-all parties, between which there are no fundamental ideological differences, and where contests for major offices (including the Presidency) are dominated by personalities and campaigns in which money for advertising is more important than ideas. Furthermore, United States Representatives are extremely vulnerable electorally, so much so that long-term strategy is subordinated to short-term pragmatics and vote-winning next time (King, 1997). This is not a model which other countries have emulated.

Nor have they adopted a number of other elements of the American style of democracy, where social exclusion from its procedures and benefits have been normal – *de facto* if not *de jure*. Americans were granted equal rights and equal protection under the laws after the Civil War, but over the next hundred years those were systematically eroded by administrative practices which made it extremely difficult for blacks to enrol as electors while gerrymandered constituencies diluted their voting strength. Participation in the American democratic process is relatively weak, as illustrated by low turnout rates at elections. The United States has what Galbraith (1992) calls a culture of contentment, divided between a contented majority – the socially and economically fortunate – and a discontented, functional underclass. The former dominate, not least in political life, where 'short-run public inaction, even if held to be alarming as to consequence, is always preferred to protective long-run action' (p. 20). The focus of government policy is 'social expenditure favorable to the fortunate' (p. 25); the favoured are 'greatly influential of voice and a majority of those who vote' (p. 144), leaving the unfavoured largely impotent within the political-electoral system. The latter's views can be largely ignored, since they are alienated from the political system and, apart from occasional periods of unrest which are usually easily stifled, have no impact on policy formation and political action. Social exclusion is not confined to the United States, of course, but Galbraith argues that it is much less extensive in Western European countries (other than the UK), whose democratic models are much more widely used around the world than are those of the UK and USA.

Conclusions

> Rhetoric aside, promotion of democracy has not been the principal objective of American foreign policy, is not now, and is not likely to be. When American policymakers have had to choose between promoting democracy and containing communism . . . 'security' has dominated 'values'.
>
> (Allison and Beschel, 1991, 102)

The Americans have done much promotional work for democracy during the twentieth century, therefore. Little of this has been altruistic: it has been strongly allied to a strategy designed to promote American economic power, focused on a deep belief in liberal, free-market regimes. Support for democracy elsewhere has been strongest where it has been perceived as the type of regime that would best advance American interests in the area: authoritarian regimes have been preferred wherever and whenever they have offered greater hope of resistance to Communism – seen as the greatest threat to the 'American world order' for much of the century – or/and greater potential for economic development along a trajectory favourable to American interests. Nor was it particularly successful: on its own continent, which Whitehead (1991, 357) describes as 'quite favorable terrain for the implantation of conventional liberal democracy', Lowenthal's (1991, 383) conclusion is that:

> From the turn of the century until the 1980s, the overall impact of US policy on Latin America's ability to achieve democratic politics was usually negligible, often counterproductive, and only occasionally positive. Although it is too soon to be sure, this general conclusion may hold true for the 1980s and 1990s as well. Despite Washington's current bipartisan enthusiasm for exporting democracy, Latin America's experience to date suggests that expectations should be modest.

But the issue remains an important discussion point in policy-making circles. The quotation at the start of this section comes from a contribution to a study commissioned by the United States Agency for International Development. Its authors not only concluded that it is desirable for the USA to continue, indeed enhance, its efforts to promote democracy but also identified 69 different actions that could be taken to that end because, as Lowenthal (1991, 402) put it in the Latin American context, 'effective democracy is the best insurance against guerrilla movements, terrorism, extremism, or anomic violence. The balanced economic social reforms that Latin America must undertake to develop and eventually to expand commercial exchange with the United States are more likely to be sustained under democratic conditions.' Democracy elsewhere best serves America's economic goals.

The outcome at the end of the century was more democracy than ever before, part of it at least a function of American pressure. But few newly-democratic regimes have adopted the American model of democracy, which implies that the United States' policy-makers were more concerned with the appearance (and the rhetoric of the appearance) than with the reality that underpinned it; or perhaps it was because they didn't want the reality of the American model subjected to close examination.

The Americans may well have contributed greatly to making the 'world safe for democracy', fulfilling President Wilson's dream some 80 years late, and democracy may be flourishing, but correlation, cause and reason should be carefully disentangled when seeking to interpret the late twentieth-century world political map. A more potent reason for the democratization trend over recent decades has been popular demand for the recognition that it brings, and which American democracy still denies to so many of its own population.

Acknowledgements

I am grateful to David Demeritt, Rita Johnston, David Slater and Peter Taylor for their detailed comments on drafts of this chapter, while absolving them from any responsibility for its final contents.

Notes

1 Bethell (1991, 67) quotes the Secretary of the Treasury urging Eisenhower to 'back strong men in Latin American governments: whenever a dictator was replaced Communists gained'.
2 Their data, taken from a study known as Polity III (Jaggers and Gurr, 1995), rates countries on five scales: competitiveness of political participation; regulation of political participation; competitiveness of executive recruitment; openness of executive recruitment; and constraints on chief executive.
3 Some of the differences between the various studies will reflect their definitions of democracy and choice of variables to represent that. Huntington employed a binary division (a country was either democratic or it was not), and his definition was that

the country's 'most powerful collective decision makers are selected through fair, honest, and periodic elections in which candidates freely compete for votes and in which virtually all the adult population is eligible to vote' (1991, 7).

4   According to Whitehead (1991, 356) 'democracy is quite frequently established by undemocratic means'.

5   Reagan believed that the Soviets had a 'master plan' to destabilize the USA through Central America (Lowenthal, 1991, 391).

6   Carothers (1991, 91) indicates the minimalist nature of much of the policy by referring to it as 'democracy by applause'!

7   Denying loans was used in 5 Latin American cases during the period 1913–33, but the main method deployed then was diplomatic non-recognition, used in 18 cases (see the table in Drake, 1991, 4).

8   Huntington's first, long, wave lasted from 1828 to 1926, and saw a net gain of 33 democratic countries. The reverse wave that followed (1922–42) saw a net reduction of 22, leaving only 11 fully fledged democracies. The second wave (1943–62) produced a net gain for democracy of 40 countries, but the following reverse wave (1958–75) saw 22 revert to non-democratic regimes, in some cases (such as Chile) assisted by the United States.

9   India exploded five devices in early May 1998, and Pakistan retaliated with five of its own two weeks later.

10   Although each State returns two Senators they are not elected in the same contest.

References

Allison, G. and Beschel, R. (1991) Can the United States promote democracy? In C. Wolf, ed., *Promoting Democracy and Free Markets in Eastern Europe*. Santa Monica: The Rand Corporation, 71–104.

Allison, G. T. and Beschel, R. (1992) Can the United States promote democracy? *Political Science Quarterly*, 107, 81–97.

Bethell, L. (1991) From the Second World War to the Cold War: 1944–1954. In A. F. Lowenthal, ed., *Exporting Democracy: The United States and Latin America*. Baltimore: Johns Hopkins University Press, 41–70.

Carothers, T. (1991) The Reagan years: the 1980s. In A. F. Lowenthal, ed., *Exporting Democracy: The United States and Latin America*. Baltimore: Johns Hopkins University Press, 90–122.

Caufield, C. (1996) *The World Bank and the Poverty of Nations*. London: Pan Books.

Charmley, J. (1993) *Churchill: The End of Glory – A Political Biography*. London: Hodder and Stoughton.

Cobbs, E. A. (1991) US business: self-interest and neutrality. In A. F. Lowenthal, ed., *Exporting Democracy: The United States and Latin America*. Baltimore: Johns Hopkins University Press, 264–95.

Dahl, R. A. (1978) Democracy as polyarchy. In R. D. Gastil, ed., *Freedom in the World: Political Rights and Civil Liberties*. Boston: G. K. Hall, 134–46.

Drake, P. (1991) From good men to good neighbours: 1912–1932. In A. F. Lowenthal, ed., *Exporting Democracy: The United States and Latin America*. Baltimore: Johns Hopkins University Press, 3–40.

Fukuyama, F. (1992) *The End of History and the Last Man*. London: Penguin Books.

Fukuyama, F. (1995) *Trust: The Social Virtues and the Creation of Prosperity*. London: Hamish Hamilton.

Galbraith, J. K. (1992) *The Culture of Contentment*. London: Penguin.

Haggard, S. and Kaufman, R. R. (1995) *The Political Economy of Democratic Transitions*. Princeton, NJ: Princeton University Press.

Hook, S. W. (1998) Building democracy through foreign aid: the limitations of US political conditionalities. *World Politics*, 5, 156–80.

Huntington, S. P. (1991) *The Third Wave: Democratization in the Late Twentieth Century*. Norman: University of Oklahoma Press.

IDEA (1997) *The International IDEA Handbook of Electoral System Design*. Stockholm: Institute for Democracy and Electoral Assistance.

Jaggers, K. and Gurr, T. R. (1995) Transitions to democracy: tracking democracy's third wave with Polity II data. *Journal of Peace Research*, 32, 469–82.

Johnston, R. J. (1999) Geography, fairness and liberal democracy. In J. D. Proctor and D. M. Smith, eds, *Geography and Ethics*. London: Routledge, 44–58.

Kennedy, P. (1988) *The Rise and Fall of the Great Powers*. New York: Random House.

King, A. (1997) *Running Scared: Why America's Politicians Campaign Too Much and Govern Too Little*. New York: The Free Press.

Kissinger, H. (1961) *The Necessity for Choice: Prospects for American Foreign Policy*. New York: Harper and Bros.

Lipset, S. M. (1959) Some social requisites of democracy: economic development and political legitimacy. *American Political Science Review*, 53, 69–105.

Lipset, S. M. (1994) The social requisites of democracy revisited. *American Sociological Review*, 59, 1–22.

Londregan, J. B. and Poole, K. T. (1996) Does high income promote democracy? *World Politics*, 49, 1–30.

Lowenthal. A. F. (1991) The United States and Latin American democracy: learning from history. In A. F. Lowenthal, ed., *Exporting Democracy: The United States and Latin America*. Baltimore: Johns Hopkins University Press, 383–405.

McDougall, W. A. (1997) *Promised Land, Crusader State: The American Encounter with the World since 1776*. Boston: Houghton Mifflin.

Maslow, A. (1954) *Motivation and Personality*. New York: Harper and Row.

Nelson, J. (1992) *Encouraging Democracy: What Role for Conditioned Aid?* Washington, DC: Overseas Development Council.

O'Loughlin, J. et al. (1998) The diffusion of democracy 1946–1994. *Annals of the Association of American Geographers*, 88, 545–74.

Owen, D. (1996) *Balkan Odyssey*. London: Indigo.

Pateman, C. (1996) Democracy and democratisation. *International Political Science Review*, 17, 5–12.

Parsons, T. (1964) Evolutionary universals in society. *American Sociological Review*, 29, 339–57.

Revel, J. F. (1983) *How Democracies Perish*. New York: Harper & Row.

Rostow, W. W. (1971) *Politics and the Stages of Growth*. Cambridge: Cambridge University Press.

Rostow, W. W. (1972) *The Diffusion of Power: An Essay in Recent History*. New York: Macmillan.

Rostow, W. W. (1985) *Eisenhower, Kennedy, and Foreign Aid*. Austin: University of Texas Press.

Rostow, W. W. (1990) *The Stages of Economic Growth: A Non-Communist Manifesto*, 3rd edn. Cambridge: Cambridge University Press.

Sheahan, J. (1991) Economic forces and US Policies. In A. F. Lowenthal, ed., *Exporting Democracy: The United States and Latin America*. Baltimore: Johns Hopkins University Press, 331–56.

Shin, D. C. (1994) On the third wave of democratization: a synthesis and evaluation of recent theory and research. *World Politics*, 47, 135–70.

Slater, D. (1998) The spatialities of democratization in global times. *Development*, 41, 23–9.

Smith, T. (1994) *America's Mission: The United States and the Worldwide Struggle for Democracy in the Twentieth Century*. Princeton, NJ: Princeton University Press.

Taylor, P. J. (1989) The error of developmentalism. In D. Gregory and R. Walford, eds, *Horizons in Human Geography*. London: Macmillan, 303–19.

Taylor, P. J. (1993) Geopolitical world orders. In P. J. Taylor, ed., *Political Geography of the Twentieth Century: A Global Analysis*. London: Belhaven, 31–62.

Vanhanen, T. (1997) *Prospects of Democracy: A Study of 172 Countries*. London: Routledge.

Whitehead, L. (1991) The imposition of democracy. In A. F. Lowenthal, ed., *Exporting Democracy: The United States and Latin America*. Baltimore: Johns Hopkins University Press, 356–82.

Chapter 11

Taking the Cold War to the Third World

Klaus Dodds

In his review of American post-war involvement in the Third World, Mi Yung Yoon concluded that:

> US intervention in Third World internal wars in the decades following World War II has been very common. In fact, the United States has been one of the most frequent interveners in such wars.
>
> (Mi Yung Yoon 1997: 580)

Such a conclusion is undoubtedly justifiable if one considers the term 'intervention' to encompass a range of activities such as the use of direct force, the supply of military weapons and expertise, the creation of development programmes such as *Alliance Through Progress* and the application of economic leverage through trade sanctions (see figure 11.1). In his study of 82 internal wars in the Third World between 1945 and 1989, Mi Yung Yoon concluded that there were 6 cases where the United States had intervened with direct force, 10 examples of economic intervention and 21 cases where various administrations had provided military arms and expertise. Examples of US intervention range from the widespread aerial bombing of Cambodia in the 1970s and the invasion of Grenada in 1983 to the support of particular 'resistance' movements in Nicaragua (Contras) and Mozambique (Renamo) during the 1980s (see Dunkerley 1994; Sidaway and Power 1998). However, there have also been occasions when the United States did not intervene in particular Third World internal wars such as the 1967 Nigerian Civil War because it was not considered critical to the overall Cold War struggle against the Soviet Union (see Mi Yung Yoon 1997).

These acts of intervention by the United States in the Third World did not, however, occur in an ideological vacuum. The events were located in a narrative drama which set the terms in which US activities in the Third World were judged (Agnew and Corbridge 1995: ix). According to most conventional interpretations of the Cold War, American interventions were motivated by a desire to constrain the virulent disease of Communism and the aggressiveness of the

Soviet Union. On the one side of this global conflict, we had the United States as the defender of freedom, liberty and capitalism and on the other side, the expansionist socialist republic of the Soviet Union. According to most variants of this moralistic interpretation, American involvement in the post-war Third World was determined by Soviet expansionism and it forced successive administrations in Washington to prepare plans to save the rest of the world from this relentless form of domination (see Agnew 1998: 105–19). In the process of trying to save the world from Communism, critics have argued that the United States often made cynical doctrinal adjustments (such as supporting anti-democratic governments) in order to support anti-Communist military regimes such as Argentina, Chile and racist *apartheid* South Africa; launched subversive campaigns in Latin America and Southern Africa and ignored acts of oppression and fraud by former dictators such as President Mobutu of Zaire (see Dijkink 1996: 59–71; Clarke 1997). Moreover, veteran critics such as Noam Chomsky have also argued that American military aid and acts of intervention in the Third World helped to bolster domestic primary and manufacturing interests and overseas market penetration rather than saving the Third World from a fate worse than capitalism (Chomsky 1992).

This chapter is initially concerned with the Cold War and the role of the United States in the Third World. Thereafter, explanations put forward for the various rounds of US intervention in the Third World between 1945 and 1989 are considered. It will be suggested that there has been no one overriding reason for US intervention in the African, Asian and Latin American continents, in spite of the arguments by some authors such as Samuel Huntington that the containment of the Soviet Union and the forces of Communism was the major motivating factor (cf. Huntington 1987 and 1993). In contrast, it is proposed that Cold War geopolitical discourses concerning the Third World were often used to legitimize US domestic and foreign interests but with considerable variations in the nature and scope of direct and indirect forms of intervention. The role of various economic, geographical, political and diplomatic interests are considered as mediating factors. Finally, the chapter concludes with a brief assessment of American interests in the Third World in the aftermath of the Cold War.

The United States, the Third World and the Cold War

The defining feature of the immediate post-war period was the creation of the United States as one of two major world powers. With the assistance of Western European nations such as Britain and France, the United States helped to establish a new international financial system based on the Bretton Woods agreement and a new international political forum called the United Nations. American economic and political hegemony, however, was challenged by the rise of the Soviet Union as the other global power in the aftermath of 1945. The eventual conflict between the two super-powers over competing conceptions of the

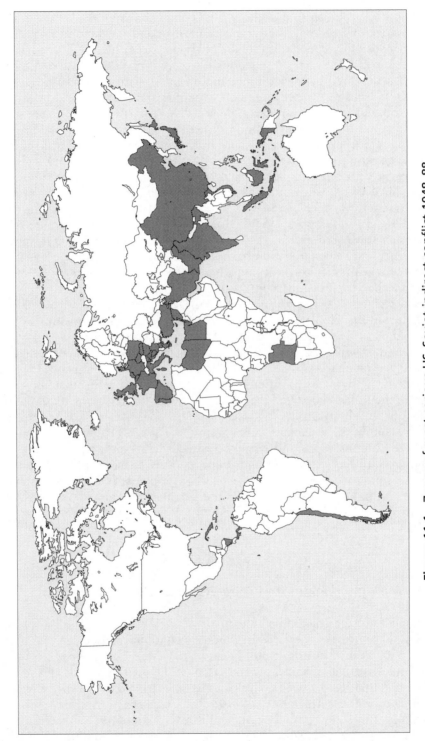

Figure 11.1 Zones of most serious US–Soviet indirect conflict 1948–88
Source: derived from Nijman 1992: 688.

international political economy and geopolitical influence had consequences for every part of the world's surface.

In a geographical vein, the Cold War originated in central Europe (epitomized by the Berlin Crisis of 1947–8), but then the centre of conflict moved abruptly to Asia in terms of events in China and Korea. The ending of the Chinese Civil War in 1949 and the onset of the Korean War in 1950 cemented, therefore, a particular geopolitical view in Washington. In both cases, the emergence of a Chinese Communist regime led by Mao Zedong and the invasion by North Korean forces of South Korea apparently confirmed the view that the Soviets and their allies were intent on geopolitical and ideological expansion. Consequently, Presidents Truman and Eisenhower argued that the US would have to play a key part in restraining this Communist offensive. With the support of the United Nations, American and other allied troops were sent to Korea for the explicit purpose of ending the civil war. After three years of bloody conflict, the wartime division of Korea was re-established at the cost of three million lives. The immediate impact of the Korean War was to harden American resolve to defend not only western Europe from would-be Communist aggression with the deployment of more troops and bombers in NATO member states such as the UK, Germany and France, but also parts of Asia and Latin America.

A new geopolitical imagination began to emerge as the conflict between the USA and the Soviet Union spread across the globe (see Ò Tuathail and Agnew 1992; Agnew 1998). Key geographical designations such as 'First World' and 'Third World' were invented by western social scientists in an attempt to highlight the profound political and social differences between the advanced industrialized countries and the recently decolonized states in Africa, Asia and the Middle East. According to some observers:

> This term [the Third World] is itself a product of the Cold War. Although it was first coined in France in the late 1940s to refer to a possible Third Estate or Third Way, it soon began to refer to those parts of the world outside the settled spheres of influence of the 'superpowers', the First and Second Worlds, in which their global conflict would be concentrated. This logic proved to be visionary. Geopolitical space was conceptualized as a threefold partition of the world that relied upon the old distinction between traditional and modern and a new one between ideological and free. Actual places became meaningful as they were slotted into these geopolitical categories, regardless of their particular qualities.
>
> (Agnew 1998: 111–12)

In this simplified geopolitical setting, the systemic-ideological conflict with the Soviet Union (and the 'Second World') was then used by the United States to justify armed intervention in Latin America and elsewhere on the basis that successive administrations were simply seeking to protect democratic and market-based societies from the evils of Communism and totalitarianism (see Brands 1998).

The 1950–3 Korean conflict then persuaded many American political commentators and leaders that the Soviet threat (and Chinese in the case of Korea) was not only substantial but that it was also going to be enduring. American financial and military aid was increasingly directed to these western-friendly states considered vulnerable to Soviet expansionism. Within the so-called Third World, states such as Taiwan, South Korea and India were given aid packages and military support. In the 1950s, for example, 5–10 per cent of national income in Taiwan was derived from US aid, whilst Ethiopia became the largest Sub-Saharan African recipient of American military aid between 1946 and 1974 because it was feared not only that the pro-western regime was vulnerable to Communist forces (up to the overthrow of the dictator Haile Selassie in 1974) but was also located close to the Suez Canal and the oil-rich Middle East (Halliday 1982; Sidaway 1998). In contrast to the frontiers of the Cold War in Europe and Asia, American allies in Latin America were encouraged to sever all ties with the Soviet Union and its allies. The Truman administration, however, also feared that Communist-inspired trade unions in Latin America might unsettle a region that not only supplied strategic raw materials but was also considered to be the geopolitical backyard of the United States. Whilst the United States provided little direct military aid, the US military developed close relations with Latin American military forces so that the latter could be assisted in the event of any Communist or pro-Soviet governments coming to power in the region (Parkinson 1990; Philips 1985).

By the 1950s, the US administration had instituted a series of policies to ensure that American interests in the Third World were protected from the threat of Soviet expansion. These ranged from the creation of security treaties such as the Rio Pact of 1947 (which involved mutual defence in the Americas) to the provision of financial and military aid. Direct intervention in the affairs of Third World states was epitomized by the involvement of the US Central Intelligence Agency providing Honduran-based rebels with funds, arms and combat training so that they could overthrow the reformist Jacobo Arbenz Guzman government of Guatemala in 1954 which had threatened to national-ize American tropical fruit interests (Immerman 1982). The emergence of the socialist leader Fidel Castro in 1959 after the Cuban revolution precipitated US military and moral support for Cuban rebels who were intent on ousting Castro during the ill-fated Bay of Pigs venture in 1961 (Johnson 1980: 5). Thereafter, Americans interests in Latin America became progressively more intervention-ist in the form of offering financial support through programmes such as the *Alliance for Progress* (which provided financial and commercial assistance to Latin American countries) to providing military training for Latin American officers in anti-Communist subversion at the School of the Americas in Panama City. It has been argued that American paranoia about Communist expansion-ism contributed to the establishment of authoritarian regimes in the region and regular intervention by Latin American militaries in domestic politics. There is no doubt that American administrations implicitly supported various military interventions in Latin America such as Brazil in 1964 and Argentina in 1962 and 1966 in the face of concerns about Communist-inspired trade unions and socialist political forces (see Philips 1985; Berger 1995). None the less, it would be wrong to assume that domestic factors in Brazil and Argentina were over-whelmed by American foreign policies and Cold War influences.

During the 1950s and 1960s, US–Soviet relations were a curious mixture of compromise and confrontation. With the emergence of the new Soviet leader Nikita Khrushchev in 1953, superpower relations were initially tense as the Soviet suppression of a reformist government in Hungary and the American curtailment of the Anglo-French attack on Egypt in the same year (1956) gave rise to fears that these conflicts might escalate into a wider conflict. The subsequent diplomatic drama over the concrete division of Berlin in 1961 and the missile crisis in Cuba in 1962 signalled the most dangerous phase of US–Soviet relations in the Cold War period. It was not uncommon for commentators at the time to be warning of a Third World War when the Americans refused to tolerate the presence of Soviet missiles on Cuban soil even though they had located similar missiles in Turkey (a member of NATO) but without taking equivalent action. After a tense stand-off, the Soviets withdrew their weapons from Cuba and then later negotiated with the American government new restrictions on nuclear testing in response to fears that a nuclear Armageddon could have occurred in October 1962 (Scott 1997: 77–8).

Unsurprisingly, the 'hotting up' of the Cold War provoked some members of the Third World to create a Non-Aligned Movement (NAM) in 1961 after an earlier 1955 anti-colonial conference in Bandung, Indonesia (Willetts 1978; Scott 1997). The latter was intended to be an international organization composed, in the main, of recently decolonized states such as India and Indonesia which sought not only to avoid the dangerous bloc politics of the Cold War but also to articulate alternative paths to modern development. One immediate consequence of events in 1961–2 (including the creation of NAM) was that the Kennedy administration became increasingly preoccupied with building better relations with potential allies in the Third World. In places such as Taiwan and South Korea, US influence (through the promotion of aid packages) was considerable given that both states had been wrecked by war and were surrounded by hostile neighbours. By the 1960s, the US persuaded these governments to adopt certain land and industrial reforms based on opening up their markets in return for US financial and military aid. Whilst larger economies such as Brazil and Chile were able to resist American advice concerning industrial policies and trading and diplomatic contacts with the Soviet Union and Cuba in the 1960s, a number of Third World states created the Group 77 in order to pressurize 'Northern' states to acknowledge the linkages between welfare, war, debt burden, environmental degradation and poverty (see Scott 1997). Other states such as the Dominican Republic were even less fortunate, as the Johnson administration ordered 20,000 US marines to land in the country and prevent a 'Communist coup' in 1965 (Dunkerley 1988). While some recent accounts have sought to play down the influence of the CIA in the overthrow of the socialist government of Salvador Allende in 1973 (see Martinez and Diaz 1996), the emergence of General Pinochet as the new military leader of Chile was also a product of domestic middle-class and business support (and the prevailing economic circumstances) for the military intervention, rather than the realization of American plans for crushing socialism in Latin America (Martinez and Diaz 1996; Ward 1997).

Overt American intervention in regions such as Latin America during the 1960s expanded further when President Johnson began to actively support anti-Communist forces in Vietnam and other neighbouring countries via a massive airlift of American troops and equipment from the mid-1960s onwards. It had been determined by US strategic thinkers and policy advisers that Vietnam was of great geopolitical significance because of its geographical location in a region (South-east Asia) proximate to China and the Soviet Union. In the period between 1965 and 1975, the United States launched numerous bombing campaigns against resistance movements in Vietnam and used chemical weapons such as napalm in their quest to clear the jungles of Communist forces. It is estimated that over 600,000 local people died in the South-east Asian region between 1969 and 1975. In spite of this massive bombardment, the US was eventually forced to retreat from Vietnam in 1973 with over 50,000 war dead because of the military stalemate.

The subsequent development of the so-called 'Vietnam syndrome' had a great impact on future American plans for direct intervention. Televisual pictures of dead American servicemen had apparently created a political culture anxious to avoid further direct entanglement with the Third World (Cummings 1992: 83–102). Furthermore, the events in Vietnam coincided with an awkward period of US–Soviet *détente* coupled with renewed tension in various parts of the Third World. On the one hand, the Nixon administration had begun a process of *rapprochement* with the USSR and China during this period culminating in restrictions on the arms race and improved sporting and cultural links with the socialist bloc. On the other hand, superpower involvement in the Third World led to the creation of the number of dangerous situations ranging from the 1973 Arab–Israeli War to support for various revolutionary movements in Southern Africa. In 1975, for example, civil war in Angola following Portuguese decolonization led to Soviet–Cuban troops supporting one faction and South African–US arms and military advice backing a rival group (Hanlon 1986; Sidaway and Simon 1993). As a consequence, the Ford and then the Carter administrations (1974–7 and 1977–80 respectively) became increasingly preoccupied with Soviet intervention in other geopolitical regions such as Southern Africa and Central America (see, for example, Wright 1997).

The ending of *détente* had devastating consequences for Europe and parts of the Third World as superpower relations plunged into the so-called Second Cold War (1979–86). In combination with the humiliation of the American forces during the hostage crisis at the American Embassy in Iran during 1978–9, American power seemed to have been exposed after the disasterous rescue attempt of the hostage which resulted in US helicopters crashing in the desert. The period between 1979 and 1981 was the most tense since the Cuban missile crisis of 1962 and coincided with a number of key developments ranging from the election of new political leaders in the USA and the UK to the Soviet invasion of Afghanistan in 1979. In the case of the latter, President Reagan and Prime Minister Thatcher used the Soviet invasion to justify the deployment of NATO nuclear missiles in Western Europe and to adopt a more confrontational approach towards the Soviet Union.

From the perspective of the Third World, President Reagan's confrontational approach had direct consequences in regions such as Central America (Ò Tuathail 1986; Dunkerley 1994; Barton 1997). Massive arms build-ups by the United States were combined with direct support for resistance movements which were perceived to be hostile to Communist and socialist governments and groupings (see Achar 1998: 98). In Nicaragua, for example, the Americans provided arms and military advice to the Contra rebels in their struggle against the socialist inspired Daniel Ortega government. In 1981–2, Reagan proposed a military budget of US$222.8 billion which was later used to support the invasion of Grenada, the military bombing of Libya and the funding of the Contra rebels in Nicaragua (Achar 1998). George Schultz, the then American Secretary of State, told the Senate Foreign Relations Committee in 1983 that the Soviet Union was engaged in 'unconstructive involvement, direct and indirect, in unstable areas of the Third World in its efforts to impose an alien Soviet model' (cited in Fisher 1997: 478).

In the early 1980s, the renewed Soviet–American confrontation had resulted, therefore, in a series of invasions, bombings and covert attacks on Third World states such as in Grenada, Nicaragua and Libya. In 1983, the downing of a South Korean airliner over Soviet airspace gave rise to new fears that a nuclear confrontation was on the near horizon. At the same time, President Reagan ordered the funding for a new programme called the Strategic Defense Initiative (Star Wars) which was supposed to involve the construction of a space-based missile protection system designed to defend the United States from intercontinental missile attack from the Soviet Union (see Clarke 1997: 176; Scott 1997: 83–4). In contrast, peace and human rights movements in Western and Central Europe were campaigning against the build-up of nuclear missile programmes and Third World states and movements were protesting against superpower military intervention and persistent economic inequalities (Thompson 1985; Kaldor 1990).

The election of President Mikhael Gorbachev to the Soviet Presidency appeared to herald a dramatic change in the Cold War. For many historians of the Cold War, Gorbachev's foreign policy coupled with domestic reforms such as *perestroika* and *glasnost* were critical in bringing the Second Cold War to an end in 1986 (see Mason 1991). His intention was to restructure relations with the West and in particular the United State by using nuclear and conventional arms reduction agreements as the cornerstone of this policy of *rapprochement*. Whilst the achievements of Gorbachev have perhaps been exaggerated in terms of his domestic reforms, there is no doubt that his emergence paved the way for an improved relationship with western powers (MccGwire 1991). The 1987 Intermediate Nuclear Forces Treaty was critical in terms of reducing the number of missile systems on the European continent. Later agreements such as the Strategic Arms Reduction Treaty in 1990 led to further cuts in long-range missiles and bombers.

In a similiar vein, the thawing of the Cold War was giving rise to further reflection over the differences between North and South rather than First, Second, and Third Worlds. The economic condition of the Third World *en masse* in the mid to late 1980s, however, was not greatly altered by the improvement of US–Soviet relations (see Halliday 1986). The demands for a NIEO had not precipitated substantial aid and technology transfers from North to South. Moreover, countries such as Nicaragua, Mozambique and Angola had been devastated by protracted civil wars, failed industrial development, rising debt burdens, environmental degradation and destroyed agricultural production systems (see Barton 1997; Dunkerly 1994; Sidaway and Power 1998). Fundamental reforms to the world economy, the international financial system and North–South trade relations were postponed by the renewal of US–Soviet competition (Clark 1997: 166–7). The disempowerment of the Third World during the 1980s was further complicated when the so-called oil shocks of the earlier decade revealed that some oil-producing Third World states were benefiting from the economic recessions of the 1980s (see Elliot 1998). As the Cold War thawed in the mid-1980s, Third World coalitions such as NAM lost political direction and began to splinter as regions such as East Asia and the Middle East developed and others such as Sub-Saharan Africa and Central America became poorer (Simon et al. 1995).

Explanations of American Involvement in the Third World (1945–89)

There has been much historical and political debate as to the rationale behind American activities in the Third World. For most of the post-war period, conservative analysts have argued that the American presence in Europe and occasional intervention in the Third World was motivated by a desire to constrain the Soviet Union as a military threat and economic competitor. The Truman Doctrine was underwritten by the geopolitical policy of containment which stated that the Soviets could only be constrained through a series of political and economic alliances as well as by the US government committing itself to a military presence in various parts of the world. For many conservative writers, therefore, the location of responsibility for the Cold War lies with the Soviets and their behaviour in Eastern Europe in the aftermath of the Second World War. From 1945 to 1950, this view of the post-war world was reinforced by 'a crisis here, a Soviet move there, and an analysis of the protagonists [in the US] which insisted that Moscow was impelled to expand and that only the United States could prevent it from achieving world domination' (Cox 1990: 30).

The division of global political space into friendly and unfriendly spaces was assisted through the deployment of two assumptions: the first, that the Soviet Union was an expansionist power hell-bent on world domination, and second, that the vulnerable regions such as South-east Asia were like dominoes and that the fall of one state to the forces of Communism would leave other geographically proximate states vulnerable. Successive Presidents from Truman to Reagan argued that friendly states in the Third World would be undermined if the US did not seek to prevent the spread of Communism from the heartland of the so-called Second World. Metaphors of disease (such as equating Communism to a virus or even a cancer) were frequently employed to convey the ineffective and slightly sinister sense of Communism (see Ò Tuathail 1992). Over time, the domino theory ensured that any local conflict in the Third World would be transformed into an expression of a global conflict between the forces of democracy and communism (see Dalby 1990).

For critics of this so-called orthodox interpretation of the Cold War, American intervention in the Third World is effectively legitimized through the employment of the Soviet Union as a monolithic and all-encompassing 'threat' against a defensive US and 'helpless' Third World (Agnew and Corbridge 1995: 76). The Cold War then became a self-fulfilling prophecy which effectively ensured that the United States had to respond to the advances of the Soviet Union by either intervening in various parts of the world or engaging in a nuclear arms race for the benefit of the 'free world'. At a discursive level, David Campbell has argued that such a viewpoint underestimated the significance of Cold War doctrines such as containment in constructing particular American identities, which then played a role in not only ensuring the promotion of 'America' as a bastion of liberty and democracy but also participated in the disciplining of sections of the American population (such as trade unions, dissenters, activists) who were labelled un-American or pro-Soviet (Campbell 1992). The discourses of the Cold War and the practices of foreign policy linked the internal life of the United States to the external behaviour of other states. Slogans such as 'reds under the bed' and high-profile anti-Communism campaigns often via television and cinema legitimized the internalization of the dangers of the Cold War (see Whitfield 1991). In the United States, therefore, foreign intervention in the Third World was often justified on the basis that this was contributing to internal defence and civil security.

The veteran critic of American foreign policy, Noam Chomsky, has argued that American intervention in the Third World was justified by a number of political strategies such as over-exaggerating the military power of the Soviet Union, promoting a massive arms build-up by the US armed forces and by depicting the Third World as vulnerable to the unrelenting advances of the Soviets. According to Chomsky, the purpose of the Cold War and American intervention in places such as the Dominican Republic and South Vietnam was to ensure American efforts to protect their economic interests and political ambitions regarding the management of a US-led global economic order. In Latin America, for example, successive governments were concerned that US supplies of raw materials and export markets would not be compromised by other American governments which might unsettle US corporate interests through radical land reforms, trade union activity and acts of nationalization.

It has also been argued that, within the field of geopolitical economy, American interventions in the Third World can only be explained in the context of the evolving post-war global political economy. According to this interpretation, the American government of 1945 was determined to resist the Soviet Union by, *inter alia*, propagating a new liberal international order based on global capitalism and open trade. The rationale for this position lay in the assumption that the expansion of the American economy was beneficial to the world-economy and vice versa. In this context, scholars such as Arturo Escobar and Wolfang Sachs have argued that the development projects imposed on the Third World in the post-war period were intimately involved with the deepening of economic and political liberalism based on policies of industrialization, limited state involvement in national economies, secular political cultures and institutional linkages with international bodies such as the World Bank and the International Monetary Fund (Escobar 1995; Sachs 1993).

In these politico-economic circumstances, the concept of intervention is far broader than just a consideration of the direct or indirect interference of the US in the Third World. US-sponsored projects based on development and modernization could be seen as an intervention in the sense that they imposed a series of policies, practices and institutions on Third World states. American ideas concerning trade, economic development and mass production/consumption played a part in the expansion of American influence within the post-war world. The struggle with the Second World served to tie American geopolitical and economic interests within a range of large economic powers such as Germany and a host of Third World industralizing states such as South Korea, Argentina and Brazil. This does not, however, imply that Third World states were simply helpless victims during this period; rather it suggests that at the very least the expansion of the world-economy was intimately tied to the geopolitical contours of the Cold War and US patterns of economic and political intervention.

Conclusions

Since the collapse of the Soviet Union and the ending of the Cold War, debates over the meaning and significance of the United States during the Cold War have persisted (see Gaddis 1997). Other scholars have pointed to the intellectual difficulties in talking about the current state of the world given the rapid political changes of the 1990s and the conceptual weaknesses inherent in Cold War categories such as First and Third Worlds (Haynes 1996). It is evident that the Third World is far too diverse and fragmented to be described in terms of one overarching geographical region. Moreover, the description of the US as the sole remaining military superpower needs to be counter-balanced by the fact that the US became the largest debtor nation in the 1980s. As two geographers have concluded, 'the break-up of the Soviet Union was not the only sign of an old order in demise: the Cold War geopolitical economy was also in disarray as mounting stagflation, indebtedness and balance of payments disequilibria clearly and successively indicated' (Agnew and Corbridge 1995: 44). (For a summary of US defence spending see figure 11.2.)

In contrast to scholars who claim to 'know' about the Cold War, most analyses of the last fifty years accept that American involvement in a global struggle with the Soviet Union was probably motivated by a mixture of reasons ranging from a commitment to containing anti-Communism to a desire to extend American economic and commercial interests in Western Europe and the Third World (cf. Gaddis 1997). Assessing American involvement during the Cold War in the Third World appears all the more difficult in a world characterized by new state

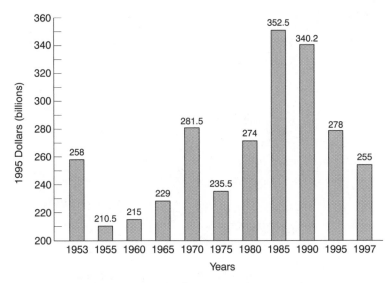

Figure 11.2   US defence spending, 1953–97

formation, new relations between processes such as globalization and regional-ism, new institutional arrangements such as intergovernmental organizations like the G8, new approaches to knowledge creation (postmodernism) and new issues such as environment and the flows of refugees (see Simon 1998; Herod, Ò Tuathail and Roberts 1998). In a world characterized by greater political and informational interconnection and transnational flows, new discourses con-cerning a forthcoming 'clash of civilizations' seem ill-equipped for understand-ing either the Cold War or the aftermath (Huntington 1993). In that sense, there is considerable room for improved dialogue between North and South over the political, economic and cultural implications of America's role in the Third World during the Cold War (see Slater 1997).

Third World writers have argued that whilst the military dimension of the Cold War has disappeared, American diplomatic, economic and political influ-ence on the affairs of Southern states remains considerable through the domi-nation of major financial institutions such as the International Monetary Fund and the implementation of Structural Adjustment Programmes (see Ayoob 1995). In his analysis of American defence spending in the 1990s, Gilbert Achar concluded that 'the current level of the US defence budget corresponds rationally to imperial expansion and exclusive global hegemony' (Achar 1998: 126). There has been no post-Cold War peace dividend according to official US Department of Defense figures which suggest that the United States military budget accounts for a third of world defence spending. Indeed, the data on defence spending suggests a percentage *increase* in the 1990s in contrast to the Second Cold War period (1980–5) under the Reagan administration. Unlike the earlier optimism about new forms of economic and political justice between North and South, many writers have concluded that the ending of the Cold War will not lead to a new international economic and political order in the twenty-first century (see Peet and Watts 1993; Sachs 1993; Simon 1998; Slater 1997).

Acknowledgements

Thanks to the editors and David Simon for their very helpful comments regarding this chapter. The usual disclaimers apply.

References

Achar, G. (1998) 'The strategic triad: the United States, Russia and China', *New Left Review* 228: 91–126.

Agnew, J. (1998) *Geopolitics: Re-visioning World Politics*. London: Routledge.

Agnew, J. and S. Corbridge (1995) *Mastering Space*. London: Routledge.

Ayoob, M. (1995) *The Third World Security Predicament: State Making, Regional Conflict and the International Sytem*. Boulder: Lynne Rienner.

Barton, J. (1997) *A Political Geography of Latin America*. London: Routledge.

Berger, M. (1995) *Under Northern Eyes: Latin American Studies and US Hegemony*. Bloomington: University of Indiana Press.

Brands, H. (1998) *What America Owes the World*. Cambridge: Cambridge University Press.

Campbell, D. (1992) *Writing Security*. Manchester: Manchester University Press.

Clarke, I. (1997) *Globalization and Fragmentation*. Oxford: Oxford University Press.

Chomsky, N. (1992) *Deterring Democracy*. London: Verso.

Clarke, M. (1997) *Globalization and Fragmentation*. Cambridge: Polity.

Corbridge, S. (1993) 'Colonialism, post-colonialism and the political geography of the Third World', in P. Taylor, ed., *Political Geography of the Twentieth Century: A Global Analysis*. London: Belhaven: 171–206.

Cox, M. (1990) 'From the Truman Doctrine to the second superpower detente: the rise and fall of the Cold War', *Jorunal of Peace Research* 27: 25–41.

Cummings, B. (1992) *War and Television*. London: Verso.

Dalby, S. (1990) *Creating the Second Cold War: The Discourse of Politics*. London: Pinter.

Dijkink, G. (1996) *National Identity and Geopolitical Visions*. London: Routledge.

Dunkerley, J. (1988) *Power in the Isthmus*. London: Verso.

Dunkerley, J. (1994) *The Pacification of Central America*. London: Verso.

Escobar, A. (1995) *Encountering Development*. Princeton: Princeton University Press.

Fisher, B. (1997) 'Toeing the hardline? The Reagan administration and the ending of the Cold War', *Political Science Quarterly* 112: 477–96.

Gaddis, J. (1982) *Strategies of Containment*. Oxford: Oxford University Press.

Gaddis, J. (1997) *We Now Know: Cold War History*. Oxford: Clarendon.

Halliday, F. (1982) *Threat from the East?* Harmondsworth: Penguin.

Halliday, F. (1986) *Making of the Second Cold War*. London: Verso.

Haynes, J. (1996) *Third World Politics*. Oxford: Blackwell.

Herod, A., G. Ò Tuathail and S. Roberts, eds (1998) *Unruly Worlds?* London: Routledge.

Huntington, S. (1987) 'Patterns of intervention: America and the Soviets in the Third World', *National Interest* 7: 39–47.

Huntington, S. (1993) 'The clash of civilizations', *Foreign Affairs* 72: 22–49.

Immerman, R. (1982) *The CIA in Guatemala*. Austin: University of Texas Press.

Johnson, J. (1980) *Latin America in Caricature*. Austin: University of Texas Press.

Kaldor, M. (1990) *The Imaginary War*. Oxford: Blackwell.

Kolko, G. (1988) *Confronting the Third World*. New York: Rantheon.

Martinez, J. and A. Diaz (1996) *Chile: The Great Transformation*. Washington, DC: Brookings Institution.

Mason, J. (1991) *The Cold War*. London: Routledge.

MccGwire, M. (1991) *Perestroika and Soviet National Security*. Washington, DC: Brookings Institution.

Morales, W. (1994) 'US intervention and the New World Order: lessons from the Cold War and post-Cold War cases', *Third World Quarterly* 15: 77–102.

Nijman, J. (1992) *The Geopolitics of Power and Conflict*. Chichester: John Wiley.

Ò Tuathail, G. (1986) 'The language and nature of the new geopolitics: the case of US–El Salvador relations', *Political Geography* 5: 73–85.

Ò Tuathail, G. (1992) 'Foreign policy and the hyper-real: The Reagan administration and the scripting of "South Africa"', in T. Barnes and J. Duncan, eds, *Writing Worlds*. London: Routledge: 155–75.

Ò Tuathail, G. (1996) *Critical Geopolitics*. London: Routledge.

Ò Tuathail, G. and J. Agnew (1992) 'Geopolitics and discourse: practical geopolitical reasoning in American foreign policy', *Political Geography* 11: 190–204.

Ò Tuathail, G., S. Dalby and P. Routledge, eds (1998) *The Geopolitics Reader*. London: Routledge.

Parkinson, F. (1990) 'South America, the great powers and the global system', in M. Morris, ed., *Great Power Relations in Argentina, Chile and Antarctica*. London: St. Martins Press: 176–96.

Peet, R. and M. Watts (1993) 'Development theory and environment in an age of market triumphalism', *Economic Geography* 69: 227–53.

Philips, G. (1985) *The Military in South American Politics*. Beckenham: Croom Helm.

Sachs, W. ed. (1993) *Global Ecology: A New Arena for Political Conflict*. London: Zed.

Scott, L. (1997) 'International History 1945–1990', in J. Baylis and S. Smith, eds, *The Globalization of World Politics*. Oxford: Oxford University Press: 71–88.

Sidaway, J. (1998) 'What is in a Gulf? From the "Arc of Crisis" to the Gulf War', in G. Ò Tuathail and S. Dalby, eds, *Rethinking Geopoltics*. London: Routledge: 224–39.

Sidaway, J. and M. Power (1998) 'Sex and violence on the wild frontiers', in J. Pickles and A. Smith, eds, *Theorising Transition: the Politcal Economy of Post-Communist Transfromations*. London: Routledge: 408–27.

Simon, D. (1997) 'Development reconsidered: new directions in development thinking', *Geografiska Annaler* 79: 183–201.

Simon, D. (1998) 'Rethinking (post)modernism, postcolonialism and posttraditionalism: South–North perspectives', *Environment and Planning D: Society and Space* 16: 219–45.

Simon, D., W. Van Spegen, C. Dixon and A. Narman, eds (1995) *Structurally-Adjusted Africa*. London: Pluto.

Slater, D. (1997) 'Geopolitical imaginations across the North–South divide: issues of difference, development and power', *Political Geography* 16: 631–53.

Smith, J. (1997) *The Cold War.* Oxford: Blackwell.

Thompson, E. (1985) *The Heavy Dancers.* New York: Pantheon.

Thompson, W. (1997) *Communism Since 1945.* Oxford: Blackwell.

Ward, J. (1997) *Latin America.* London: Routledge.

Whitfield, S. (1991) *The Culture of the Cold War.* Baltimore: Johns Hopkins University Press.

Willetts, P. (1978) *The Non-Aligned Movement.* London: Pinter.

Wolf-Philips, L. (1987) 'Why Third World? Origins, definitions and usage', *Third World Quarterly* 9: 1311–28.

Wright, G. (1997) *The Destruction of a Nation: United States Policy towards Angola since 1945.* London: Pluto.

Wolfe, A. (1984) *The Rise and Fall of the 'Soviet Threat'.* Boston: South End Press.

Yung Yoon, M. (1997) 'Explaining US intervention in Third World internal wars 1945–1989', *Journal of Conflict Resolution* 41: 580–602.

Chapter 12

US Influence in the Making of the Contemporary Amazon Heartland

Berta K. Becker and Roberto S. Bartholo, Jr

'the Amazon has been our undertaking ... The civic sense of our presence, however, will not be sufficient, in current times, to impede the aggressive presence of those who would dispute it, convinced that we are not prepared to retain it as a part of our political being. The danger must be met by executing a policy that serves as an answer to that negative conclusion.'

Arthur Cezar Ferreira Reis

An Environmental Heartland?

The rescue of the concept of *heartland*, proposed by Sir Halford Mackinder a hundred years ago, to designate the Eurasian continental mass as a foundation for the power of the Imperial State, may seem paradoxical at the dawning of the twenty-first century. Is there sense in using a concept relative to territorial attributes, in terms of their extension, texture and geographical position, as a condition for economic self-sufficiency and political autonomy of the state, in the era of globalization, contraction of space and time, and of cyberspace?

The answer is yes, considering that the concepts must be situated in the historical contexts in which they are developed. And this chapter assumes that the South-American Amazon has become an *environmental heartland.*

As was well-emphasized by Mackinder, technological advance is inherent to the concept of heartland. It was the development of steamships and railways which, while deeply altering social relationships, added strategic value to the largest landmass on the planet. Endowed with the advantage of far-flung internal circulation which permitted reaching the shores of the World island and the sea, it was, at the same time, a natural fortress, walled by mountain chains and glacial seas which rendered it inaccessible to maritime power. Thence, his famous assertion: whoever dominates the East of Europe, will dominate the heartland; whoever dominates the heartland will dominate the World island, and whoever dominates the World Island, will dominate the world.

Today, relationships and concepts are redefined by the joint action of two elements:

1. the scientific-technological revolution, which creates a new form of production, the raw materials for which are information and knowledge, transformed into sources of economic productivity and political power (Castells 1985; von Weizsäcker 1979; Bartholo 1992); and
2. the environmental crisis, perhaps the major limitation to the expansion of capitalism in conventional forms of production (Daly 1991; Taylor 1997), which imposes new patterns of relationship with the economic resources (Jonas 1979; Bartholo 1986).

Within this transformational context, nature is reassessed. On one hand, the new form of production tries to free itself of the resource base, using a lower volume of raw materials and energy, and producing new materials. The valuation of natural elements is effected on another level, conditioned by new technologies. This is especially the case of the use of nature as a source of information, vital to biotechnology, based on the decoding, reading and instrumentalization of biodiversity. But it is also the case of the as yet unsolved theoretical possibility of the use of hydrogen isotopes as an input for energy. In other words, nature is valued as capital for present or future realization and as a source of power for contemporary science (Becker 1997).

In terms of geographical, territorial space, the strategic valuation of the Amazon as a heartland, is based:

1. on territorial extension, as proposed by Mackinder, as well as the self-defence thereof, represented, in this case, by the 'Amazonian factor' comprised of the immense distances and by the coverage by the forest mass which, until now, have hindered occupation;
2. the new significance acquired by territorial extension, as a double asset: the land itself, and an immense natural capital, the maximum expression of which is biodiversity. The December 1997 Conservation International report indicates that Brazil is the country possessing the largest megadiversity on the planet. Brazilian biodiversity is the largest in the world in terms of plants, fresh-water fish and mammals, second in terms of amphibians, third in birdlife and fifth in reptiles. Biological biodiversity is a source of new genetic resources and biological active principles of major market interest and social relevance;
3. the new strategic geographical position, as a crossroads of the new North American, Asian and European power blocs;
4. its cultural identity which, based on social diversity, constitutes an inestimable source of local knowledge of tropical nature, the alternative practices of which generate new forms of use and protection of biodiversity;
5. the potential and opportunity for the world, represented by the development of alternative uses of natural resources;

6. *last but not least*, the expansion of communications and circulation (of information, money, business) permitted by telecommunications networks connecting points within the territory horizontally, and vertically, in national and transnational space.

In a cultural-symbolic representation, production from the Amazon heartland is conditioned by the centrality of biodiversity and sustainability in the world today. Since the 1970s, the question of the limits of economic development has metamorphosed into the sustainability of Earth as the *locus* of life. The Amazon has become the greater symbol of a double-faced ecological challenge – the valuation of natural capital and human survival. Several environmental movements and non-governmental organizations (NGOs) have spread their nets in the heartland thanks to telecommunications, decisively penetrating the planetary image.

The Amazon has been the subject of an ancient dispute. Independent Brazilians had to face with dialogue and diplomacy the accusations of the North American press that Brazil damaged the interests of humanity by developing a closed-door policy for the Amazonian space. And the contemporary Amazon, already an urbanized forest, has a socio-environmental web, in a large measure directed from abroad, without full knowledge of the democratic Brazilian government and society.

In this new situation the influence of the United States combines a two-faced strategy, employing both veiled and explicit coercive forces. The first face of the American strategy aims at an appropriation of the decison-making process of economic use of natural capital (e.g. non-governmental organizations acting as apparently independent agents, but in fact implementing the American strategy). The second face aims at a direct intervention in territory (e.g. the installation of military bases in South America as part of the War on Drugs).

The final outcome of this process depends on the Brazilian capacity for negotiation in the face of the contemporary North American strategy in a world where the ecological factor becomes part of the global geopolitical agenda. Various are the states and organizations ready to 'defend' the environmental heartland, reflecting the scenario of contemporary globalization. But this does not invalidate paraphrasing Mackinder: whoever dominates the environmental heartland will dominate the natural capital of the future.

An Ancient Dispute

The beginnings of settlement

Efforts to integrate the Amazon in the European civilizational field began with Francisco Orellana's adventurous excursion in 1541. In accordance with the Treaty of Tordesillas, the region was almost entirely Spanish, the limiting meridian passing through what is today the city of Belém. The first exploratory incursions were British and Dutch. Trusting in the Iberian Union sealed by a common

Royal House, the Spaniards ignored the titles conferred by the Treaty of Torde-sillas. Portuguese and Portuguese-Brazilians, arriving from the north-east, having settled in Presépio, the original nucleus of Belém, situated at the mouth of the great valley, began fighting the British and Dutch and initiated territo-rial occupation along its tributaries. Although the Council of Portugal, meeting on 6 April 1615, advised Philip III of the threat, the Spanish Council of State left the defence of the gateway and course of the Amazon exclusively to the Portuguese. With the restoration of Portugal as an independent nation, the Spaniards were obliged to take bitter notice of the consequences of such negligence.

Explorers, soldiers, civil and ecclesiastical authorities flowed from Presépio, spreading north, west and south. An entire network of fortifications was built. Spices from India began to be replaced by the 'drugs of the wilderness', typical of the forest (cocoa, clove, piassava, annatto, cinnamon, oleaginous seeds, timber, etc.), as well as sugar, tobacco, cotton and coffee.

The door left ajar

The Portuguese implemented a policy combining arms and diplomacy, with a view to guaranteeing them sovereignty over the Amazon Basin. Their great efforts to develop systemized information on the territory throughout the sev-enteenth and eighteenth centuries are characteristic. Missionaries, soldiers and explorers participated in this 'grand inquiry'. In parallel, Portugal closed the Amazon to foreigners. This attitude was not mere obscurantism on the part of a 'retrograde' country feeling threatened by Illuminism. It was part of an Imperial strategy for guaranteeing sovereignty.

In respect of the use of the Amazonian rivers, the Portuguese Empire took drastic measures, prohibiting the free circulation of any vessels. After Brazilian independence, the question acquired new outlines: It was the era of steamships and, in the North American press, Brazil is accused of damaging the interests of humanity by developing a closed-door policy for the Amazonian space. The Brazilian solution was the incorporation of a steamship company with Brazilian capital, 'for purposes of mercantile circulation and the civilization, opening and settlement of the Amazonian wilderness' (Ferreira Reis 1960, p. 91).

Independent Brazil had to dose its 'closed-door' policy with dialogue and diplomacy. In a revolutionary century, in which a pragmatic science was emerg-ing, committed to bringing human well-being to its full potential, the curiosity of foreign scientists and researchers fell heavily on the Amazon.

The great cycle of scientific expeditions opened with J. B. von Spix and K. F. Ph. Von Martius in 1820, and J. von Natterer between 1829 and 1835. Special mention should be made of the relationship between the scientific expeditions and the transplant, by the British, of rubber tree seedlings to plantations in the east, the root of the economic decadence of the Brazilian Amazon. Without the support activities of scientific expeditionaries such as Dr R. Spruce (1849–64), it is doubtful that the enterprise would have been successful.

In short, the cycle of great scientific expeditions is intertwined with economic and geopolitical interests. This permitted a biographer of R. Spruce to affirm that the travelling naturalists were precursors 'who placed a guide-book of the riches of the Amazon at the disposal of big business' (Ferreira Reis 1960, p. 110).

The 'Hiléia Amazônica' (Amazon Rainforest) Institute

In the period following the Second World War, the project for the creation of the 'International Institute for the Hiléia Amazônica' attained great notoriety. Its history is exemplary. The preparatory seminar, held in London by UNESCO, awakened a great interest in the scientific community. In August 1947 the 'preliminary conference' was installed in Belém, and prepared a detailed plan of the studies to be entrusted to the Institute, involving a broad spectrum of programmes which, in the language of the time, focused on the economic utilization of biodiversity, human development, traditional populations, and indigenous populations in particular.

The international agreement for the creation of the Institute was signed at Iquitos in April 1948, *ad referendum* of the national governments. Submitted to Congress as a presidential message, the 'Iquitos Convention' encountered resistance, which saw in it an instrument for the 'internationalisation of the Amazon'. Specific doubts were raised as to the Institute interfering with national sovereignty, and a deliberation for seeking an opinion from the General Staff of the Armed Forces was passed. The General Staff declared that the scientific objectives pursued by Brazil 'could be fully realised without risk to our sovereignty had care been taken to leave the executive functions, within each country, in the hands of a national agency' (Ferreira Reis 1960, p. 211).

In view of the new situation, the Brazilian Ministry of Foreign Affairs obtained, from the signatory nations, an Additional Protocol with explicit safeguards referring to 'the objectives of pure and uncompromised research of the Institute, within the most scrupulous respect for the sovereignty of each of the countries in whose territory it is authorised to carry out its activities, in accordance with their laws and regulations' (Ferreira Reis 1960, p. 213).

The fears of the nationalists were well-founded, in that the hidden agenda of the Institute was the internationalization of the Amazon, which emphasized that once constituted, it could 'acquire, possess, and raise assets, contract and undertake obligations, receive donations and contributions, move funds, create and manage scientific centres and, in general, undertake all legal acts necessary to both purposes and functions' (Ferreira Reis 1960, p. 215). In these attributions, the nationalists indentified foreign colonization under the rhetorical guise of utilizing the Amazon for the universal good.

The fact is that abroad, the 'Iquitos Agreement' was received as an opportunity for expansion of capital and populations, raising the curtain on possibilities heretofore unknown not only to other governments interested in the

·settlement of the Amazon, but also to an entire new host of consortia, organizations, partnerships and even individual persons.

Congress, after being informed of the General Staff's position and of the Additional Protocol of the Foreign Ministry, opted, due to the seriousness of the situation, for the suspension of any further decision.

The programmed network

Between 1965 and 1980 the military regime brought the 'frontier economy' to its peak. This is a paradigm under which progress is deemed to be infinite growth and prosperity based on the exploitation of natural resources perceived as being equally infinite (Boulding 1966). To this end, they developed a spatial technology of double (technical and political) control, the 'programmed network', comprised of an integrated complex of programmes and plans. Multiple frontiers were opened, the largest being the Amazon, the assimilation of which was considered a maximum priority (Becker, 1992).

The modernization process being under its direct control, the federal government implemented extensive spatial integration networks – roads, telecommunications, urbanity, hydroelectricity and railway – promoted huge mining projects, subsidized the flow of capital for private acquisition of land, and induced migration (Becker, 1992). This strategy, strongly supported by financial resources provided by the IDB and, especially, the IBRD, resulted in generalized conflict – involving land and environmental issues – which were later exacerbated by the emerging financial and political crises of the state and the mounting environmentalist pressure. However, in spite of all the turbulence, the Amazon was no longer the same: it became an urbanized forest, with 60 per cent of its 20 million inhabitants concentrated in urban nuclei situated along the major river and land circulation routes.

At the end of the 1980s, two major regional transformation vectors were already emerging, expressing the transitional structure of the region and the state (Becker 1995):

1. The techno-industrial vector (TIV), comprising projects of groups interested in the mobilization of natural resources (especially minerals and timber) and in new markets for the electrical-electronic industry, located in Manaus, as well as in the implementation of appropriate infrastructure;
2. The techno-ecological vector (TEV), involving two types of projects. The conservationist, communitary 'from the bottom up' alternatives, each one being realized in a given ecosystem, by populations of distinct ethnic and/or geographic origins, with different social, productive and political structures, and each, as a survival strategy, allying itself with transnational networks in diverse partnerships.

The TEV created a new socio-environmental web, envolving indigenous territories (20 per cent of the region's total area), conservation units (6 per cent) and isolated community experiments. The web, in a large measure directed from abroad, without full knowledge of the Brazilian government and society, fuels the risk of fragmentation of the heartland. But the outcome of this process depends on the capacity for dialogue and diplomacy which may come to be established by Brazil, transforming the pressure from the TEV into an inducement to change to a pattern of sustainable and nationally sovereign development, the definition of which is to be established during the process of negotiation itself (Becker 1997).

The Two Faces of the Contemporary Strategy of the American Government

The production of the environmental heartland is not exclusively the work of the United States. Europe in general – particularly Germany – actively participates in this action. The discourse in favour of the defence of the heartland includes: protection of biodiversity, containment of deforestation, and 'aid' for the traditional populations, with the significant omission of the necessity for mechanisms designed to protect their 'cultural capital'. The age-old heritage of knowledge and technical procedures for the instrumentalization of Amazonian biodiversity, as well as the very genetic heritage of these populations, may thus be easy booty for contemporary 'biopiracy', as indicates the Final Report of the Committee Investigating Accusations of Biopiracy in the Amazon, presided over by congresswoman Socorro Gomes (PROBEM 1997).

The influence of the United States in the Amazon heartland combines both veiled and explicit coercive forces.

In the issue of protection of biodiversity, forms of veiled coercion predominate. These forms of coercion aim at the control of the use of territory through non-production, or minimal production. They are veiled in the sense that they do not intervene directly in the territory, but they mean an appropriation of the decision-making process of economic use of natural capital. It is an extremely selective strategy of inclusion/exclusion, focused on the richer, preserved, forest areas and traditional and indigenous populations, ignoring the core character of the urban issue in the contemporary Amazon. More recently it privileges partnerships with municipal and state governments, sidelining, as much as possible, the central government.

The 'War on Drugs' and Brazil's answer: the SIVAM project

In the specific conditions of the heartland, the most explicit form of coercion, of a paramilitary nature, is the attempt to intervene directly in the territory by means of the 'War on Drugs', through action by the Drug Enforcement Administration (DEA) governmental agency. In South America, the United States has vigorously pursued destroying production of marijuana and cocaine and impeding their trafficking. US pressure on the Amazonian countries does not conceal

its desire for direct intervention in the heartland. To this end, the United States
has supported the installation of a series of military bases in neighbouring countries, in a veritable siege on the Brazilian Amazon. Why not in Brazil? Because
the Brazilian military formulated an innovative response, creating a gigantic
control project based on modern technology: the SIVAM (Amazonian Surveillance System) project. This is a strategy designed to counter territorial
intervention in the name of the War on Drugs, as well as in defence of the
environment.

The fact that a US firm – Raytheon – was the winning bidder was
greatly influenced by the funding facilities offered: for the first time in 15
years, the Eximbank once again offered Brazil a loan, and with enormous
concessions. The fact is, that the heartland will enter the twenty-first century
under the direction of one of the most sophisticated information systems in the
world.

The 'modesty' of USAID

There is only modest action by the US government on the environmental issue,
through USAID. Appearances are misleading, inasmuch as American strategy is
implemented through an intricate network of apparently independent agents,
who act under its hegemony. These are agencies, universities, firms and especially NGOs and the World Bank, constituting *de facto* civilian arms of
the American government. Every effort to evaluate the relative significance
of American investments must include investments made by these multiple
agents.

USAID's environmental investment policy has a clear-cut anti-Brazilian state
outline, projects being executed through their agents, in agreement especially
with NGOs, and avoiding the funnelling of resources to the Brazilian Government through the appropriate channels of the Ministry of Foreign Affairs.
USAID presents two justifications for this procedure: 1) the cancellation of
the Brazil–USA military agreement, which led the Carter administration to end
USAID activities in Brazil, and 2) the fact that Brazil's situation has changed to
that of an *advanced developing country*, leading the American government to
shift its aid to poorer countries in Latin America.

In 1990, USAID's actions underwent a major strategic change: the Congress
of the United States authorized USAID to implement a programme to address
important global climate change issues in 'key' countries, including Brazil.
USAID, therefore, launched the Global Climate Change Programme (GCC) in
Brazil. The primary goal of this programme is to reduce Brazil's contribution to
the global emission of greenhouse gases by reducing deforestation in Brazil's
Amazon region. Towards this end, the GCC Programme has promoted the development of ecological and economically sustainable policies and activities to
manage forest resources in the Amazonian states. In 1996, the new Integrated
Environmental Management Project opened new possibilities in the field of biodiversity conservation and renewable energy.

The most important arm of USAID is, undoubtedly, the WWF, the powerful NGO whose projects, supported by USAID, are only a small part of its action in Brazil.

The Protection of Biodiversity by Non-governmental Agents

The WWF

This is number one of the world's leading conservation organizations, present in 120 countries, and acting significantly and autonomously in Brazil. Created over 35 years ago, the WWF has over 5,000,000 members today, with 1,500,000 Americans occupying first place and over 300,000 Britons, and approximately 30 per cent of the population of the Netherlands. The organization collects 180 million dollars per year, between contributions from members, foundations, corporations and governments. Significantly, notwithstanding the fact that it is a global organization with headquarters in Switzerland, its Department for Latin America and the Caribbean is headquartered in Washington.

The WWF's current line of action is associated with a redefinition of strategy, dividing the planet into 800 regions, defined by the density of ecological interactions, prioritizing the most important so-called Global 200.

Brazil's elevation, in 1996, to the status of a WWF national organization, represented the transfer of a good number of projects to the country, in addition to greater projection and representativity and a reorientation of fundamental posture: not simply to denounce, but to present alternatives.

The environmental face of the World Bank

In Brazil, the 'programmed network' was made possible by the World Bank's loans, such as in the case of the *Polonoroeste* (Project for the Integrated Development of the Brazilian North-west). During the eighties, it ceased to support projects in the country and then began to finance environmental projects, and is today the largest lender to developing countries for environmental conservation and management. Their loans, at below-market rates, with a Brazilian government participation of approximately 10 per cent, constitute, undoubtedly, an inducement for the building of the environmental heartland.

In 1992, two major programmes were instituted in the Western Amazon, with the objective of recuperating the perverse effects caused by the *Polonoroeste*:

a) the *Prodeagro* (Programme for the Development of Agriculture and Enviroment) – in the State of Mato Grosso – amounting to US$205 million. However, inasmuch as the state has been showing its own economic dynamism, this loan, however big, is not a determining factor in its transformation;

b) the *Planafloro* (Plan for the Development of the Agriculture, Cattle
    and Resources of the Forest) – in the state of Rondônia – amounting
    to US$228 million. Here, on the contrary, in view of the state's depend-
    ence on colonization by small landowners, and its loss of dynamism due
    to the stagnation of traditional agriculture and a mining crisis, the
    programme performs a crucial role in its trajectory. Specific objectives
    of the *Planafloro* were conservation of biodiversity, protection and
    reinforcement of the boundaries of conservation units, indian reserva-
    tions, public forests and extraction reserves, prevention of illegal defor-
    esting and clearing by burning, development of sustainable exploration
    systems in agro-forestry areas, and the improvement of the technical-
    operational state institutions, particularly those active in the environmen-
    tal arena.

The World Bank is also the manager of the Pilot Plan for Protection of
Brazilian Rain Forests (G-7), established in 1991, the largest environmental programme implemented in a single country, on a scale as large as the Amazon
Heartland itself. Fruit of a partnership between the Brazilian Government and
the Group of Seven (G-7), the European Union, non-governmental organizations and the World Bank, it constitutes one of the major expressions of the
new form of international cooperation. Investments from donations were forecast at US$280 million, approximately 20 per cent of which through a Fund –
the Rain Forest Trust Fund – and the remainder through co-financing by the
donor countries, in addition to Brazilian participation. The explicit objective of
the PPG7 is the protection of biodiversity and containment of deforestation. Its
strategy is the selective funding of projects favourable to environmental preservation and the strengthening of non-governmental organizations held to be
mediators capable of overseeing the implementation of the projects, including
their financial aspects.

The projects which have received, to date, the largest donations are: Natural
Resource Policy, Demonstrative Projects, and Centres of Excellence. The first
of them, which acts in conjunction with the state governments, has as its core
elements economic-ecological zoning, inspection and control. The second
focuses directly on communities, usually in partnership with NGOs. The third
is aimed at strengthening science and technology in prioritized theme areas.
Other projects are in different stages of preparation, with special emphasis on
the Ecological Corridors Project, with US$45.62 million earmarked for it
alone.

The World Bank is not only the manager, but also the great intellectual
mentor of the G-7. During the first five years of the programme, it practically
dominated the decision-making process. The creation of The Ministry of the
Environment and its Secretariat for the Coordination of The Affairs of the Legal
Amazon, in 1995, assured Brazil's participation in the management of the
Programme. Through not always harmonious dialogue and adjustments, the
Programme shifted its initial conservationist goal to a consensus on sustainable
development.

Globalization of Research and the Heartland as an 'Experimental Paradise'

A new scale of biodiversity protection is being outlined, accentuating the role of the heartland as an 'experimental paradise' similar to tax havens. These large-scale 'top-down' projects, not justified by local social demand, reveal the priority of scientific research in their implementation. Some are already under way. The new ones forecast are on a gigantic scale, coherent with the new tendencies towards globalization of research based on far-reaching partnerships. And American scientists perform a major role in these megaprojects.

Exemplary are the Ecological Corridors, a true revolution in biodiversity protection strategy, and part of the PPG7. It proposes the installation of a web of diverse protected areas, composing a mosaic of immense territorial extent, so as to permit the greatest possible flow of species and genetic material, with the goal of preserving biodiversity *in situ*. In the Amazon, five ecological corridors have been selected. A great innovation in this context is the partnership with the private sector, in exchange for tax exemption and special credits and loans.

The ICSU (International Council of Scientific Unions), which stimulates the globalization of research through an intricate network of projects centred on global change, is also present in the heartland. This is thoroughly interdisciplinary and international research, connected to consultancies, governmental decision-making, international agencies and NGOs.

The IGBP (International Geosphere–Biosphere Programme) is one of the major nodes in this network, and the LBA (Large Scale Biosphere–Atmosphere Experiment in the Amazon) project is the first to be supported by the three largest global change research programmes: the IGBP; the WCRP (World Climate Research Programme); and the IHDP (International Human Dimensions Programme).

The USA exerts a strong influence on the global programmes, both through the Scientific Unions and individual scientists. In the case of the LBA, American interests are revealed by the direct participation of NASA in the LBA Scientific Planning Group itself, between 1992 and 1995.

Realizing the 'Natural Capital'

Biodiversity's industrial use is already beginning, as aimed at by PROBEM (the Brazilian Programme of Molecular Ecology for the Sustainable Use of the Amazonian Biodiversity). Its main objectives are to improve research and development of biotechnology and the use of natural products.

PROBEM's main projects are (PROBEM 1997):

1. the construction of an Amazonian Biotechnology Centre (ABC), a world-class laboratory complex to be built in the Industrial District of the free zone of the city of Manaus;

2. the coordination of a network of laboratories with 80 national research groups belonging to 30 national institutions, for the bio-prospecting of the flora and fauna of the heartland. Ten top foreign laboratories are already committed to this network, half of them from the USA: Cornell, Stanford, UCLA, Berkeley, Columbia;
3. the certification of natural products with a PROBEM seal of quality, the patenting and control of industrial property, and the marketing of (including trademarked) products, services and technologies;
4. the central drive of ABC is to establish a Biotechnological Industrial Centre, concentrating on the areas of pharmaceutical products, cosmetic materials, food, bio-insecticides, enzymes, essential oils, antioxidants and natural dyes and fragrances. To this end, the ABC will develop a Pre-Industrial Activities Sector.

PROBEM aims at integrating laboratory research with a broad Company Incubation sector, and a specialized consultancy service focusing on technological research and development, industrial and intellectual patents, management and transfer of technology, quality certification and control of biotechnology products.

Conclusion

The recent past reveals that international cooperation is an instrument of veiled coercion. This means that there is a risk that the Brazilian national state's sovereignty may be ignored by a foreign appropriation of the decison-making process regarding economic use of the natural capital of the Brazilian territory.

But it is also true that dialogue and diplomacy and, in particular, the strengthening of communication channels between the democratic Brazilian state and civilian society, can transform veiled coercion into an instrument of change. In this context, emphasis should be given to the significance of a very recent experiment: the establishment of the 'Agenda Úmidas' for the sustainable development of the state of Rondônia, a pilot projet financed by the World Bank. The 'Agenda Úmidas' aimed at the setting of strategic guidelines for the sustainable development, involving a wide participative process of local stakeholders which went far beyond the horizons of mere conservationism. This could well be an exemplary case of the possibility of transforming veiled coercion into a policy for development.

Among the megadiverse countries, Brazil stands out due to a singular endogenous capacity for the development of scientific and technological research, notwithstanding the fact that integration between industry and endogenous capacity for science and technology is still at a relatively low level of realization. PROBEM constitutes a Brazilian strategic answer: recovery of decision-making power concerning the transformation of 'natural capital' into support for sustainable development.

But this is only one facet of the question, for if Brazilian diplomacy does not wish to abdicate from the defence of national sovereignty, it must undertake every effort to enforce the concept of *national consent* in favour of certain measures of international interest in its international relationships. As indicated by Meira Mattos (1993), in this respect the core issue is the adjustment of consent between sovereign nations. How to make of this principle of international right a normative value of the highest order is not a simple task. Brazil must make its best effort to meet the American strategy of veiled and explicit coercion with the powers of dialogue and diplomacy.

Seeing the Amazon as an 'area of attraction' is nothing new. From green 'hell' to green 'paradise', according to the evolution of ideologies, the region has been criss-crossed by scientists and disputed for centuries by pioneers and adventurers. As an area of economic possibilities, the region has long been seen as a gigantic 'storehouse of raw materials', even though the economic utilization of this immense asset has remained at a relatively low level of profitability.

Today, however, we are at the dawning of a great transformation: new experimental spaces and horizons of expectations are emerging for the realization of 'natural capital', based on the advances of molecular biology and biotechnology. In this new scenario, Brazil faces the potential challenge of providing the terms for an institutional-creative answer, capable of updating the words of Arthur Cezar Ferreira Reis, and going beyond *maintaining the environmental heartland as a part of our political being.*

References

Bartholo, R. S., Jr, 1986. *Os Labirintos do Silêncio*, Ed. Marco Zero, Rio de Janeiro.

Bartholo, R. S., Jr, 1992. *A Dor de Fausto*, Ed. Revan, Rio de Janeiro.

Becker, B. K. 1992. *Amazônia*. Ed. Ática, São Paulo, 1995–7.

Becker, B. K. 1995. Redefinindo a Amazônia: o vetor técno-ecológico, in *Brasil. Questões Atuais da Reorganização do Território*, Castro, I. E., Costa Gomes, P. C. and Corrêa, R. L. (eds), Bertram Brasil, Rio de Janeiro.

Becker, B. K. 1997. *A Amazônia em Ação*, Secretaria da Amazônia Legal, mimeo, Brasília.

Boulding, K. 1966. The Economics of the Coming Spaceship Earth, in H. E. Jarret (ed.), *Environment Quality in a Growing Economy*, Johns Hopkins, Baltimore.

Castells, M. 1985. High Technology Economic Restructuring and the Urban-Regional Process in the United States, in *High Technology, Space and Society*, Castells, M. (ed.), *Urban Affairs Annual Review*, 28.

Daly, H. E. 1991. From Empty-world Economics to Full-world Economics. In *Environmentally Sustainable Economic Development*. UNESCO.

Ferreira Reis, A. C. 1960. *A Amazônia e a Cobiça Internacional*, Companhia Editora Nacional, São Paulo.

The LBA Scientific Planning Group. 1996. *O Experimento de Grande Escala da Biosfera – Atmosfera na Amazônia* (LBA). Brasil, Holanda e EUA.

ICSU. 1996. *Understanding Our Planet*. Paris, France.

Jonas, H. 1979. *Das Prinzip Verantwortung*, Insel Verlag, Frankfurt am Main.

Meira Mattos, C. 1993. *Amazônia e a Dissuasão Estratégica*, Revista da Escola Superior de Guerra, 26, ano IX, Nov., pp. 85–8.

PROBEM – *Amazônia*. 1997. SCA/MMA, Brasília.

SIVAM. 1998. *Sistema de Vigilância Começa a Ser Implantado*. Assessoria de Comunicações – CCSIVAM, Rio de Janeiro.

Taylor, P. J. 1997. As Raízes Geo-históricas do Desafio da Sustentabilidade e uma Alternativa Política Pós-Moderna, in Becker, B. K. and Miranda, M. (eds), *A Geografia Política do Desenvolvimento Sustentável*, Editora da UFRJ, Rio de Janeiro.

Von Weizsäcker, C. F. 1979. *Die Einheit der Natur*, Carl Hanser Verlag, München.

Chapter 13

American Power and the Portuguese Empire

James Sidaway

The United States of the Cold War period, like ancient Rome, was concerned with all political problems in the world. The loss of even one country to Communism therefore, while not in itself a threat to American physical security, carried implications that officials in Washington found highly disturbing. They became greatly concerned with the appearance as well as the reality of events, and there was much talk of dominoes . . . the attitude that 'we have to prove that wars of national liberation don't work' (a curious attitude for the children of the American revolution to hold) did carry the day . . .

(Ambrose, 1985, xvii)

East and West, North and South: Cold War, Containment and Decolonization

Writing about the vision of American power and destiny presented to the readers of *Time*, (the popular weekly American news magazine) in the first years of the Cold War, Hinds and Windt (1991, 157) note how the magazine had:

> shared the president's view of the new divisions of the world, the antagonists, and the ideological content of the worldwide struggle. Equally important, the magazine blithely noted that if people did not like the way that some described the policy (imperialism), then another more congenial description (containment) should be adopted. It was as simple as that. Change the language and one changes the way people view the policy.

But it was not always to be so simple. This chapter will examine some of the ways that the United States sense of global mission of anti-Communist containment was rendered more complex by its articulation with another sense of global mission, that of Portuguese imperialism. This is only one small aspect of a much bigger story about how the United States variously contested, co-existed with or subsumed the older European overseas empires during the epoch of Cold War. In general accounts of this, the American relationship with Portuguese colonialism is usually but a side story to the course of events in, for example, the Middle East, Cuba or Vietnam. Yet the relationship with Portugal was symptomatic of the profound contradictions of American imperial power and the way that what is still most commonly re-told as a story of East versus West division (the Cold War) was much more than that and had, *amongst other things*, a significant North–South dimension.

The projection of American power has been distinctive not only in terms of its scale (the first global superpower), but also in terms of having developed in an age of formal decolonization and rhetorical anti-imperialism. At the end of the Second World War, the United States had a global network of strategic commitments (bases, troops, military hardware, defensive agreements) and economic interests and was formulating a doctrine of anti-Communist 'containment' directed at the Soviet Union and its allies. Barnet (1973, 18) noted how therefore in the post-Second World War era:

> All the elements of a powerful imperial creed are present: a sense of mission, historical necessity, and evangelical fervor. Perhaps, most important, the American imperial creed was fitted with particularly effective rhetorical devices to clothe greed in the language of charity and to obscure the national will to win by calling it the burden of responsibility.

At this time, when American commentators and politicians were declaring that the US would be supporting *Freedom* (which was equated with anti-Communism and capitalism) the remaining formal territorial overseas empires of the old European imperial powers (notably, Britain, France, Belgium, the Netherlands and Portugal) were coming under pressure from national liberation movements seeking independence and charting what were frequently rather different invocations of 'Freedom' (as ending capitalist exploitation and foreign domination).

In summary, the United States came to its moment of greatest imperial expansion just as the era of formal decolonization and independence unfolded. The European powers were weakened by the Second World War, in both material terms as well as ideologically. Not only had millions of colonial subjects been mobilized and borne arms during the war, but the possibility of defeating European colonial domination had been demonstrated (for example by the rapid Japanese advances in Asia). A wave of radical demands for freedom, for economic and political decolonization swept Africa and Asia. European responses varied. Sometimes independence was granted quickly, in the face of massive popular demands (e.g. India). In other cases however, the colonial powers (often bolstered or supplanted by an armed and determined settler population) dug in for what often became extremely violent and protracted struggles (e.g. Algeria and Rhodesia). Given the diversity of paths (sometimes violent, sometimes peaceful, sometimes revolutionary, sometimes 'constitutional') to decolonization, the complex courses of national liberation and the birth of what became known as the 'Third World' would be uneven and contain a mixture of violence, heroism and deceit. Inevitably as the greatest power on earth, with a strong sense of its global mission, the United States was drawn in. The format

of US involvement would reflect these complexities combined with the contradictions evident in the US claims to be *both* a beacon of Freedom, itself a society
based on an historic anti-colonial revolt, *and* the leader of a global strategy of
containing Communism.

The Ambivalent Place of Portugal

Portugal had been officially neutral in the Second World War. After 1945 the
anti-democratic regime run by Antonio de Olivieira Salazar, which had come
to power in the late 1920s following a military coup, was still firmly in place.[1]
Moreover central to its vision of greatness and destiny was continued possession of a colonial empire in Africa and Asia. The most significant territories
were in Southern (Angola and Mozambique) and West (Portuguese Guinea)
Africa. But Portugal also held a series of Atlantic islands. Nearest to the metro-
pole were Madeira and the Azores (which had no indigenous populations) and
further south populated mostly by creole populations were Cape Verde and São
Tome and Principe. There were also tiny enclaves in the Gold Coast (later to be
absorbed by Ghana) as well as territories on the west coast of India (Goa,
Damão and Diu) and further east, Macão (next to Hong Kong) and East Timor
(at the eastern end of Indonesia).

Portugal was not about to relinquish its empire. Not only was it profitable
for the Portuguese elites, but it provided a safety-valve (of emigration) for
peasant and working-class Portuguese, displacing pressure for land reform and
social justice in a Portugal dominated by a landed and commercial clique with
close ties to the right-wing regime. The empire was also central to the regime's
self-image as a beacon of civilization, Catholicism, and progress with a global
mission.

De Figueiredo (1975, 74) notes how:

> The empire, some twenty-two times the size of Portugal and almost as big as
> Western Europe, gave a sense of purpose and a new dimension to the 'Revolução
> Nacional' ['National Revolution'] which had emerged from what Salazar himself
> had defined as the 'dictatorship of reason and intelligence'. Schoolchildren, the
> main target of the grand re-education designs of the regime, were proudly shown
> by their schoolmasters the map of the empire superimposed over the map of
> Europe to illustrate the slogan: '*Portugal não é um pais pequeno*' ('Portugal is not
> a small country'). When one turned to Salazar's analysis of the Portuguese eco-
> nomic situation at the time one sees that he too had been developing his own ver-
> sions of the theories of expansionism and the need for 'living space' that were a
> dominant feature of Nazi Germany's 'New Order'.

Of course, in the 1930s such imperialist dreams were hardly distinctive. The
Axis powers, Belgium, Holland, Britain and France all had colonial strategies,
vast pretences, and racist mythologies of spreading 'civilization' to 'backward'
peoples. After 1945 however, such programmes looked increasingly anachronistic with widening pressures and movements for decolonization and the

relative eclipse of European powers by the new (US and Soviet) Superpowers. Portugal, though, still led by Salazar, was determined to hold its empire, which, amongst its other functions for the Portuguese elite, was *the* key to official claims of Portuguese greatness.

Meanwhile, the Portuguese Atlantic territory of the Azores became strategically important to the US during the Second World War. The Atlantic archipelago, nine islands extending over 500 miles in the Atlantic Ocean, 'a third of the way from Lisbon to New York' as a sympathetic foreign observer of Salazar's regime described the Azores (Kay, 1970, 160), were to be used as a vital base by the allies during the war. Despite official Portuguese neutrality, an ancient treaty of alliance (dating back to 1373) between Portugal and England was invoked by the British government. Salazar had little choice but to agree, for without Portuguese cooperation, the US and Britain would likely have taken possession of the islands (with or *without* any Portuguese permission). The limited range of 1940s aircraft made a refuelling stop essential in the Atlantic crossing. The Azores were well placed for this and they could also serve as a useful allied naval base, of enormous significance in the battle for the Atlantic.

Following a long negotiation (and a British ultimatum to Portugal citing the old alliance), the Azores were made available to the allies from 18 August 1943. Portugal, Britain and the US signed an agreement permitting some allied access to Azorean ports and airfields in 'exchange' for guarantees of Portuguese sovereignty over the islands and promises that such recognition-guarantees would be extended to the entire Portuguese empire. At first, Salazar argued that only British forces could use the facilities, but the impossibility of separating British from American forces and the value of the islands to the US as a staging post on eastward routes to South-east Asia led the Americans to pressure Portugal for a further concession. For a while, Salazar held out and refused, supported by the Portuguese press. But with the pro-Axis Vichy government of southern France controlling French West African colonies and negotiating with Hitler to offer Dakar in Senegal as an Axis base, and with influential strands of Congress and the media behind him, US President Roosevelt authorized his chargé d'affaires in Lisbon to negotiate with Salazar. In the early 1940s, this was George Kennan (later to become famous, in July 1947, as the author of a key Cold War text published in the influential American journal *Foreign Affairs*,[2] which purported to reveal 'The Sources of Soviet Conduct'). As part of his negotiations, Kennan conceded US recognition of Portuguese sovereignty throughout the Portuguese empire. With this concession on the table and the prospect of German retaliation fading as Hitler was forced on the defensive, Salazar felt able to sign a formal agreement with the US on 28 November 1944, allowing US access to Azores facilities for five years.

So after initial equivocation, Portugal had successfully established itself as a link in the 'Western system'. This transatlantic framework, soon to be codified in the form of the North Atlantic Treaty Organization (NATO) resulted in a certain space of movement and flow (at first overwhelmingly from the USA) of capital, diplomatic exchange, cultural products (such as music, cinema and fashion), influence, power and military force. It was about to become hegemonic in Western Europe and beyond, as America built alliances and systems of 'containment' across the Atlantic and the Pacific. Portugal had orientated itself towards this developing 'system' and integrated itself into what American's would often call 'The Free World', at little cost (remaining officially neutral to the end of the war) and receiving guarantees that the overseas empire would be confirmed as 'Portuguese' lands in the eyes of the US and Britain. Portugal was even to receive US military support in the act of regaining Japanese-occupied East Timor, the most distant segment of Portugal's global empire.

When the trans-Atlantic alliance system of NATO was being negotiated in 1947, with Portugal as a party to the discussions, Salazar unsuccessfully sought to extend the area included in the provisions of the treaty to include all of the Portuguese empire. Whilst this was resisted by the United States, Portuguese membership in NATO and American aid did permit a technical mobilization of the Portuguese armed forces. In addition, the Azores agreement was renegotiated in 1948 and 1951. In the latter year, NATO access to the islands in the event of war was conceded without conditions or future time limit.

Arms and other military hardware provided to Portugal by virtue of its NATO membership were to become increasingly significant in the 1960s. In the immediate post-Second World War years and through the 1950s anti-colonial resistance in the Portuguese empire was not coherently organized. For example in the most significant of the Portuguese colonies, of Angola, Guinea and Mozambique, it was only in the 1960s that national liberation movements began to organize. In the early 1960s, inspired by the example of successful anti-colonial mobilization elsewhere on the continent, and with a degree of material support (mostly of weapons and military training) from China and the Soviet Union, the *Frente para a Libertação de Moçambique* (Frelimo), the *Movimento Popular de Libertação de Angola* (MPLA) and *Partido Africano de Independência de Guiné e Cabo Verde* (PAIGC) all began to wage (what were envisaged as protracted) armed struggles in pursuit of independence for Mozambique, Angola and Guinea. All these movements were broadly Marxist in ideological orientation and although they were all characterized by splits, only in Angola did other significant non-Marxist movements develop, the heterodox *Frente Nacional de Libertação de Angola* (FNLA) and the *União Nacional para a Independência Total de Angola* (UNITA). These splits within Angolan nationalism would later become subjects of Cold War intervention as the MPLA received Soviet aid and Cuban troops whilst UNITA was backed by the USA (and South African troops).

In the meantime, with Portugal as a loyal (albeit anti-democratic) anti-Communist ally, the US would inevitably be drawn in contradictory ways in respect of the unfolding anti-colonial insurgencies in the Portuguese African territories and the question of the defence and future of the Portuguese overseas empire. Although the Indian invasion and occupation of Goa and associated Portuguese Indian territories in 1961 (which Portugal waged a polemical campaign against) certainly troubled the visions of Salazar and emboldened those elsewhere in Portugal and its 'overseas provinces' who sought change, it was in

Africa that the most substantial challenges arose. Following the wave of African decolonization at the start of the 1960s, the Portuguese empire was looking increasingly anachronistic and vulnerable. In the 1950s the issue could still be more or less ignored. As long as most of Africa was still enduring European colonial domination, the immediate status and future of the African Portuguese empire was hardly an issue for the United States and other 'great powers'. But this quickly began to change in the 1960s. As the African continent asserted demands for independence and (black African) majority rule and as 'new' post-colonial states emerged (starting with Ghana in 1958), so Africa, and with it the Portuguese empire, would increasingly become domains of Cold War confrontation. The utter unwillingness of Salazar to consider the end of Portuguese colonial rule meant that the struggle for national liberation in the Portuguese colonies would become an armed insurgency against a country which now insisted that its African colonies were simply 'provinces'[3] integral to the geography of Portugal, which in turn was allied with the US and armed by NATO.[4] In this context, the strategic gaze of the US was turned on Angola, Mozambique and Guinea at the start of the 1960s. US concern had been reinforced by the closure of the Suez Canal in 1957 as a result of hostilities between Egypt and Israel. The closure of Suez made the shipping route around the Cape of Good Hope at the tip of Southern Africa more important. Soon it was being scripted as a vital Western lifeline (not least for oil supplies, en route from the Persian Gulf to the West). The relative proximity of Angola and Mozambique to these routes rendered them 'strategic' in the US global vision, particularly as concern grew about the future of South Africa. Furthermore, ports in Angola and Mozambique could be used by the US Navy without the considerable political difficulty that could arise from the similar use of ports in apartheid South Africa.

When John F. Kennedy became US President in 1960, at the very moment when African decolonization was accelerating (a record number of African countries achieved independence at this time), the US sought to accommodate something of the aspirations of an increasingly vocal, organized and self-confident fraternity of 'post-colonial' African, Asian, Caribbean and Latin American countries, which were increasingly adopting the designation of Third (not East, nor West) World. Such accommodation was part of a strategy to defuse revolutionary potential. Particularly after the shock of the Cuban revolution (which rapidly took a pro-Soviet direction in the 1960s) Kennedy sought to respond to the Third World with a mixture of enhanced capabilities for US intervention combined with a rhetoric of accommodation, aid and progress. The idea of development as (capitalist) modernization was reiterated and elaborated, for example by Kennedy's adviser, Walter Rostow (1960), and his schema of 'The Stages of Economic Growth' which were embodied in Kennedy's programmes for the recovery of the American initiative and (self-)confidence in the Third World. Although the Portuguese government was also to some extent affected by the ideas of 'development' and 'modernization' that were evident in Kennedy's renewal of US Third World policy, this was tempered in Portugal by the continued wariness of foreign (especially American or democratic) influence and the deeply reactionary politics of the Salazar regime which made practically no concessions to political or cultural opposition. Indeed, as the anti-colonial revolts swept Portuguese colonies in Africa and as Portugal found itself the object of condemnation from the Communist and increasingly numerous and vocal Third World states at the United Nations and meetings of the Non-Aligned Movement, so spokesmen for Portugal and the empire (including Salazar) took an increasingly hysterical line of condemnation, denial and insistence of Portuguese uniqueness, rights and privileges.

A developed geopolitical discourse about Portugal's vital and central place in the Christian, anti-Communist West was invoked and continually reiterated in statements by Salazar, his deputy (and after 1968, successor) Caetano, and an array of greater and lesser military and colonial officials. Setting the tone and disassociating himself from its invocation of 'democracy' Salazar himself had described the Atlantic Pact of 1949 which established NATO as:

> the symbol and the expression of a new crusade: that of the defense of the Western and Christian civilization.
>
> (quoted in Teixeira, 1992, 121)

By the 1950s, Portuguese colonial and military figures had reappropriated the geopolitical terminology of heartland, choke-points, Eur-Africa that had been common in French and German geopolitical discourse earlier in the century and which articulated with broader Western and particularly United States codifications of 'Containment'.[5] In Portuguese discourses, however, it tended to take on romantic, poetic, even metaphysical (semi-religious) overtones. For example, the Atlantic would become the 'Ocean of Western civilization' ('discovered' of course by the 'pioneers of European and Christian civilization', the Portuguese).

This was mixed with ideas of strategic 'choke-points' and western 'life-lines' and of Portuguese stability, vitality and destiny lending the Portuguese empire a vital role; 'a position of incontestable value' in the 'defence of the Free World' (Comprido, 1956, 268). Notably for Portuguese writers, this 'Free World' (usually contrasted with an atheistic-oriental-Communist world) could not be defined in terms of its democratic credentials (as it frequently was in US and British discourses of the time). Instead it was constituted by such things as a 'spirit of Western civilization', 'anticommunism' and 'a unity of origin and thought that resulted in a historic process . . .' (Comprido, 1956, 229; my translation).

In slightly less poetic but no less extravagant forms, de Carvalho (who continues to write geopolitics today) visualized Portugal's place in the Atlantic Alliance as part of its world-destiny. For in the event of 'an attack from the East' (directed at NATO):

> there would be a 'first line' where the clash would inevitably have to take place and a 'second line' sheltered by the Atlantic and indispensable support to the first line in men, equipment, raw material and provisions. The ocean therefore appears as a hinge which one must inevitably dominate so that each group of nations will be able to carry out its part of the common task. And this immediately gives a

dominant position to two countries: Iceland in the Arctic Polar Circle and Portugal in the Central Axis of transcontinental communications. Since the northern
route presents obvious difficulties, our country is therefore the first and the best
spear-head that the reinforcement zone possess in the operational zone. In other
words: Portugal is the European country with which the allies of the other continent are most easily in liaison. The air bases of the Azores, in the middle of the
Atlantic, are an indispensable pillar of the airbridge connecting the friendly
margins of the great blue river.

                    (de Cravalho, 1964, 91–2, translation in Crollen, 1973, 77)

Meanwhile in the USA, the Azores bases were being represented as of vital
strategic significance, as in a 1953 article in *Christian Science Monitor*. Next to
the title 'Uncle Sam's Biggest Gas [fuel] Station', in a map the Azores are shown
at the centre of American bases in the West. Through these kinds of mappings
(and similar documents were common in Portuguese publications), the ideas
about strategic spaces and linkages which were characteristic of both American
and Portuguese geopolitical discourses were made more 'powerful' and 'real' to
wide audiences.

However, despite such rhetorics, and the continued significance of the Azores
in the activities and contingencies of the US armed forces and 'strategic planners', for a short period in the early 1960s the relationship between Salazar's
Portugal and the US was far from 'special'. During Kennedy's presidency from
1960 (cut short by his assasination in 1963) the Portuguese–American relationship briefly soured, with disputes about the future of the Portuguese empire
the main issue of contention.

The Kennedy administration was the moment at which the US rhetorical commitment to the end of (other formal) empires was permitted to resurface, albeit
in a context still overwhelmingly shaped by Cold War and containment preoccupations. Defusing and co-opting the radical pressures emerging in the Third
World was a significant theme at the start of the Kennedy presidency. For Africa,
the Caribbean, Latin America and Asia, this meant a renewed stress on US-
directed counter-insurgency, combined with aid and a cultural and economic
offensive which sought to regain the American and capitalist initiative. The
latter anticipated what later in the context of American strategy in Vietnam
would be termed the 'battle for hearts and minds'. This was also a renewal of
the heroic story of American soldiers and agents, an image of reinvigorated,
potent American leadership and brave and strong men, held to embodied in
Kennedy himself (see Dean, 1998).

America could also reiterate the claim of being moral beacon to the post-
colonial world whilst proclaiming its difference from the European colonial
empires. Hence, as Serapião and El-Khawas (1979, 173) noted, when armed
revolt directed against Portuguese settlers broke out in Angola in 1961:

G. Mennen Williams, Assistant Secretary for African Affairs, pointed out that
American colonial experience and revolutionary struggle for independence put her
in a position to understand and share the aspirations of people who struggle for
independence. He criticized the bad living conditions of Africans in Mozambique

and other territories and concluded that we do not think it possible that small groups of people can continue to determine the affairs of large populations.

In the context of the US voting for a UN security council resolution that condemned repressive measures in the Portuguese colonies, plus the imposition of an embargo on military equipment destined for use in 'Portuguese Africa', Salazar reacted with furious and hysterical statements. He told his foreign minister Franco Nogueira that:

> The present American government is a case where the lunatics have taken over the asylum. . . . American policy is even more revoltingly imperialist than that of the Soviet Union.
>
> (cited in Maxwell, 1995, 51)

But Salazar had little to fear from the US and other NATO allies, as the overdetermining Cold War containment logic reasserted itself (see Mahoney, 1983). Even before Kennedy's assassination in 1963, the US had quietly toned down its criticisms of the Portuguese. The renegotiation of the Azores base deal had much to do with this. But the continued, albeit qualified US support for Portugal ran deeper than this narrow strategic concern. Kennedy's successors Lyndon B. Johnson (1963–9) and Richard Nixon (1969–74) not only stopped critical comments but actively bolstered Portuguese military capacity to wage counter-insurgency.

Therefore, as the wars of national liberation in Guinea, Angola and Mozambique were stepped-up in the 1960s and early 1970s, so US support (together with that from other NATO allies) became a decisive factor in Portuguese counter-measures. What became known in 1965 as the 'Anderson Plan', in which the State Department and the US Ambassador in Portugal (Admiral George Anderson) devised a scheme whereby the national liberation movements and the Portuguese regime could move to negotiated agreement and a plebiscite on the status of the overseas provinces (to be overseen and guaranteed by the United States) was rejected by an ever-intransigent and nervous Portugal (see Samuels and Haykin, 1979). With its rejection by the Portuguese (who feared that events could quickly escape their 'control') went, as Macqueen (1997, 54) notes, 'the last serious opportunity for the regime to extricate itself from Africa and the wars by negotiation'.

The Anderson Plan therefore could be no more successful than Kennedy's statements and short-lived arms embargo at forcing a Portuguese concession. As the anti-colonial wars became more violent and widespread, the US became a key Portuguese ally. Under President Nixon and his (geopolitics-obsessed) Secretary of State, Henry Kissinger, the United States provided ample diplomatic, economic and military support, including aircraft, defoliants (of the highly toxic kind also used by the US in Laos and Vietnam to destroy forests so that Communist insurgents lost cover), and advisers.[6] Not only did the Azores base remain useful to the US (now as a transhipment point *en route* to Indochina as well as on route to the 'Middle East', a place whose centrality to US strategic dramas grew in the late 1960s and early 1970s), but Southern Africa in particular became scripted as strategic in US geopolitical discourses.[7] Salazar's incapacitation by a stroke in 1968 and his replacement by his deputy Marcello Caetano changed virtually nothing in respect of Portuguese policy. The wars went on – with continued tacit American support for Portugal. But even with military aid from America and other NATO allies (notably West Germany), they were ultimately unwinnable. Although the national liberation movements were not strong enough to defeat the Portuguese, nor could the Portuguese forces eliminate the insurgents by military means. Disquiet grew in the junior ranks of the armed forces. The regime, however, was unyielding. The situation represented an apparent stalemate: unwinnable wars and an intransigent regime which could not contemplate decolonization without undermining its own sense of mission. Portugal could not decolonize without a wider democratization and vice versa.

The end was to come unexpectedly, for most in Portugal and its allies including the US. The Americans had few links with opponents to the Portuguese regime and no real intelligence contacts amongst the lower-ranking military officers who were actively conspiring against the old regime. Therefore the Americans were taken by surprise when disaffected lower-ranking officers staged a coup on 25 April 1974. With little will to fight on, the authority structures of the old regime quickly collapsed. The fact that the officers who staged the coup were left-leaning sent alarm bells ringing in Washington. Recalling this moment, Maxwell (1995, 67) notes how:

> During the early 1970s . . . Portugal had been the perfect ally for the United States. It was docile, dependent and had nowhere else to go. The United States' embassy in Lisbon became a quiet pasture for conservatives who had fallen temporarily from grace in Washington, including William Tapley Bennett, Jr., a former US ambassador to the Dominican Republic and Admiral George W. Anderson, Jr., a former chief of naval operations and organizer of the Cuban blockade of 1961. In Portugal they were surrounded by an exceptionally congenial expatriate community. Ensconced among the mimosa groves of Estoril and Cascais or the almond blossoms of the Algarve were fallen dictators (the late Fulgencio Batista), would-be or former monarchs (Don Juan of Spain, ex-King Umberto of Italy), arthritic British colonels, a handful of ex-Nazis, and Elliot Roosevelt who owned a stud farm and had a stake in the Portuguese real estate company Torralta. They all enjoyed Portugal: Admiral Anderson bought a villa in the south. Like so many others rehabilitated by Nixon's election, such men were not without influence. William Tapley Bennett, Jr. was appointed assistant United States ambassador to the United Nations. Admiral Anderson became chairman of President Nixon's foreign intelligence advisory board. They were sure sources of 'unbiased' views on the scruffy soldiers and assorted 'reds' who emerged from the woodwork of old Portugal to ruin 'their' paradise in 1974.

Rapid and chaotic decolonization ensued. Within eighteen months, all the Portuguese colonies were independent, except East Timor, which was invaded and occupied by Indonesia,[8] and Macão, which China refused to reabsorb.

Portugal was no longer a quiet, loyal ally and its former African territories were now being ruled by Communist regimes. Portuguese troops and millions of Portuguese settlers returned to metropolitan Portugal, abandoning property and businesses in the former empire. The fall of the Portuguese empire was one of the events that produced an enhanced sense of vulnerability-threat-danger in the USA.

The repeated sense of events slipping out of (US) control in Portugal and its former empire, and more widely in the Third World (see Halliday, 1986, 1989) subsequently fed into the renewal of American 'resolve' represented by President Ronald Reagan (1981–9) and his projects of supporting 'anti-Communist' rebels in the Third World. In the 1980s, leading beneficiaries of this policy were the Afghan Mujahidin (one of the most faction-ridden and re-actionary movements in the world) fighting the then Soviet-backed regime in Kabul and the Nicaraguan Contras, mostly made up of members of the armed forces of the former (US-backed) Nicaraguan dictator Anastasio Somoza, who had been deposed by a popular revolution in 1979–80. A reinforcement of a will to intervene (most usually via such proxy forces) would also be worked out in the spaces of the former Portuguese empire, the US backing the brutal Indonesian occupation of East Timor and supporting UNITA, the most reactionary (but anti-Communist) of the nationalist movements in Angola. In the 1980s, Angola became the scene of violent protracted war between UNITA and the MPLA government, the latter bolstered by Soviet aid and Cuban troops and the former receiving ample US and South African aid and fighting alongside South African troops. Hundreds of thousands of lives would be lost. This engagement, however, is another story of US power, of 'containment' and the US will to reconstruct 'order' – an enduring theme in American alliances and strategies – often (as in Angola and East Timor) at tremendous human cost.

Conclusions

In charting some of the key themes and directions of US engagement with Portugal in the period between the rise to American global strategy (the 1940s) and the fracture of the Portuguese empire in 1974–5, this chapter has brought out just how complex and contradictory the development and exercise of American imperial power would become. The enterprise was inevitably contradictory, reflecting the paradox (present at its 'founding moment') of the United States as simultaneously a post-colonial state, beyond European empire, and an imperial force in its own right. The account has also shown something of how the Cold War would be marked by and express this. As this chapter has indicated, the Cold War therefore encompassed much more than a simple tale of East–West confrontation.

It was also a story of decolonization and of conflict between North–South, and below the surface appearances of a straightforward confrontation between East and West, or a competition between capitalism and Communism, were debates about civilization, 'race' and destiny, all evident not only in Portugal, but also in the ways that the Third World was conceptualized in the USA, with deep precedents in US history and wider notions of its manifest destiny (for primers see Bills, 1990; Kaplan and Pease, 1993; Mahoney, 1983). Although it has not been foregrounded here, this chapter has also touched upon the way that an American *masculine* will to control and to order found expression in the Cold War. This was perhaps most visible in Kennedy, but it has been evident in all US presidents. In the latter sense, the Cold War therefore relied, as Sylvester (1996, 406) has shown, on things beyond the remits of usual 'International Relations' analysis:

> on militarized gender, on constructions of 'men' and 'women' that could be mobilized for the propaganda 'war' . . . 'Men' on the frontlines of freedom and 'women' cultivating a conservative conformism at home.

For the USA, the Cold War was additionally a way, as Campbell (1992) has demonstrated, to secure a sense of self with a common (Communist) enemy.[9] Above all, and as this chapter has traced through the case study of the Portuguese empire (which self-consciously traced its origins to the fifteenth century and continually celebrated this historical 'primacy'), American policies in the Cold War articulated with other histories that had different tempos and logics.

With these points in mind, we can join Stephanson (1998), who has recently emphasized the complexity and diversity of histories that were folded into and interpreted through the Cold War contest. As he notes, not only was the Cold War a much more intricate story than has often been assumed, but it is not something that has been excised from US psyches. And therfore the longings for control, for order, for the right 'information' to perfect it, and for a world made in its own image (of which the Cold War was also in part an expression) continue to haunt, possess and contain American strategy, as it goes marching on.

Notes

1   There is some debate amongst historians as to the validity of the term 'Fascist' to describe the regime led by Salazar (1932–68), who quickly worked up to the position of Portuguese Prime Minister after he was invited by the military (who had staged a coup in 1926) to become Finance Minster in a government seeking to restore order after more than a decade of weak governments and economic chaos.

   Certainly, throughout its long trajectory, Salazar's regime was associated with a messianic idea of Portuguese destiny, a contempt for democracy, a celebration of hierarchy and deference, a deeply reactionary form of Catholicism, and in the earlier years some Nazi-style mobilization and dressing-up in brown shirts and long boots. After 1945, the latter was played down, but the outpourings about Portuguese destiny and civilization did not diminish.

2   The monthly half-academic, half popular *Foreign Affairs* became one of the key forums for semi-official writing about American foreign policy. Kennan's anonymous

article on 'The Sources of Soviet Conduct' was to be its most cited text. Ambrose (1985, 99–100) notes how: 'Its author was soon widely known; its reception nothing short of spectacular. It quickly became the quasi-official statement of American foreign policy . . . The sentence . . . that was most frequently quoted . . . and the one that became the touchstone of American policy, declared that what was needed was "the adroit and vigilant application of counter-force at a series of constantly shifting geographical and political points, corresponding to the shifts and manoeuvres of Soviet policy." This implied . . . that crisis would follow crisis around the world, as the Soviet-masterminded conspiracy used its agents to accelerate the flow of Communist power into "every nook and cranny." ' For more on Kennan, including his role in Portugal, see Stephanson (1989).

3 In 1951, the colonies were redesignated as overseas provinces. Salazar became fond of declaring that they were constitutionally just like other provinces of Portugal, such as the Algarve or Lisbon. The logic of this was an attempt to defuse calls for decolonization. For if the colonies were simply 'provinces', then Portugal was not a colonial power. Like other cases where colonies were redesignated as 'provinces' of the colonial power (such as Britain in Ireland and France in Algeria), it also meant that the path to decolonization lay through metropolitan politics.

4 In addition to US aid, advice and military supplies, Portugal received military equipment from West Germany, Holland, Belgium and the UK. The full story of this is beyond our scope here, but can be pieced together from Crollen (1973), Minter (1972), Serapião and El-Khawas (1979) and Telo (1996). The West German role was particularly significant, related no doubt to the fact that East Germany offered support to the national liberation movements.

5 On geopolitical discourses of 'heartland', 'choke-points', etc., see Agnew (1998) or Ó Tuathail (1992, 1996). A very good primer is also the collection of texts and commentary assembled in Dalby, Routledge and Ó Tuathail (1998). For more on Portuguese geopolitics, see Sidaway (1999), the edited collection by Atkinson and Dodds (1999) of which it forms a part will also provide a fairly comprehensive guide to the histories, evolution and dissemination of geopolitical discourses.

6 In addition to Serapião and El-Khawas (1979) which covers Kennedy and Johnson, Atunes (1986) is a comprehensive source on Nixon's engagement with Portugal. On the geopolitics of his secretary of state Henry Kissinger, see Hepple (1986) and Ó Tuathail (1994). The Portuguese dependence of the US grew considerably towards the end of the Salazar–Caetano regime. Kissinger comments (in one of his volumes of memoirs) how Portugal became the only European country to offer US aircraft refuelling rights during the American resupply operation to Israel during the October 1973 war between Israel and Egypt and Syria: 'Meanwhile we had obtained Portuguese permission to use Lajes airfield in the Azores for refuelling. When the first approach had been made on Friday October 12, the Portuguese government had stalled. It had no interest in antagonizing the Arab nations. It sought to extract some military equipment for its colonial wars in Mozambique and Angola. To this we were not prepared to agree. I had therefore drafted a Presidential letter of unusual abruptness to Portuguese Prime Minister Marcello Caetano that refused military equipment and threatened to leave Portugal to its fate in a hostile world. By the middle of Saturday afternoon, the Portuguese gave us unconditional transit rights at Lajes airbase' (Kissinger, 1982, 520).

7 The US was drawn into the complex politics of Southern Africa, always having to keep a certain distance from South Africa, but in reality working to preserve the

status quo in the region if the only alternative was revolution. An account is beyond the scope of this chapter; however, see Lake (1976), Minter (1972), Ó Tuathail (1992) and Schraeder (1994).

8   The Indonesian occupation of East Timor was conducted (in defiance of the UN and the local population) with American backing. The US was concerned at the Marxist leanings of Timorese nationalism and preferred that its anti-Communist ally Indonesia (itself ruled by a military-dominated regime installed with US backing in 1965) should take charge.

9   Barnet (1973, 252–7) is a wonderful primer on this: 'What characterized America was now its power, and the citizens' sense of belonging was somehow related to the vicarious exercise of national power. More and more an American came to mean someone who identified with the strugle against America's enemies. Americanism became defined in terms of un-Americanism. Through his American identity the citizen could try to recapture the spiritual bonds and kinships with his (sic) fellow citizens even as he was becoming more and more isolated from them in his day-to-day existence. Anticommunism came to serve as a kind of glue to hold a rapidly fragmenting society together.'

References

Agnew, J. 1988: *Geopolitics: Re-visioning World Politics* (London: Routledge).

Ambrose, S. 1985: *The Rise to Globalism: American Foreign Policy since 1938* (Harmondsworth: Penguin).

Atunes, J. F. 1986: *Os Americanos e Portugal vol 1: Os anos de Richard Nixon (1969–74)* (Lisbon: Dom Quixote).

Barnet, R. J. 1973: *Roots of War* (Harmondsworth: Penguin).

Bills, S. L. 1990: *Empire and Cold War: The Roots of US–Third World Antagonism, 1945–7* (Basingstoke: Macmillan).

Campbell, D. 1992: *Writing Security: United States Foreign Policy and the Politics of Identity* (Minneapolis: University of Minnesota Press).

Comprido, J. B. 1956: 'Importância Geopolítica de Portugal para a estratégia do Mundo Livre'. *Anais do Club Militar Naval*, LXXXVI 7–9, pp. 223–68.

Crollen, L. 1973: *Portugal, the US and NATO* (Leuvan: Leuvan University Press).

Dalby, S., Routledge, P. and Ó Tuathail, G. (eds) 1998: *The Geopolitics Reader* (London and New York, Routledge).

Dean, R. D. 1998: 'Masculinity as ideology: John F. Kennedy and the domestic politics of foreign policy'. *Diplomatic History*, 22, 1, pp. 29–62.

de Carvalho, H. M. 1964: *Portugal no Pacto do Atlantico, politica externa Portuguesa, 70* (Lisbon: Junta de Investigações do Ultramar).

de Figueiredo, A. 1975: *Portugal: Fifty Years of Dictatorship* (Harmondsworth: Penguin).

Halliday, F. 1986: *The Making of the Second Cold War* (London: Verso).

Halliday, F. 1989: *Cold War, Third World: An Essay on Soviet–US Relations* (London: Hutchinson Radius).

Hepple, L. 1986: 'The revival of geopolitics'. *Political Geography Quarterly*, 5, pp. 21–36.

Hinds, L. O. and Windt, T. O. 1991: *The Cold War as Rhetoric: The Beginnings, 1945–1950* (New York: Praeger).

Kaplan, A. and Pease, D. E. (eds) 1993: *Cultures of United States Imperialism* (Durham and London: Duke University Press).

Kay, H. 1970: *Salazar and Modern Portugal* (London: Eyre and Spottiswoode).

Kissinger, H. A. 1982: *Years of Upheaval* (Boston and Toronto: Little, Brown and Co.).

Lake, A. 1976: *The 'Tar Baby' Option: American Policy toward Southern Rhodesia* (New York: Columbia University Press).

Macqueen, N. 1997: *The Decolonization of Portuguese Africa: Metropolitan Revolution and the Dissolution of Empire* (Harlow: Longman).

Mahoney, R. D. 1983: *JFK: Ordeal in Africa* (New York and Oxford: Oxford University Press).

Maxwell, K. 1995: *The Making of Portuguese Democracy* (Cambridge: Cambridge University Press).

Minter, W. 1972: *Portuguese Africa and the West* (Harmondsworth: Penguin).

Ó Tuathail, G. 1992: 'Foreign policy and the hyperreal: the Reagan administration and the scripting of South Africa'. In T. Barnes and J. Duncan (eds), *Writing Worlds: Discourse, Text and Metaphor in the Representation of Landscape* (London and New York: Routledge), pp. 155–75.

Ó Tuathail, G. 1994: 'Problematising geopolitics: survey, statesmanship and strategy'. *Transactions of the Institute of British Geographers NS*, 19, 3, pp. 259–72.

Ó Tuathail, G. 1996: *Critical Geopolitics* (Minneapolis: University of Minnesota Press).

Rostow, W. W. 1960: *The Stages of Economic Growth: A Non-Communist Manifesto* (Cambridge: Cambridge University Press).

Samuels, M. A. and Haykin, S. M. 1979: 'The Anderson plan: an American attempt to seduce Portugal out of Africa'. *Orbis*, 23, 3, pp. 649–69.

Schraeder, P. J. 1994: *United States Foreign Policy toward Africa: Incrementalism, Crisis and Change* (Cambridge: Cambridge University Press).

Serapião, L. B. and El-Khawas, M. 1979: *Mozambique in the Twentieth Century: From Colonialism to Independence* (Washington, DC: University Press of America).

Sidaway, J. D. 1999: 'Iberian geopolitics'. In D. Atkinson and K. Dodds (eds), *Geopolitical Traditions* (London and New York: Routledge).

Stephanson, A. 1989: *Kennan and the Art of Foreign Policy* (Cambridge, MA: Harvard University Press).

Stephanson, A. 1998: 'Fourteen notes on the very concept of the Cold War'. In G. Ó Tuathail and S. Dalby (eds), *Rethinking Geopolitics* (London and New York: Routledge), pp. 62–85.

Sylvester, C. 1996: 'Picturing the Cold War: an art graft/eye graft'. *Alternatives*, 21, pp. 393–418.

Teixeira, N. S. 1992: 'From neutrality to alignment: Portugal in the foundation of the Atlantic Pact'. *Luso-Brazilian Review*, 29, 2, pp. 113–26.

Telo, J. A. 1996: *Portugal e a NATO: o reencontro da tradição atlântica* (Lisbon: Edições Cosmos).

Chapter 14

The Reality of American Idealism

Gillian Youngs

Introduction

There are many ways of thinking about American idealism and this chapter presents a particular perspective on the issue. Its focus is transitions in American international theory over the twentieth century. In the realm of international affairs the US has been the great actor of this century. In many ways it has defined fundamental notions of what can be understood as *action* on the international stage. This chapter seeks to illustrate the contribution political theory has made to the achievement and communication of this. The arguments reflect on mainstream US international theory as primarily oriented towards understanding the changing nature of international relations with direct regard to the American position within them. The story of this theory parallels the major transformations in that position through the century, from an isolationist power to the leading military, economic, political and cultural player in global terms.

The key historical moments in this story include the two world wars, where the decisive impact of growing US involvement in international affairs was demonstrated. The post-1945 era has been characterized by the growth of the international economy, world trade and transnational corporate activity. The decade of the 1970s was marked by the end of the Vietnam war, the oil crises and the collapse of the dollar–gold standard. The bipolar framework (US–Soviet/West–East) dominated perspectives on power during the Cold War period and in post-Cold War times focus has moved do the US as sole superpower and leader of the expanding capitalist world. The reach of capitalism can now be considered truly global in terms of the major regions, for as well as the former Soviet bloc countries as spheres for capitalist activity there is the Communist giant China whose state-steered modernization process has attracted the lion's share of foreign direct investment to developing countries in recent years (Youngs 1997b and c).

In this new 'global age' (Youngs 1999) focus is placed on integrated social analysis of power which incorporates cultural as well as economic and political forces. Processes of globalization – frequently labelled Americanization or Westernization (see, for example, Darby 1997) – are a new prime concern for international relationists wishing to identify current forces for change (Clark 1997). One of the most extreme visions in this respect remains Francis Fukuyama's (1992) 'end of history' thesis which could well be renamed 'a theory of the American century'. This thesis is one of the most complex theoretical explanations and justifications of the irresistibility of the American dream in its modernizing, technological, capitalistic and, of course, liberal forms (Youngs 1997a). In its emphasis on political *and* economic liberalism it follows the dominant trend in US international theory, albeit in its own distinctive vein. It illustrates a major tendency in discourses of globalization to offer an allegedly universally applicable explanation of the contemporary world (Youngs 1996).

This chapter traces major developments in US international theory from idealism through realism to neorealism and neoliberalism. It begins with a discussion of discourse as an analytical basis for critically examining political theory. The following sections assess mainstream strands of American international theory for their preoccupation with America's changing role in the world. The main arguments illustrate how theory can usefully be understood as an expression of an evolving US identity. The last section discusses the post-Cold War era and examines the Fukuyama thesis as a discourse of globalization.

Theory as Discourse

Treating theory as discourse opens up complex possibilities for doing much more than assessing theory in its own terms. The fundamental myth that theory stands apart from reality, that it is somehow separate from other forms of practice, is overturned. Instead theory is considered as just one, albeit specific, form of practice which is necessarily informed by other forms of practice (Youngs 1999). Thus there is a desire to contextualize theory, to ground it, rather than treating it as an abstract given which somehow floats free from socio-historical conditions. The poststructural turn in critical international relations theory has explicitly revealed the importance of negotiating theory as discourse, and thus theory as an implicated element of social practices (Der Derian and Shapiro 1989; George 1994; see also Peterson 1992). The use of the term 'implicated' here illustrates that theory is intensely related to other forms of practice. One of the major reasons for adopting such an approach is to include theory in our understanding of power, to remove theory from any abstract and innocent status in this respect. The abstract nature of theory tends to endow it with timeless qualities, to loosen it from its historical and institutional associations, to veil those, to disconnect them from considerations of it.

There are many ways in which investigating theory as discourse is an exercise in recontextualization and social relocation: seeking out and identifying the social connections which have been lost from sight through the *abstraction* of theory from them over time. This incorporates an overt recognition that theories, notably dominant theories, are part of the historical processes which shape the world we live in, and, importantly, the ways in which we understand it. Furthermore they contribute directly to what we might think is possible within it and the manner in which this can be achieved. This requires us to look at how theoretical discourses are constructed and how theoretical developments are interrelated in order to understand the workings of their power.

International theory is by its very nature a grand theory. It aims to encapsulate the widest sphere of human relations: the diplomatic sphere, traditionally referred to as high politics, which tops the hierarchy of political domains. It may seem an awful truism to say that the most influential of international theory is likely to be produced by the most influential in international relations. And thus in the twentieth century it is not surprising that US international theory has played an increasingly influential role. In many senses it could be argued to have defined the field, or at least significant parameters of that field. What's more, and this is equally important, it has steered to a substantial degree the patterns of change within the field. Like most social science, international theory is addressing a moving, dynamic target. International relations are subject to constant change on smaller and larger scales. But that change can undoubtedly be viewed and understood from a myriad perspectives. Theory is part of the process that narrows down which perspectives are prioritized and which left to one side. US international theory has had its impact, as I will explain further below, in prioritizing perspectives which frame international relations and transformations within them in American terms. It is possible to view such theory as an expression of the American experience in international relations as much as it may be an attempt to explain or understand them. This is not to suggest that this in any way sums up such theory; rather to recognize that this is a part of its character.

Attention to theory as discourse encourages us to recognize the historical relevance of theory in very broad ways (Foucault 1966, 1969 and 1971; Burchell, Gordon and Miller 1991; Hindess 1996). This is a direct challenge to ahistorical abstract senses of theory and a claim that theory is part of the means for reflecting on and understanding history.

From Isolationist Starting Points: Idealism to Realism

The major transition in US international theory in the first half of the twentieth century was from idealism to realism. The idealist phase is most closely associated with America's move through its different stages of relative isolationism of the pre-First World War and inter-war years, and its Democratic president of 1913–21, Woodrow Wilson. The Wilsonian foreign policy, as one commentator has put it, was characterized by 'his complete faith in America's liberal-exceptionalism', his belief that 'the United States represented a new departure among the nations in both a moral and political sense' (Levin 1968: 2). This liberal-internationalism was encapsulated in his vision of the League of Nations which ultimately failed to gain the support of Congress and thus US membership. Assessments of Wilsonian values indicate the prioritization of a combined political and economic vision of liberal-capitalist expansionism (p. 14). The vision identified this expansionism as a universal good for international relations in the form of 'a stable American-inspired world order' (p. 8).

> The heart of the matter is that Wilson's conception of America's exceptional *mission* (my emphasis) made it possible for him to reconcile the rapid growth of the economic and military power of the United States with what he conceived to be America's unselfish service to humanity.
>
> (ibid.; see also Link 1971: 72–87)

The security and financial crises of the inter-war years, including the global shockwaves around the 1929 Wall Street crash, and the horrors of the Second World War in Europe and Asia, severely tested idealistic approaches to international affairs. With the US emerging as the military, political and economic superpower at the end to the war, the framework for the liberal-capitalist expansionist vision of Wilson could be put firmly in place. The United Nations system and the Bretton Woods financial institutions (International Monetary Fund and World Bank) founded in the immediate post-war years remain today as foundational to efforts to maintain an American-led world security order and increasingly open international economy guided by the liberal principles of free trade. These principles have been reflected in the General Agreement on Tariffs and Trade (GATT) and now the World Trade Organization (WTO) (Hoekman and Kostecki 1995). But the human cost of the Second World War, and the unrelenting nature of German expansionism in Europe under Hitler which paved the way for it, called for a new focus on *power* in the analysis of world affairs.

E. H. Carr's (1946 [1939]) *Twenty Years' Crisis: 1919–1939* and Hans J. Morgenthau's (1985 [1948]) *Politics Among Nations: The Struggle for Power and Peace* can be identified as two texts explaining this development in international theory from idealism to realism (Vasquez 1983: 15–17; George 1994). They represent between them a transatlantic fix on this development, Carr being located in the UK and Morgenthau in US academia (see also Carr 1942). They also in varying ways affirmed the growing importance of social *science*, that is, the application of scientific principles to the study of society (Hollis and Smith 1991). Carr's (1946: 65–7) discussion reminds us that belief in progress is common to both utopian and realist approaches, but the latter pursues the 'scientific' hypothesis that 'There can be no reality outside the historical process.' A realist perspective, he explained, contested notions of Wilsonian utopianism determining policy. 'The most cursory examination shows that the principles were deduced from the policies, not the policies from the principles' (p. 73).

In general terms realists treat politics as a pragmatic realm, hence Morgenthau's (1985: 5) enduring image of looking over the shoulder of the statesman as he writes his dispatches. The scientific claims of Morgenthau's (p. 4) 'theory of international politics' are incorporated into 'six principles of political realism', which include the belief that politics 'is governed by *objective laws* that have their roots in human nature' (my emphasis) and that it is possible to develop '*a rational theory*' (my emphasis) that reflects them. What primarily defines politics for Morgenthau is 'the concept of interest defined in terms of power' (p. 5). His framework recognized tensions between the *ideal* principle of sovereign equality between states and the *real* dominance of international relations by the two superpowers of the postwar era, the US and the Soviet Union (p. 8).

From Morgenthau's standpoint the possibility of change in the international system can be derived only from the real rather than the ideal. 'The realist cannot be persuaded that we can bring about . . . transformation by confronting a political reality that has its own laws with an abstract ideal that refuses to take those laws into account' (p. 12). Part of this reality is the understanding of international politics as a means towards ends on the basis of 'a struggle for power' (p. 31). There has been rich debate about the metaphysical and foundational overtones of Morgenthau's central commitment to the momentum of power in international relations, 'his reification of necessity in the form of power' (Griffiths 1992: 76). Among the many critical concerns about realist theory are questions about the status of its very own claims to be a *scientific* theory (Hollis and Smith 1991).

The realist perspective asserts the primacy of politics in international relations and the definitive influence of those states with the most power in those relations. It is therefore not suprising that it endured as a dominant theory in Cold War times. Commentators like Jim George (1994: 94) urge us to take this setting seriously in its broadest ideological terms and to recognize the links between realism's claims to scientific characteristics and the 'Enlightenment progressivist' principles of the US.

> In this discursive context, the emerging discipline of International Relations began to speak more confidently of its (rational-scientific) knowledge and of the correct means/ends method by which the enduring Realist wisdom of Western history must be transposed to the world of nuclear weapons, deterrence, and the Third Word.
>
> (ibid.)

Such critiques of realist theory as discourse, as theoretical discourse making claims to the power and status of science, take us deep into the world of theory *as* practice. They locate realist theory within the wider historical process of Enlightenment progress and rationality, of modernization and development, and recognize the US as the most powerful expression of those processes and the claims of scientific principles of rationality which underpin them. Thus they implicate theory as a powerful part of the social articulation and application of these principles.

In many ways it is possible to view the power-politics emphasis of realism, and its articulation of the definitive role of American power in all its modern forms, as an expression of idealism in another guise. To put it simply, realism can be understood as a particular and further *idealization* and thus celebration of American power and its multiple political and economic roles for change in the world. For all realism's claims to rationalistic pragmatics, echoes of Wilsonian idealism live on within it. And both idealism and realism as phases of US international theory have been concerned with American *action* in international relations and rationales for it. This is the case whether we are thinking of American involvement in the two world wars or its interventionism of the Cold War years, notably in Korea, Vietnam and Grenada. Henry Kissinger (US Secretary of State 1973–7), whose writings and experience fuse the theoretical and practical worlds of contemporary international relations, articulates the enduring and ambivalent qualities of the American mission in world affairs.

> The singularities that America has ascribed to itself throughout its history have produced two contradictory attitudes toward foreign policy. The first is that America serves its values best by perfecting democracy at home, thereby acting as a beacon for the rest of mankind; the second, that America's values impose on it *an obligation to crusade for them around the world* [my emphasis]. Torn between nostalgia for a pristine past and yearning for a perfect future, American thought has oscillated between isolationism and commitment, though, since the end of the Second World War, the realities of interdependence have predominated.
>
> (Kissinger 1994: 18)

Realism Transformed

As the world changed, realism changed too. The next major development in dominant US international theory was neorealism. Kenneth Waltz's (1979) *Theory of International Politics* is generally regarded as the founding text of this school of thought, but it is useful to contextualize it within earlier material on transnationalism (Keohane and Nye 1972) most closely associated with the school of neoliberalism. In recent times the 'debate' between neorealism and neoliberalism has occupied the foreground of mainstream preoccupations within US international theory (Baldwin 1993). The extent to which we can distinguish between the two schools of thought is indeed debatable, as I and others have argued (Powell 1994; Youngs 1999). As I will sketch out further below, what they share can be considered more important than the things which set them apart from one another.

Neorealism developed in the changed conditions of a bipolar world and directly addressed those developments from the US position. It held firmly to key founding elements of realism (as has neoliberalism in the main), including: theoretical claims to scientific rationality; the primacy of power politics; the prime role of the major powers in international relations; and an associated state-centric framework for interpreting those relations (see Ashley 1984). But by the 1980s the international conditions for the expression of US power had changed significantly from its point of supreme influence at the end of the Second World War. With European recovery and integration and the growth of international production, trade and financial exchange, there were now three nodes of economic power referred to as the triad: the US, Europe and Japan. Interdependence became central to considerations of US hegemony in the world

(Keohane 1984; Keohane and Nye 1989), not least because of the growing expression of American power through processes of transnationalism, notably the activities of its leading transnational corporations. Major events in the 1970s which focused concern upon the exact status of American hegemony included the final stages of the Vietnam war and US defeat, the collapse of the gold–dollar standard and the oil crises.

Neorealism is a status quo, 'problem-solving' theory (Cox 1981). It seeks to address questions and solve problems on the basis of the existing order. Developed in the Cold War era, neorealism clearly addressed US hegemony in the context of bipolar power. In realist vein, Waltz's theory adheres to the principle that the major powers have definitive roles on international relations. On this basis he judged the notion of the world as interdependent as having limited relevance.

> The United States and the Soviet Union are economically less dependent on each other and on other countries than great powers were in earlier days. If one is thinking of the international-political world, it is odd in the extreme that 'interdependence' has become the word commonly used to describe it.
>
> (Waltz 1979: 144)

Waltz's theory is unequivocal when it comes to America's expanding transnational corporate activity. Its international lead in this sphere gave it a 'special position' and one not shared by the Soviet Union (p. 151). 'Some American firms may be vulnerable; America as a nation is not. Someone who has a lot can afford to lose quite a bit of it' (p. 148). The US-centric nature of neorealism is substantially encapsulated in its concerns with the changing nature of American hegemony. The universal relevance of concern with this issue is communicated through the status quo orientation of neorealism, which identifies the established order as 'by definition, a "public good" (George 1994: 128).

As Jim George points out when commenting on Robert Gilpin's (1987) *The Political Economy of International Relations*:

> The end result, inevitably, is an anarchic scenario in which the focus of attention is order and the given interests of all (rational) states are calculated as support for institutional mechanisms that produce this order (e.g. World Bank/IMF) established by the United States, the hegemonic orderer.
>
> (ibid).

The Post-Cold War Setting: Neorealism Unleashed

In post-Cold War international relations the US has regained its immediate postwar stature as sole superpower but in very changed global conditions. Despite the Cold War, the last half of the twentieth century can be interpreted in many respects as a fulfilment of the Wilsonian vision of an American-led international order based on liberal capitalist expansionism. Post-Cold War, the global potential of that expansionism seems limitless. The last Communist giant, China, may be holding fast in terms of its state ideology, but its political economy must increasingly be understood in terms of its partial modernization, decentralization and growing integration into the global system of liberal capitalism (Youngs 1997c and 1998).

The unipolar (rather than bipolar) setting for American hegemony has opened the way for unprecedented embedding of the neorealist articulation of the established order as 'by definition, a "public good"'. But in a recent interview Kenneth Waltz (1998: 378) also suggested that the competitive power politics element of neorealism should not be ruled out of consideration for the future relevance of this theory because of the possibility of strategic opposition to the US. In the neoliberal camp where 'institutionalist theory . . . emphasizes the role of international institutions in changing conceptions of self-interest' (Keohane 1993: 271) there has been, not surprisingly, optimism about the growing potential for cooperation in international relations.

> Institutionalism . . . sees increased economic and ecological interdependence as secular trends, and therefore expects that as long as technological change prompts increased economic interdependence, and as long as threats to the global environment grow in severity, we will observe a continuing increase in the number and complexity of international institutions, and in the scope of their regulation.
>
> (p. 285)

As we head towards the new millennium in conditions of unipolarity, the tendency for new *global* theories is strong. One of the most well-known is Samuel Huntington's (1996) 'clash of civilizations' thesis in which he emphasizes civilizational 'fault lines' of contemporary and potential conflict (p. 252), and asserts 'the fundamental cultural gap between Asian and American societies' (p. 307). Processes of 'globalization and fragmentation' (Clark 1997) are attracting close attention in efforts to frame international relations anew in full recognition of the political, economic and cultural cross-border structures and interactions which have come increasingly to characterize them (see for example: Harvey 1990; Giddens 1991; Kofman and Youngs 1996; Zloch-Christy 1998). I want to focus here for illustrative purposes on one global theory explicitly the field of international relations which is probably the best known of all such theories. Francis Fukuyama's (1992) 'end of history' thesis is a testament to idealistic beliefs in the ultimate global triumph of America's liberal political economy. Its popularity has extended well beyond the academic boundaries of much of its argument. An elaborate theoretical formulation, it is also a celebratory universalistic articulation of the American way.

There are many senses in which if Wilsonian values are taken as an introduction to America's potential leadership role in twentieth-century international relations, then the Fukuyama end of history thesis can be understood as a conclusion to that role. Common to both, it can be argued, is an idealism about liberal political economy as the ultimate social end. As I have argued elsewhere:

> A distinctive element of Fukuyama's explanation of global relations is the stress on the *combined* role of liberal economics and politics in meeting human needs,

material and non-material. His claim that we can talk in terms of 'universal' history at all is highly dependent on his arguments concerning the attractions of a certain form of state organisation, i.e. liberal democracy, coupled with liberal economic principles. Fukuyama's position rests on idealised notions of liberal politics and economics. It views them very much as open frameworks offering seemingly endless opportunities for individual material gain and a social sense of self. Thus the 'global culture' which results is a technologically driven liberal political economy (Fukuyama 1992: 126). This idea of a global culture signals the transnational triumph of the technological imperative. The implicit suggestion is that 'technical rationality' (Ashley 1980) as championed by the politics and economics of liberal capitalism *is* culture. The assertion erases the importance of any other understanding of culture.

(Youngs 1997a: 31)

Integrated within this approach to culture is a sense of social homogeneity defined significantly by the scientific/technical rationality which is such a key characteristic of US political economy and power. I have argued that 'the Fukuyama perspective is neo-colonial in turn, a kind of post-imperialist dream or vision of the ultimate triumph of the *West* (ibid.; see also Little 1995). We return here to the topic of modernization mentioned earlier in relation to rationalistic Enlightenment principles of *progress*.

The universal destiny which the Fukuyama thesis posits is framed in terms of liberal progress as a given social good. The global culture described is defined on that basis. Technology and culture are related conceptually but not in an open way. Technology is reduced to the technological imperative, 'i.e. the evolutionary understanding of technological progress or advance' (p. 32). Culture is reduced to the expression of that progress. In its explicit focus on history the thesis is conceptualizing time in the reductive terms of an ideology of progress, capturing the world in a crude oppositional framework: 'divided between a post-historical [modernized] part, and a part that is still stuck in history' (Fukuyama 1992: 276, 385).

Continuities within Theory: The Future is American?

There has been a continuity of futurism in mainstream American international theory. This chapter has charted its major transitions over the century but it has also revealed embedded continuities. These prompt us to think about this form of theory and its *idealization* of the American way. It is theory which has projected the American state and its liberal modernism as an ideal model for humanity. This has been variously achieved, but the discontinuities should not deter us from considering underlying continuities. These are contained in the emphasis on different aspects of the American model of modernity according to the times. The main components are: a progressive, expanding capitalist economy; a rationalistic, scientifically organized and inspired society; a fusion of political with economic values to produce a holistic form of liberalism.

In all its different guises dominant US international theory this century has been outward-looking. It has reflected a view of America's place in the world, its potential and actual power in the world. This discussion has indicated the links of theoretical transitions to substantive developments. In this respect mainstream American international theory can be perceived as one mirror in which to view the US as part of a changing world this century. But at all times, whether more or less explicitly, this theory has been communicating the values which underpin the American worldview as a view of America *in the world*. The internationalism of the idealist phase expressed an overt missionary zeal about American leadership as a dynamic liberal force for potential universal harmony. The emphasis on rationality and power in the realist phase both reflected the traumas of world conflicts and failed efforts to maintain peace, and the hope which scientific ordering principles might bring to human society. Neorealism has continued the realist tradition but adapted it to a more complex international political economy where the factors underpinning American hegemony include institutional and transnational processes. Nevertheless neorealism remains a state-centred explanation of international politics, and the state at the centre of it all is clearly America. In recent post-Cold War times neoliberalism's emphasis on the potential for cooperation has been restated, but so have there been fresh claims for the pertinence of realist-style power politics perspectives to the possibilities of strategic opposition to the US. Huntington's 'clash of civilizations' approach offers a dark scenario of alternatives in this regard.

As we move towards the new millennium under conditions of American sole superpower status and globalization, space has been created for the revival of the triumphalism of the idealist phase in the early part of the century. I have used Fukuyama's familiar end of history thesis to illustrate this point. While it has its own distinctions, it in some ways also brings together the different stands idealizing the American model of modernity in mainstream US international theory over the century. The thesis celebrates the progressive nature of the technological imperative and posits it as the new global culture. But I have not drawn on Fukuyama's arguments to suggest simply that the story of mainstream American international theory has come full circle this century. I have done so rather as an affirmation of the continuity of idealism about the American model of modernity, differently expressed, through the contrasting stages of this theory. This assessment suggests that if the twentieth century is the American century then this theory is part of the explanation.

References

Ashley, R. K. 1980: *The Political Economy of War and Peace. The Sino–Soviet–American Triangle and the Modern Security Problematique.* London: Pinter.

Ashley, R. K. 1984: The Poverty of Neorealism. *International Organization,* 38, 225–86.

Baldwin, D. (ed.) 1993: *Neorealism and Neoliberalism: The Contemporary Debate.* New York: Columbia University Press.

Burchell, G., Gordon, C. and Miller, P. (eds) 1991: *The Foucault Effect: Studies in Governmentality.* Hemel Hempstead: Harvester Wheatsheaf.

Carr, E. H. 1942: *Conditions of Peace.* London: Macmillan.

Carr, E. H. 1946: *The Twenty Year's Crisis 1919–1939: An Introduction to the Study of International Relations*, 2nd edn. 1st edn published 1939. London: Macmillan.

Clark, I. 1997: *Globalization and Fragmentation: International Relations in the Twentieth Century*. Oxford: Oxford University Press.

Cox, R. W. 1981: Social Forces, States and World Orders: Beyond International Relations Theory. *Millennium: Journal of International Studies*, 10, 127–55.

Darby, P. (ed.) 1997: *At the Edge of International Relations: Postcolonialism, Gender and Dependency*. London: Pinter.

Der Derian, J. and Shapiro, M. (eds) 1989: *International/Intertextual Relations. Postmodern Readings of World Politics*. New York: Lexington.

Foucault, M. 1966: *Les mots et les choses*, Paris: Éditions Gallimard. Tr. by A. Sheridan Smith as *The Order of Things*, London: Tavistock Publications, 1970.

Foucault, M. 1969: *L'archéologie du savoir*. Paris: Éditions Gallimard. Tr. by A. Sheridan Smith as *The Archaeology of Knowledge*, London: Tavistock Publications, 1970.

Foucault, M. 1971: *L'ordre du discours*. Paris: Éditions Gallimard. Tr. by I. McLeod as The Order of Discourse, in M. Shapiro (ed.), *Language and Politics*, Oxford: Blackwell, 1984.

Fukuyama, F. 1992: *The End of History and the Last Man*. London: Penguin.

George, J. 1994: *Discourses of Global Politics: A Critical (Re)Introduction to International Relations*. Boulder, CO: Lynne Rienner.

Giddens, A. 1991: *The Consequences of Modernity*. Cambridge: Polity.

Gilpin, R. 1987: *The Political Economy of International Relations*. Princeton, NJ: Princeton University Press.

Griffiths, M. 1992: *Realism, Idealism and International Politics*, London: Routledge.

Harvey, D. 1990: *The Condition of Postmodernity: An Enquiry into the Origins of Cultural Change*. Oxford: Blackwell.

Hindess, B. 1996: *Discourses of Power: From Hobbes to Foucault*. Oxford: Blackwell.

Hoekman, B. M. and Kostecki, M. M. 1995: *The Political Economy of the World Trading System: From GATT to WTO*. Oxford: Oxford University Press.

Hollis, M. and Smith, S. 1991: *Explaining and Understanding International Relations*. Oxford: Clarendon Press.

Huntington, S. P. 1996: *The Clash of Civilizations and the Remaking of World Order*. New York: Simon and Schuster.

Keohane, R. O. 1984: *After Hegemony: Cooperation and Discord in the World Political Economy*. Princeton. NJ: Princeton University Press.

Keohane, R. O. 1993: Institutional Theory and the Realist Challenge After the Cold War. In Baldwin 1993, 269–300.

Keohane, R. O. and Nye, J. S. (eds) 1972: *Transnational Relations and World Politics*. Cambridge, MA: Harvard University Press.

Keohane, R. O. and Nye, J. S. 1989: *Power and Interdependence: World Politics in Transition*, 2nd edn. 1st edn published 1977. Boston: Little, Brown.

Kissinger, H. 1994: *Diplomacy*. New York: Simon and Schuster.

Kofman, E. and Youngs, G. (eds) 1996: *Globalization: Theory and Practice*. London: Pinter.

Levin, N. G. 1968: *Woodrow Wilson and World Politics: America's Response to War and Revolution*. New York: Oxford University Press.

Link, A. S. 1971: *The Higher Realism of Woodrow Wilson and Other Essays*. Nashville: Vanderbilt University Press.

Little, R. 1995: International Relations and the Triumph of Capitalism. In Booth, K. and Smith, S. (eds), *International Relations Theory Today*. Cambridge: Polity, 1995, 62–89.

Morgenthau, H. J. and Thompson K. W. 1985: *Politics Among Nations*, 6th edn. 1st edn published 1948. New York: Knopf.

Peterson, V. S. 1992: Transgressing Boundaries: Theories of Knowledge, Gender, and International Relations. *Millennium: Journal of International Studies*, 21, 183–206.

Powell, R. 1994: Anarchy in International Relations Theory: the Neorealist–Neoliberal Debate. *International Organization*, 48, 313–44.

Vasquez, J. A. 1983: *The Power of Power Politics: A Critique*. London: Pinter.

Waltz, K. N. 1979: *Theory of International Politics*. New York: McGraw-Hill.

Waltz, K. N. 1998: (Interviewed by F. Halliday and J. Rosenberg) *Review of International Studies*, 24, 371–86.

Youngs, G. 1996: Dangers of Discourse: The Case of Globalization. In Kofman and Youngs 1996, 58–71.

Youngs, G. 1997a: Culture and the Technological Imperative: Missing Dimensions. In Talalay, M., Farrands, C. and Tooze, R. (eds), *Technology, Culture and Competitiveness and the World Political Economy*. London: Routledge, 1996, 27–40.

Youngs, G. 1997b: Globalized Lives, Bounded Identities: Rethinking Inequality in Transnational Contexts. *Development*, 40(3), 15–21.

Youngs, G. 1997c: Political Economy, Sovereignty and Borders in Global Contexts. In Brace, L. and Hoffman, J. (eds), *Reclaiming Sovereignty*. London: Pinter, 1997, 117–33.

Youngs, G. 1998: The Political Economy of Transition: The Case of Hong Kong. In Zloch-Christy 1998, 262–78.

Youngs, G. 1999: *International Relations in a Global Age*. Cambridge: Polity.

Zloch-Christy, I. (ed.) 1998: *Eastern Europe and the World Economy: Challenges of Transition and Globalization*. Cheltenham: Edward Elgar.

Chapter 15

Contradictions of a Lone Superpower

David Campbell

Saving Captain O'Grady

For an understanding of United States foreign policy in the post-Cold War world, the story of Captain Scott O'Grady is instructive. As part of NATO's Operation Deny Flight, the enforcement of the no-fly zone over Bosnia instituted in 1993, US aircraft patrolled 'the Balkan skies' to ensure that none of the parties to the conflict enjoyed air support.[1] Few if any encountered hostile forces, until one day in June 1995. On the second of the month, a Bosnian Serb surface-to-air-missile downed a US Air Force F-16 fighter, forcing its pilot, Captain Scott O'Grady, to eject.

After parachuting safely to the ground, O'Grady endured six days and nights in 'enemy territory'. While Bosnian Serb forces searched for the missing airman, US forces planned a search and rescue mission for the first aviator shot down since the 1991 Gulf War. Unsure of whether O'Grady had survived, those hoping to rescue him began to lose hope as no evidence of his whereabouts were forthcoming. That is, until a fellow pilot from the base at Aviano in Italy prolonged a reconnaissance flight and picked up a faint radio message from O'Grady. Within hours, a force of more than 60 marines and 40 aircraft (one of the largest 'packages' deployed since the Gulf War) swept into north-western Bosnia to get O'Grady. Although there was no resistance on the ground, the rescue force encountered some fire as they flew towards the Croatian coast and onto a US aircraft carrier in the Adriatic Sea (O'Grady 1996; BBC 1997).

Back home, a week of media coverage which overtook even the O. J. Simpson saga, turned the incident into a foreign policy test for the Clinton administration. Focused on O'Grady's family as they awaited news, this coverage culminated in Ted Koppel, the prominent host of ABC News *Nightline*, broadcasting from the family living room, and magazine accounts of how O'Grady had been plucked 'out of the Balkan depths' in a manner that recalled 'the right stuff' (O'Grady 1996: 164; *Time* 1995; *US News and World Report* 1995). In response, O'Grady's return became a series of public celebrations stretching all the way from the USS *Kearsarge* to Aviano (where Neil Diamond's 'America' serenaded the military at a reception) and on to the White House. With the Captain cast as an American icon and fêted as a hero, his story became a fable of America in the post-Cold War world.

During a ceremony at Andrews Air Force Base in Washington, one general told the assembled admirers that O'Grady's experience in Bosnia

> reminds us that while the Cold War is behind us, we live in an uncertain and dangerous world. The readiness of our armed forces, and the quality of our people, are essential to our nation's ability to project its power and defend its interests ... [this rescue] exemplifies both the strength of America and the commitment of its people.
>
> (O'Grady 1996: 181)

The President, speaking at a Pentagon reception after hosting a White House luncheon for the O'Grady family, chimed in with similar thoughts. Clinton declared that the rescue 'said more about what we stand for as a country, what our values are and what our commitments are than any words the rest of us could utter.' O'Grady – who had told the President in an earlier phone conversation that 'the United States is the greatest country in the world' – responded first by thanking 'God for His love' and then dedicated the attention heaped upon him to all members of the armed forces who had gone unrecognized in previous conflicts (The White House 1995). Those military personnel present at the ceremony left feeling that this was 'a great day to be an American' (Air Force News Service 1995).

In many of its themes, O'Grady's story invokes dimensions of American identity which take it beyond the confines of a strategic account. Laced with religious imagery, declarations of faith and accounts of revelations – O'Grady (1996: 184) writes that his 'six days in Bosnia were a religious retreat for me, a total spiritual renewal' which resulted in his being 'born again' – the story recalls the way in which America's foreign ventures have often moralized about the country's mission in a faithless world (Campbell 1998). With its dedication to prisoners of war and those 'missing in action', it is a tale that demonstrates how the country can overcome the failures of past interventions and 'return with honour'.[2] Concluding with O'Grady's (1996: 188) 'pilgrimage' to the Tomb of the Unknown Soldier at Arlington National Cemetery in Washington, the story of this rescue can be read as a parable about the place faith, family and the military should be accorded in America's post-Cold War identity.[3]

Many of the details of O'Grady's survival and rescue reveal much about America in the post-Cold War world, though perhaps in ways unintended by either O'Grady or his adulators. Reminiscent of earlier captivity narratives – stories dealing with, for example, the fate of American colonists at the hands of the 'uncivilized' Indians (Drinnon 1990) – O'Grady's apparently heroic efforts, in a place which repeated intelligence briefings had informed him contained 'no safe areas' and 'no friendly forces', mask the fact that for an airman trained in survival techniques, avoiding people in civilian clothes carrying hunting rifles should not be the greatest challenge of his military career (O'Grady 1996: 38, 78, 80).

More significantly, although Operation Deny Flight was said to be a mission designed to keep the Bosnian skies free from violence (BBC 1997), the enforcement of the no-fly zone by O'Grady and others deflected attention from the war on the ground and the reluctance of NATO countries to commit forces to combat the violence there. This illustrated all too vividly the contradiction that is at the heart of the O'Grady story: the willingness of the US military to deploy a substantial force to save one man, in contrast to its reluctance to deploy any forces to respond to the deaths of more than 200,000 Bosnians. Encapsulated in this contrast, this paper will argue, is the uncertain meaning of US foreign policy in the post-Cold War period.

While America's leaders proclaim their country's might and its superiority, they remain unwilling to back their assertions of primacy with the military resources at their disposal. Adamant that they lead the sole remaining superpower, America's politicians are reticent about matching that claim with the currency of superpower. This is *not* to say that the United States *should* embark on a global strategy of military adventurism consistent with its constant reiteration of the idea that US leadership is essential, *or* to claim that military intervention is either the first or best option in all circumstances. It is, in contrast, to foreground the fact that the United States is in something of a strategic grey zone, caught between the universalism of its political appeals and the particularism of its limited capabilities (Warner 1996). While it might be thought that the uncertainty of this situation would be eased by either scaling back the rhetorical flourishes or increasing the willingness to use a superpower's resources until the two dimensions were correlated with one another, the nature of American foreign policy, especially its pivotal role in the (re)generation of American identity, means that this gap and the problems to which it gives rise is likely to persist. An appearance of incoherence and inconsistency is thus likely to be a structural feature of US foreign policy for some time to come.

The purpose of this chapter is to explore the dilemmas of US foreign policy in the post-Cold War world by focusing on how this strategic grey zone developed, and what its implications are, with respect to both strategic policy and political identity. To achieve this, the argument uses O'Grady's rescue, and the larger context of US policy with respect to the Bosnian war, as the launch pad for some of the themes which have come to dominate strategic thinking about America's role as the lone superpower in a dangerous world. As all contemporary strategic thinking is compared and contrasted to the Cold War experience, this chapter will highlight aspects of that period which need to be rethought in order to understand the present.

Bosnia and the Narcissism of US Foreign Policy

Bosnia was, arguably, the second crisis of the post-Cold War world, following shortly after the allied success (at least from the official perspective) of the Gulf War. But in contrast to the decisive action and clearly articulated objectives of that conflict, the war in Bosnia saw only military inaction and political confusion on the part of Europe and the United States (Bert 1997, Gow 1997). While the images of concentration camps and the details of ethnic cleansing were widely evident, and despite the post-Second World War conviction that such things should 'never again' happen, few governments regarded the conflict a sufficient threat to their vital interests to warrant the commitment of men, women or materiel. When a NATO plan (OpPlan 40–104) for military deployment was drawn up, its aim was to provide cover for a possible UN withdrawal from Bosnia. This indicated that the alliance was willing to commit forces to implement a failure – to provide cover for a retreat from even the hitherto limited 'humanitarian' involvement – but not to address the politics of the problem (Holbrooke 1998: 65–7).

Inaction with regard to Bosnia was a product neither of inattentivness nor lack of discussion on the part of officials in the United States. The Bush administration was aware of the forthcoming break-up of Yugoslavia and its likely consequences, but concluded it knew of no way to prevent it without reinforcing the idea that the US should be the world's policeman (Gompert 1996). In the Clinton administration, Bosnia was much discussed, but 'most high-level meetings on Bosnia had a dispirited, inconclusive quality . . . Although no one could ignore the crisis, there was little enthusiasm for any proposal of action, no matter what it was. The result was often inaction or half-measures instead of a clear strategy' (Holbrooke 1998: 81).

Part of the reason for this 'collective spinelessness' (Weiss 1996) was that the policies that were implemented for Bosnia were not directed at Bosnia *per se*. Instead, as the second crisis of the post-Cold War period, Bosnia became a symbol of US foreign policy, and, most significantly, a site upon which the United States could trial aspects of its strategies for this new era in which the Cold War guidelines for intervention were no longer operative (Danner 1997; Ó Tuathail forthcoming). Indeed, at various times during the four years of the Bosnian war the United States and its European allies declared that what was important was not so much a specific event or issue in Bosnia, but what that event or issue meant for the alliance between Europe and the United States, and especially the 'credibility' of one or other of the partners. If the inaction of allied forces meant that the United States looked weak and unable to lead, then this was a problem – not because it meant little was being done for Bosnia, but because the image of the United States suffered. The pivot for US policy towards Bosnia was therefore more often than not United States concern about itself and its allies, rather than a concern for Bosnia (Danner 1997; Holbrooke 1998: 84).

By the summer of 1995 the feeling that the on-going war in Bosnia was, regardless of US reluctance, an affront to the US was particularly acute in the highest reaches of the US administration. This egotism of policy combined with developments in the war to produce a more robust American commitment of diplomacy in sync with force designed to end the war. As the chief negotiator of the Dayton accords, Richard Holbrooke (1998: 359–60) argued:

> Strategic considerations were vital to our involvement, but the motives that finally pushed the United States into action were also moral and humanitarian. After

Srebrenica and Mount Igman the United States could no longer escape the terrible truth of what was happening in Bosnia . . . Within the Administration, the loss of three friends on Mount Igman carried a special weight; the war had, in effect, come home.

For Holbrooke, the massacre of thousands of Bosnian/Muslims at Srebrenica in July 1995, and the death in a road accident on Mount Igman above Sarajevo of three of his negotiating team, finally meant the war had 'come home'. Of course, for millions of Bosnians the war had been at home for more than three years, with disastrous effects. But although the carnage of this period had been much evident to the world during that time, it took something more immediate to the self to force action.

The Cold War Prism

That concerns about America's sense of self would condition its response to a crisis like Bosnia is hardly novel, as much about the Cold War itself can be characterized in this manner.[4] Rethinking this aspect of the Cold War is important for the argument here, because the consensus view that the post-Cold War period betokens new challenges stems from a (misplaced) surety about the nature of the Cold War.

While conventional wisdom dictates that the Cold War was a situation induced externally by the behaviour and beliefs of the Soviet Union, the majority of the national security bureaucracy's internal and secret assessments of the early Cold War emphasized that, although the threat to the United States and western Europe was most easily represented by the activity of Communist forces and the Soviet Union, the danger being faced was neither synonymous with nor caused by them. Moreover, when the nature of the East–West struggle was considered, its terms were as much cultural and ideological as they were geopolitical.

Indeed, the post-war texts of United States foreign policy always acknowledged that their initial concern was the absence of order, the potential for anarchy, and the fear of totalitarian forces or other negative elements which would exploit or foster such conditions. It was the (in)famous National Security Council Memorandum 68 (NSC-68 1978), after all, which declared 'even if there were no Soviet threat' the United States would still pursue a policy designed to cope with the 'absence of order among nations' (NSC-68 1978: 401, 390). In effect, then, a document as important as NSC-68 recognized that the interpretation of the Soviet Union as the pre-eminent danger to the United States involved more than a simple recognition of an external reality.

Moreover, assessments of the threat to America and its interests since the Second World War have identified a wide range of concerns. Despite considerable differences in the order of magnitude of each, over the years policy-makers have cited world Communism, the economic disintegration of Europe, Red China, North Vietnam, Cuba, Nicaragua, Libya, 'terrorists', drug smugglers, and assorted Third World dictators. In the post-Cold War period, this parade of possible enemies has been extended to include the prevailing atmosphere of ambiguity and uncertainty; countries such as Iran, Iraq, North Korea and other 'rogue states'; conditions such as environmental degradation, population growth, migration flows, and refugee movements; divergent belief systems such as 'fundamentalist Islam'; the culture clash of incommensurable civilizations; and even previously unimaginable possibilities such as 'killer asteroids', all represented in ways similar to the identity politics of the Cold War (Huntington 1993; Klein 1994; Miller 1993; *New York Times* 1992, 1992a).

None of these sources posed a threat in terms of a traditional calculus of (military) power, and none of them could be reduced solely to the Soviet Union. All of them, however, were (and are) understood in terms of their proclivity for anarchy and disorder, and the danger they posed for a particular American sense of self.[5] In this sense, we can regard many of the foreign policy texts which warn of threats as akin to the seventeenth-century literary genre of the jeremiad, or political sermon. In these rites, Puritan preachers combined searing critiques with appeals for spiritual renewal (something which Captain O'Grady claimed to have received amidst the dangers of Bosnia). These exhortations drew upon a European tradition of preaching the omnipresence of sin so as to instil the desire for order (Bercovitch 1978).

Thinking about the texts of foreign policy as jeremiad-like articulations of danger alters our understanding of the Cold War. As a struggle which exceeded the military threat of the Soviet Union, and a struggle into which any number of potential candidates – regardless of their strategic capacity to be a threat – were slotted as a danger, what is understood under the rubric of *the* Cold War takes on a different cast. If we recall that the phrase 'cold war' was coined by a fourteenth-century Spanish writer to represent the persistent rivalry between Christians and Arabs (Halliday 1990: 7), we come to recognize that the sort of struggle the phrase denotes is a struggle over identity: a struggle that is not context-specific and thus, for the United States, a struggle not rooted in the existence of a particular kind of enemy like the Soviet Union, but which reveals much about the self.

Although the global inscription of danger in United States foreign policy was something that long preceded the Cold War (e.g. the strategies of 'manifest destiny' in the nineteenth century), it was in the post-WWII period, when numerous overseas obligations were constructed, that the identity of the United States became even more deeply implicated in the military capacity and external reach of the state. In this sense, the Cold War needs to be understood as a disciplinary strategy that was global in scope but national in design. As a result, the Cold War can be understood as an ensemble of political practices and interpretative dispositions associated with the (re)production of political identity. In these terms, 'cold war' signifies not a discrete historical period the meaning of which can be contained, but an orientation towards difference in which those acting on behalf of an assumed but never fixed identity are tempted by the lure of otherness to interpret all dangers as fundamental threats which require the mobilization of a population. Illustrative of this was the so-called first crisis of the post-Cold War period, the United States-led war against Iraq (Campbell 1993).

This means that the collapse, overcoming, or surrender of one of the protagonists in the Cold War (as conventionally understood) does not mean that the orientation to the world identified as cold war has been rendered obsolete. In consequence, although we might term the period after the fall of the Berlin Wall and the demise of the Soviet Union as the *post*-Cold War period – though it is perhaps better understood as the 'post-Cold War yet pre-epithet new era' (Liotta 1997: 95) – it is marked as much by the persistence of well-established frames of reference for interpreting and representing danger as it is by new and allegedly incomprehensible phenomena.

Overcoming the Syndromes of Vietnam and Somalia

However, to highlight the persistence of Cold War ways of thinking into the post-Cold War period is not to suggest that the more things changed the more they remained the same. Different crises did provoke different responses which revealed that an ambivalence about America's place and purpose in the world were far from resolved in the triumphant firepower of Operation Desert Storm. Indeed, whereas the Gulf War was celebrated for its regeneration of American pride through violence (a not uncommon feature of American history; see Slotkin 1973), Bosnia was a different matter altogether.

In many ways, Bosnia broke the mould of the Cold War compact. Seemingly liberated from the restraints imposed by a global competitor, the US appeared to be in a position to exercise unchallenged force. At home, however, Bosnia reversed the traditional position whereby conservatives favoured military assertiveness and liberals did not (Muravchik 1992). While freed from the Cold War's geopolitical strictures – and encouraged by talk from the then UN Secretary General Javier Perez de Cuellar about an emerging right of humanitarian intervention which would override the principle of sovereignty (Steadman 1993) – the United States, in concert with the Europeans, failed to develop or implement a coherent strategy that would enable an effective response to the violence in Bosnia.

Operation Deny Flight, the mission on which Captain O'Grady served, was a good illustration of the half-measures the United States and its allies deployed with regard to Bosnia. It was instituted because although the new era was marked by fewer limits, the US military none the less seemed burdened by the legacy of previous US interventions. As Holbrooke (1998: 8) wondered, would the United States, in its 'post-Vietnam, post-Somalia mood', be willing to confront what his colleague Robert Frasure (in a typically deprecating way) called 'the "junkyard dogs and skunks of the Balkans"'?[6]

Colouring American strategic thought above all else has been the experience in Vietnam. Defeat at the hands of the North Vietnamese in 1975, after more than a decade of war, left 58,000 American soldiers dead and United States society split, produced what foreign policy observers have called – in a manner that suggests it is a sickness pervading the American body politic – a 'Vietnam syndrome'. This is the conviction that the military should not deploy its resources unless political objectives were crystal clear and the forces could be sent on such a scale as to make victory almost certain. At its heart is the belief that rarely if ever are American lives to be spent in pursuit of international goals. Its influence is reflected in transformations of American strategic doctrine during the 1980s spurred by subsequent disasters.

The October 1983 bombing of the US Marine barracks in Beirut, which left more than 200 service personnel dead, prompted a revision of when, where and how US military force might be used. Casper Weinberger, Secretary of Defense in the Reagan administration, detailed a doctrine which set forth restrictive criteria before troops could be sent abroad. The 'Weinberger Doctrine' called for deployment only when vital interests were at stake, diplomacy had failed, and a domestic consensus was created to support deployment of such a magnitude as to make military victory more than likely (Hoffman 1995: 23–4).

The basic thrust of this doctrine was reinforced by the idea of 'decisive force', which was central to the 1992 National Military Strategy (NMS) and advocated by the Chairman of the Joint Chiefs of Staff (JCS), General Colin Powell. According to the NMS, the idea was to ensure 'the ability to rapidly assemble the forces needed to win – the concept of applying decisive force to overwhelm our adversaries and thereby terminate conflicts swiftly with minimum loss of life'. Embodying the military's approach to the Gulf War, the thinking behind the idea of decisive force was to avoid in future the 'half measures and confused objectives' that were said to contribute to the Beirut bombing, as well as the 'protracted conflict' and 'divided nation at home' which most believed characterized the failure of Vietnam (Hoffman 1995: 24).

The concept of decisive force has become 'firmly rooted in the US military lexicon and culture' (Hoffman 1995: 26). Given that it was instituted so as to overcome the implications of defeat in Vietnam, its use means the ghost of Vietnam, though bloodied in the Gulf, continues to reappear every time military deployment is debated. This was evident with regard to Bosnia, where discussions about possible military options were dominated by the idea that anything other than (though perhaps also including) a replay of Operation Desert Storm would inevitably lead the US into a 'quagmire' reminiscent of the Mekong delta (Danner 1997; Ó Tuathail 1996: ch. 6).

Discussions about the use of force in Bosnia were enjoined by another spectre to accompany that of Vietnam. Although the images of concentration camps and ethnic cleansing had dominated the world's media throughout the summer of 1992, footage of an unfolding disaster in Somalia soon offered competition for political attention. In December 1992 US forces were dispatched, but not to the killing fields of Bosnia. Instead, they went ashore under the blaze of television lights near Mogadishu to facilitate the distribution of aid. Operation Restore Hope, it was judged, offered fewer costs and greater benefits – for the United States (Danner 1997).

This limited and non-traditional use of the military – agreed to by General Powell on condition it did not become a precedent for Bosnia – soon encountered, however, some of the dilemmas strategic planners responsible for the idea of decisive force had worked hard to prevent. After a UN peacekeeping force had taken charge in Somalia, American commanders unilaterally decided to put a bounty on the head of the 'warlord' they held responsible for attacks on UN forces, General Mohammed Farah Aidid. Almost a decade to the day after the Beirut barracks bombing, a party of US special forces pursued Aidid's forces in central Mogadishu. The largest fire-fight since Vietnam ensued, and it was reported that 18 US soldiers and some 200 Somalis had been killed. The latter were forgotten as America was enraged by the sight of Somalis dragging the body of a downed helicopter pilot through the streets of the city. Subsequent investigations reveal that more than 1,000 Somalis – few of them Aidid's militia – were indiscriminately massacred by panicked US troops caught in a series of ambushes (*The Observer* 1998).[7]

In the aftermath of this carnage, the idea of 'mission creep', a change in objectives brought on by the circumstances on the ground without prior planning, gained currency. As a term which suggests that a 'quagmire' will be the end result of a limited or non-traditional use of force, it was argued that US involvement in multilateral and multinational operations could not be tolerated. What was obscured in the fallout from Mogadishu was that the hunt for Aidid was a unilateral US operation of which UN commanders were unaware. None the less, the UN was held responsible for a peacekeeping mission which had gone beyond its objectives and cost American lives, and 'crossing the Mogadishu line' entered the strategic jargon as a scenario to avoid.

The fiasco in Mogadishu, and the blame wrongly ascribed to the UN, presaged an immediate shift in American policy. It killed off talk of 'assertive multilateralism' as a new strategy for the US (Urquhart 1998), and was taken to support the notion that the US was diminishing its sovereignty by subordinating itself to UN authority. After Republicans captured control of the US Congress in the 1994 mid-term elections, President Clinton responded to this sentiment by issuing Presidential Decision Directive 25, which limited US involvement in collective security operations, declared that the US was opposed to a UN army, and indicated that there would be no US military units earmarked for peacekeeping operations (Schlesinger 1995).

Somalia thus joined Vietnam as a ghost of things gone wrong, leading Holbrooke (1998: 216–17) to observe that a new 'Vietmalia' syndrome governed American thinking about the use of force. It fomented a significant debate about whether or not US forces would join the NATO allies on the ground in Bosnia, the outcome of which was a decision to send troops after a peace deal had been agreed and not before (Holbrooke 1998: 218–23). The defining characteristic of a 'Vietmalia'-governed strategy was an aversion to incurring casualties and an insistence on an exit strategy.

While the minimization of casualties is a self-evidently prudent strategy, ever since the development of 'decisive force' it has been a primary requirement for the US. It is a condition, however, not shared (at least to the same extent) by some of America's NATO allies. Canadian forces were present in Mogadishu, and suffered losses while engaging in violence against civilians. Dutch forces were implicated in the massacre at Srebrenica and put themselves in danger. Yet in neither country was the motto of 'not a single body bag' installed as one of the bases for considering future international duties. The Dutch defence minister observed after the disaster of Srebrenica that putting his country's armed forces in harm's way was a necessary price to pay when an international order in which human rights and social justice are priorities had to be defended (Weiss 1996: 90). The American preoccupation with the number of military personnel who might be endangered in a deployment suggests the United States' priorities lie elsewhere. The reaction to Captain Scott O'Grady's downing is evidence in abundance of this.

This well-developed aversion to casualties, however, does not come into play if the other criteria of 'decisive force' – clear objectives and deployment of forces sufficient to win – are deemed to have been met (Conversino 1997). According to Huntington (1997: 35), 'a national interest is a public good of concern to all or most Americans; a vital national interest is one which they are willing to expend blood and treasure to defend'. Rather than being distinct, this means that interests and casualties are almost inseparable as criteria. Indeed, the relationship between clear objectives and sustainable casualties borders on the tautological. That is because clear objectives require a well-delineated national interest, which, in turn, is something understood via reference to the possibility of casualties. Interests are 'vital' if one is prepared to die, but no one is prepared to die unless they are vital.

The difficulty in resolving this circular logic has produced another limiting criteria for the use of US forces. This involves the centrality accorded an 'exit strategy'. In situations where interests cannot be represented as vital, and substantial casualties cannot be contemplated, planners and the public seem willing to accept deployment if, prior to that deployment, a schedule for the return of the troops is fixed. The idea is that if the dangers of 'mission creep' leading inexorably to a 'quagmire' cannot be wholly avoided in terms of casualties, then the worst excesses can be avoided by specifying when the forces deployed can get out. In the absence of yardsticks by which the interests that led to an intervention could be considered vital, an exit strategy provides a judgemental substitute (Tellis 1996; Rose 1998).

Explicitly endorsed by figures such as Clinton's Secretary of Defense, William Cohen, and National Security Adviser Anthony Lake, exit strategy doctrine is less about a strategy than a deadline (Holbrooke 1998: 210–11). It betrays a desire for quick and easy interventions, short on both costs and time, when the argument that they serve vital national interests cannot be made. Given that the construction of the national interest, vital or not, is an inescapably politically fraught exercise (Weldes 1999), advocacy of an exit strategy becomes a substitute for a debate about the political goals of foreign policy. In the context of Bosnia, talk about the need for an exit strategy is often code for a desire to see the country partitioned (Mearsheimer 1997; Pape 1997).

The impetus for enshrining the exit strategy doctrine during the first Clinton administration stemmed from the failed interventions of its time in office.

Between 1993 and 1995, the inability to obtain support for the policy of 'lift and strike' (lifting the arms embargo on the Bosnian government and providing air support to turn back Serbian aggression) in Bosnia, the deaths of US soldiers in Mogadishu, and the turning back of a US navy vessel sent to support a return to democracy in Haiti, combined to make US policy the object of criticism. Derided as pursuing a global strategy of social work, the administration was said to be concerned with the internal conditions of peripheral places rather than the traditional prerogatives of international order and security (Mandelbaum 1996).

Despite the hesitations produced by this spectacle of feckless policies, the second Clinton administration saw an emboldened attitude. Bolstered in large part by its role in bringing about the cessation of the Bosnian war – albeit via a problematic political structure imposed on that country (Campbell 1999) – the administration was aided by shifts in the strategic opinions of the military following a series of new appointments. Led by General John Shalikashvili, Chairman of the JCS, 'Clinton's warriors' have endorsed a doctrine of limited intervention (Worth 1998). Increasingly supportive of non-traditional missions, especially as a rationale for maintaining high levels of defence expenditure, this change in attitude has to some extent softened the strictures of 'decisive force'. However, although the open-ended US commitment to NATO stabilization forces in Bosnia agreed in June 1998 demonstrates a fixed exit strategy will not always be invoked, and General Shalikashvili has been publicly mindful of the unavoidable nature of casualties in military operations (Conversino 1997: 21), the continuing concerns about 'mission creep', especially in the Balkans, remain.

Although the lessons learnt from past interventions in Vietnam, Beirut and Somalia have seen strategic thought ebb and flow in its emphases, few if any of the changes have challenged the legacy of Cold War representations of danger. As a consequence, America's Cold War commitment to a massive military has continued into the post-Cold War period. Notwithstanding the dramatic changes in the world since the fall of the Berlin Wall, defence spending in the United States is within the Cold War range, albeit down from its Cold War peak. With an annual budget of more than US$250 billion, the United States outspends its closest rival by more than a factor of three (Gholz et al. 1997: 8, 12). Successive examinations of defence spending – such as the Pentagon's 1993 Bottom-Up Review and the proposed Quadrennial Defense Review – have not substantially altered the overall pattern of America's defence posture, which is principally designed to simultaneously fight and win two 'major regional conflicts' (Robb 1997). Additionally, the 1994 Nuclear Posture Review, although it codified the 60 per cent reduction in the nuclear stockpile throughout the 1990s, did not challenge the strategic basis of nuclear deterrence nor the need to maintain a force of 3,500 warheads targeted at Russia (Cox 1995: 50–1). As a result, while much is made of the novel challenges of the post-Cold War period, the US is prepared to handle them with a Cold War military infrastructure.

The Persistence of Pre-eminence

The continuance of a Cold War military machine in a post-Cold War geopolitical environment has gone hand-in-hand with strategic objectives which, despite the fluctuations in thought of the 1980s and 1990s, have remained remarkably durable. Which is not how things at first appear. Each new issue of the journals of quasi-official Washington opinion – such as *Foreign Affairs, Foreign Policy, International Security, Orbis, The Washington Quarterly*, and the like – pours forth with complaints of strategic incoherence, ruminations about the loss of an intellectual compass, and proclamations on 'how to create a true world order' (Hendrickson 1994; Binnendijk and Clawson 1995; Odom 1995). At first glance, the only consensus seems to be that there is no consensus on what role the United States should be adopting in the post-Cold War world. Yet, for all the introspection about the United States' role in the world, the underlying aim of ensuring 'an environment in which democratic capitalism can flourish in a world in which the US still remains the dominant actor' is unchanged official policy (Cox 1995: 5), and largely accepted by most analysts, even if they disagreed on the means of achieving this goal.

The preservation of the United States' global pre-eminence, which the leaked 1992 Defense Planning Guidance detailed as the need to ensure a global rival does not emerge (*New York Times* 1992), has been the core of both the Bush and Clinton administrations' policies (Mastanduno 1997). The Clinton administration simultaneously employed, however, one or more of the competing strategic visions on offer (Posen and Ross 1996–7: 44). But no matter the multiplicity of approaches or tactics, no one in authority has veered from the path of regarding the United States as an exceptional nation whose superiority in world affairs is indispensable (Christopher 1998; Maynes 1998). When Under Secretary of State Peter Tarnoff suggested in 1993 that the US had a less interventionist and more multilateral foreign policy – as a means of explaining that the ditching of the 'lift and strike' policy for Bosnia in the face of European objections was a considered position for a new era rather than the defeat it appeared – he was swiftly rebuked by his superiors for casting into doubt the need for US leadership (Warner 1996: 44–7). As Secretary of State Warren Christopher (1998: 524) opined after his retirement, 'by the end of the term, "Should the United States lead?" was no longer a serious question'.

The Problems of Pre-eminence

The American consensus on United States pre-eminence stretches beyond its own shores. As Kagan (1998: 24–5) notes, when the leadership of the lone superpower was hamstrung by domestic distractions, the editorial pages of many Asian, European and Middle Eastern newspapers fretted about the loss of direction the world might consequently suffer. This might not be evidence of the universal acclaim for the benevolent empire that Kagan observes, but it is evidence of the fact that US leadership depends on the willingness of others to be led. In the absence of a competitor for global leadership, because few other states have either the desire or the resources, America's pre-eminence should therefore be seen as an effect of the strategies of others rather than a natural product of its own worthiness. However, the fickleness of this situation – evident when countries such as India and Pakistan refuse to bow to the pleas of the US president – exposes the fragility of claims about the inevitability and durability of America's lone superpower status (*The Guardian* 1998; *The Observer* 1998a).

Even if accepted, this position of perceived pre-eminence results in two significant problems for the United States. The first is that being number one in terms of perceived power does not mean you can always achieve what you set out to do. Huntington (1997: 42) sets out the dilemma in a typically triumphalist (and somewhat problematic) fashion:

> A decade after the end of the Cold War, a paradox exists with respect to American power. On the one hand, the United States is the only superpower in the world. It has the largest economy and the highest levels of prosperity. Its political and economic principles are increasingly endorsed throughout the world. It spends more on defence than all the other major powers combined and has the only military force capable of acting effectively in almost every part of the world. It is far ahead of any other country in technology and appears certain to retain that lead in the foreseeable future. American popular culture and consumer products have swept the world, permeating the most distant and resistant societies. American economic, ideological, military, technological, and cultural primacy, in short, is overwhelming.
> American influence, on the other hand, falls far short of that.

According to Huntington (1997: 42), the incapacity of power to be automatically translated into authority and influence can be explained by recognizing that in the United States there exists a gap between a 'strong country' and a 'weak government'. This is because, with the exception of the military, American resources cannot be marshalled by the state in the service of specific strategic goals.

This analysis, however, begs a more basic question, one which probes its rendering of 'power'. If American primacy is in all categories overwhelming, yet this does not translate into influence (let alone getting your own way), what is the meaning of 'superpower'? What is 'super' about relations of power which cannot deliver the unilateral capacity for achievement they seem to promise? If, as is the case, the 'lone superpower' cannot do much if anything by itself, what meaning can the category 'lone superpower' have? And if analysts recognize that most of the dimensions of power (cultural, economic and the like) they believe are 'super' are not synonymous with the state, why does their discourse of superiority incessantly invoke them as though they were state-controlled resources?

In the post-Cold War period the incessant conversation about the United States' mission has gone beyond the logic of representation to incorporate what Weber (1994), invoking Jean Baudrillard, calls the order of simulation. Given the asymmetrical relations of power which characterize the interdependent and globalizing nature of the late-modern world, no stable referent or signified entity can be produced by this debate. No matter how much the sovereignty of the United States is proclaimed as something to be asserted and defended – and 'sovereignty' has become, in a manner not dissimilar to that of 'Eurosceptics' in Britain, the valued commodity of conservative American politics (Beinart 1997) – it is unable to be defined, demarcated or secured. It is only through this debate's iteration of a 'sovereignty' or the status of 'superpower' can these phenomena be accessed, for no such conditions are unilaterally possible (as the impact of the financial crisis which originated in East Asia in 1997 has demonstrated).[8] In this sense, the debate about America's status as the lone superpower in a unipolar moment has become a *simulation of* and *substitute for* the United States' preponderant power and pre-eminence.

That the United States' position as a lone superpower can be considered the product of the order of simulation does not mean it is without real effects. Indeed, the discourse of American pre-eminence exposes a second problem with significant consequences. Given the gap between promise and performance in the application of power, there is a basic contradiction between a country which wants to maintain its supposed pre-eminence at the same time as its population appears increasingly averse to bearing the costs of this putative global position (Mastanduno 1997: 87).[9] While 'the national interest' is an example *par excellence* of something which appears to be stable but is in actuality simulated, there does seem to be a sentiment abroad in the country that 'American national interests do not warrant extensive American involvement in most problems in most of the world' (Huntington 1997: 36). Added to the conviction that no cause seems sufficient to warrant casualties, we can observe practical limits to even the simulation of power.

This leaves the United States in a precarious situation. Not because it is at risk in an uncertain world, but because the contradictions that emerge through the order of simulated superpower will create risks at home and aboard. At home, the risk will be one of political frustration and *ressentiment* stemming from the tension between the unending stating of global goals and the limited desire to bear their costs. Abroad, the unilateral defence of sovereignty – as in the United States' determination to ensure that Americans are immune from the workings of the permanent International Criminal Court, and the cruise-missile attack against 'terrorist' sites in Sudan to avenge the deaths of its citizens (*The Observer* 1998b, 1998c) – will more often than not put the United States on the margins of global opinion even as it proclaims its leadership position.

Conclusion

The record of US policy towards the war in Bosnia, when contrasted with the rescue of Captain Scott O'Grady, illustrates the bind the US places itself in *vis-à-vis* the rest of the world. The US is a country which has the capacity, and constantly declares it is the only one which has the capability, to assert global leadership. But when faced with genocidal violence it declined to commit its resources in response to hundreds of thousands of deaths – until one of its personnel went missing for six days and nights. Clinton was right when he observed the rescue of O'Grady said more than anything about what the US stood for, what its values and commitments were. However, what many heard about the US in the rescue of O'Grady was that it would put aside the spectres of interventions past only for its own.

The debate about the United States' place in the world, although more often than not conducted on the register of interests and policies, embodies a logic of representation central to the constitution of American identity, as the discussion above on the character of the Cold War suggested (Campbell 1998). The national identity that results from this and other practices is not by any means fixed and stable. It requires reiteration, and the rescue of O'Grady was one minor but symbolic moment in which this reiteration occurred. It demonstrated how, even in the so-called post-Cold War world, the identity politics of the Cold War intersects with fears derived from previous military interventions to shape the terrain on which future decisions will be made.

Those spectres of interventions past – the so-called syndromes of Vietnam and Somalia – continue to profoundly influence American strategic policy. They have given rise to the idea that complex situations become 'quagmires' and lead to 'mission creep' and should thus be avoided, unless decisive force can be applied. They have produced an aversion to casualties at all costs, and a desire for a clearly defined exit strategy prior to the entrance of military forces. As a result, US policy has often become inward-looking, concerned with its own image, power and prestige, rather than the issue at hand.

Paradoxically, none of these limitations on what is and is not desirable and/or possible have made much difference to the material basis of US military power. Long after the demise of the Berlin Wall and the collapse of the Warsaw Pact, the US continues to spend hundreds of billions of dollars on its arsenal, far outstripping in both quantitative and qualitative terms the military of any potential competitor. All of which fuels further the contradictions of the 'lone superpower'. If the US continues to be unwilling to support its purported pre-eminence with resources (preferably exercised multilaterally so as to match the transnational character of the world), or is unable to adopt the stance of a 'normal' country without hegemonic aspirations (Bandow 1994), the contradictions of a lone superpower will mean that its symbolic capital is unlikely to be valued highly.

Acknowledgements

For research assistance, thanks are due to Erna Rijsdijk. Martin Coward, Kate Manzo and Gearóid Ó Tuathail provided helpful comments.

Notes

1 Rendering the space of their operations as 'Balkan' has been a given for NATO. As such, these activities participate in the imagining of the Balkans (Todorova 1997).

2 It is a staple of American conservatives that there remain US service personnel still held captive or unaccounted for, especially in Vietnam. For a reading of the mythology surrounding POWs and MIAs, see Franklin (1992).

3 On the significance of tombs of unknown soldiers, see Anderson (1991). O'Grady's symbolic value for neoconservatives was identified in a number of media spots, especially a March 1996 interview for the Washington-based Family Research Council (whose motto is 'Family, Faith and Freedom') radio programme 'Straight Talk'. Details on the FRC, which regards national security policy as part of the agenda of family concerns, can be found at <http://www.frc.org>.

4 This section draws upon the arguments in Campbell (1998).

5 The dangers could be multiplied, however, when American identity was (as is both ontologically and empirically the case) contingent. Conservatives, such as Samuel Huntington, claim that the advent of multicultural America has meant the nation no longer has a secure sense of identity. Given the proposition that national interests spring from national identity, Huntington (1997: 28–9) argues that this uncertainty is responsible for the country's lack of clarity about its purpose, with the result that 'subnational commercial interests and transnational and nonnational ethnic interests have come to dominate foreign policy'. In the context of policy towards Bosnia, this attitude – which linked multiculturalism with 'balkanization' – was especially important (Campbell 1998a).

6 While the response to the violence in Bosnia differed from the Gulf War, the representations of that violence invoked understandings common to the Cold War (Campbell 1998a). For example, an American diplomat referred to Bosnian Serb territory as 'Indian country' – while US units in IFOR named their bases and areas using frontier references (e.g. Fort Apache) – thereby invoking a frame of reference common to US military expeditions at home and abroad since the nineteenth century. The comment is attributed to Robert Frasure (Holbrooke 1996).

7 Little attention has been paid to the number of Somalis who were killed as a result of Operation Restore Hope. While it is said that more than 100,000 were saved by the humanitarian efforts of the US-led forces, the CIA privately concedes that between 7,000 and 10,000 Somalis were killed in hostilities with US forces, with a total of 34 Americans dead (Maynes 1995: 98; Clark and Herbst 1996).

8 This is not to suggest that lamentations about sovereignty are fictive errors. In the order of simulation, the distinction between real/unreal is collapsed so that something like 'sovereignty' is neither fact nor fiction but a simulacrum or truth effect (Weber 1994: 125–6). This means that while the polemics of the American right – such as Buchanan's (1998) tilt at the windmill of global capitalism in the name of sovereignty – appear hopelessly unrealistic, they are no more wrong or right than the Clinton administration's invocation of the defence of sovereignty as the reason for its marginalization in the July 1998 vote to establish a permanent International Criminal Court (a stance which followed the claims of the American right; see Schaefer 1998). The point is that both these iterations of sovereignty are the only way in which sovereignty can be made available.

9   With congressional opposition resulting in a failure to pay more than $1.5 billion in
    dues owing to the United Nations (Urquhart 1998), and with an aid budget equiv-
    alent to 0.15% of GDP placing it only 21st amongst industrialized nations in the
    provision of assistance (Schlesinger 1995: 6), this aversion is easily documented.
    Former Secretary of State Warren Christopher (1998: 527) decried those Americans
    who wanted to have it both ways: 'Congress tended to talk a lot about US leader-
    ship but to shirk the hard decisions to provide the necessary resources for it.'

References

Air Force News Service (1995) O'Grady: Attention 'still unreal'. Found at
    http://www.allstar.fiu.edu/aerojava/O'Grady.htm, accessed 24 Aug. 1998 at 13:32.
Anderson, B. (1991) *Imagined Communities*, revised edn. Verso, London.
Bandow, D. (1994) Keeping the Troops and the Money at Home. *Current History* 93,
    8–13.
BBC (1997) *999 Special: Missing in Action*. British Broadcasting Corporation, Bristol,
    6 May.
Bercovitch, S. (1978) *The American Jeremiad*. University of Wisconsin Press, Madison.
Bert, W. (1997) *The Reluctant Superpower: United States Policy in Bosnia, 1991–95.*
    St. Martin's Press, New York.
Beinart, P. (1997) The Nationalist Revolt. *The New Republic*, 1 Dec.
Binnendijk, H. and Clawson, P. (1995) New Strategic Priorities. *The Washington
    Quarterly* 18, 109–26.
Buchanan, P. (1998) *The Great Betrayal: How American Sovereignty and Social Justice
    are Being Sacrificed to the Gods of the Global Economy*. Little Brown, Boston.
Campbell, D. (1993) *Politics Without Principle: Sovereignty, Ethics and the Narratives
    of the Gulf War*. Lynne Rienner, Boulder.
Campbell, D. (1998) *Writing Security: United States Foreign Policy and the Politics of
    Identity*, revised edn. Manchester University Press, Manchester.
Campbell, D. (1998a) *National Deconstruction: Violence, Identity and Justice in Bosnia*.
    University of Minnesota Press, Minneapolis.
Campbell, D. (1999) Apartheid Cartography: The Political Anthropology and Spatial
    Effects of International Diplomacy in Bosnia. *Political Geography* 18 (4), 395–435.
Christopher, W. (1998) *In the Stream of History: Shaping Foreign Policy for a New Era*.
    Stanford University Press, Stanford.
Clarke, W. and Herbst, J. (1996) Somalia and the Future of Humanitarian Intervention.
    Foreign Affairs 75 (2), 70–85.
Conversino, M. J. (1997) Sawdust Superpower: Perception of US Casualty Tolerance in
    the Post-Gulf War Era. *Strategic Review* 25, 15–23.
Cox, M. (1995) *US Foreign Policy After the Cold War: Superpower Without a Mission*.
    Royal Institute of International Affairs, London.
Danner, M. (1997) The US and the Yugoslav Catastrophe. *The New York Review of
    Books*, 20 Nov.
Drinnon, R. (1990) *Facing West: The Metaphysics of Indian Hating and Empire Build-
    ing*. Schocken Books, New York.
Franklin, H. B. (1992) *M.I.A., or Mythmaking in America*. Lawrence Hill Books, New
    York.
Gholz, E., Press, D. G. and Sapolsky, H. M. (1997) Come Home, America: The
    Strategy of Restraint in the Face of Temptation. *International Security* 21, 5–48.

Gompert, D. C. (1996) The United States and Yugoslavia's Wars. In *The World and Yugoslavia's Wars*, ed. R. H. Ullman, pp. 122–44. Council on Foreign Relations, New York.

Gow, J. (1997) *The Triumph of the Lack of Will: International Diplomacy and the Yugoslav War*. Hurst & Co, London.

The Guardian (1998) Washington Goes Back to Square One, 30 May.

Halliday, F. (1990) The Ends of Cold War. *New Left Review* 180, 5–23.

Hendrickson, D. C. (1994) The Recovery of Internationalism. *Foreign Affairs* 73, 26–43.

Hoffman, F. (1995) 'Decisive Force': A New American Way of War? *Strategic Review* 23, 23–34.

Holbrooke, R. (1996) The Road to Sarajevo. *The New Yorker*, 21–28 Oct.

Holbrooke, R. (1998) *To End a War*. Random House, New York.

Huntington, S. P. (1993) The Clash of Civilizations. *Foreign Affairs* 72, 22–49.

Huntington, S. P. (1997) The Erosion of American National Interests. *Foreign Affairs* 76 (5), 28–49.

Kagan, R. (1998) The Benevolent Empire. *Foreign Policy* 111, 24–34.

Klein, B. (1994) *Strategic Studies and World Order*, Cambridge University Press, Cambridge.

Liotta, P. H. (1997) No Man's Land: US Grand Strategy in the Wake of Yugoslav Intervention. *Mediterranean Quarterly* 8, 81–101.

Mandelbaum, M. (1996) Foreign Policy as Social Work. *Foreign Affairs* 75 (1), 16–32.

Mastanduno, M. (1997) Preserving the Unipolar Moment: Realist Theories and US Grand Strategy After the Cold War. *International Security* 21 (4), 49–88.

Maynes, C. W. (1995) Relearning Intervention. *Foreign Policy* 98, 96–113.

Maynes, C. W. (1998) The Perils of (and for) an Imprial America. *Foreign Policy* 111, 36–47.

Mearsheimer, J. J. (1997) The Only Exit from Bosnia. *New York Times*, 7 Oct.

Miller, J. (1993) The Challenge of Radical Islam. *Foreign Affairs* 72, 43–56.

Muravchik, J. (1992) The Strange Debate Over Bosnia. *Commentary* 94, 30–7.

New York Times (1992) Pentagon Imagines New Enemies to Fight in Post-Cold-War Era, 17 Feb.

New York Times (1992a) Asteroid Defense: 'Risk is Real', Planners Say, 7 April.

NSC-68 (1978) United States Objectives and Programs for National Security, April 14, 1950. In *Containment: Documents on American Policy and Strategy, 1945–1950*, eds Thomas H. Etzold and John Lewis Gaddis, pp. 385–442. Columbia University Press, New York.

The Observer (1998) US Soldiers Massacred 1,000 Somalis in Panic, 22 March.

The Observer (1998a) When Washington's Writ Runs Into the Sand, 7 June.

The Observer (1998b) Nowhere to Run for War Criminals, 19 July.

The Observer (1998c) The Missile Strike, Bungling Pentagon and Nerve Gas Factory That Never Was, 30 Aug.

Odom, W. E. (1995) How to Create a True World Order. *Orbis* 39, 155–72.

O'Grady, Captain S., with Coplon, J. (1996) *Return With Honour*. Harper Paperbacks, New York.

Ó Tuathail, G. (1996) *Critical Geopolitics*. University of Minnesota Press, Minneapolis.

Ó Tuathail, G. (forthcoming) A Strategic Sign: The Geopolitical Significance of 'Bosnia' in US Foreign Policy. *Environment and Planning D: Society and Space*.

Pape, R. (1997–8) Partition: An Exit Strategy for Bosnia. *Survival* 39, 25–8.

Posen, B. R. and Ross, A. L. (1996–7) Competing Visions for US Grand Strategy. *International Security* 21 (3), 5–53.

Robb, C. S. (1997) Challenging the Assumptions of US Military Strategy. *The Washington Quarterly* 20, 115–31.

Rose, G. (1998) The Exit Strategy Delusion. *Foreign Affairs* 77, 56–67.

Schaefer, B. (1998). The International Criminal Court: Threatening US Sovereignty and Security. Executive Memorandum No. 537, The Heritage Foundation, Washington, 2 July.

Schlesinger, A. (1995) Back to the Womb? Isolationism's Renewed Threat. *Foreign Affairs* 74 (4), 2–8.

Slotkin, R. (1973) *Regeneration Through Violence: The Mythology of the American Frontier, 1600–1860*. Wesleyan University Press, Middletown.

Steadman, S. J. (1993) The New Interventionists. *Foreign Affairs* 72 (1), 1–16.

Tellis, A. J. (1996) Terminating Intervention: Understanding Exit Strategy and US Involvement in Intrastate Conflicts. *Studies in Conflict and Terrorism* 19, 117–51.

Time (1995) Out of the Balkan Depths, 19 June.

Todorova, M. (1997) *Imagining the Balkans*. Oxford University Press, New York.

Urquhart, B. (1998) Looking for the Sheriff. *The New York Review of Books*, 16 July.

US News and World Report (1995) The Right Stuff, 19 June.

Warner, D. W. (1996). *US Foreign Policy After the End of the Cold War*. PSIS Occasional Paper No. 3, Graduate Institute for International Studies, Geneva.

Weber, C. (1994) *Simulating Sovereignty: Intervention, the State and Symbolic Exchange*. Cambridge University Press, Cambridge.

Weiss, T. G. (1996) Collective Spinelessness: UN Actions in the Former Yugoslavia. In *The World and Yugoslavia's Wars*, ed. R. H. Ullman, pp. 59–96. Council on Foreign Relations, New York.

Weldes, J. (1999) *Constructing National Interests*. University of Minnesota Press, Minneapolis.

The White House (1995) Remarks by the President Thanking the Armed Forces for the Rescue of Captain Scott O'Grady. Office of the Press Secretary, 12 June. Found at http://whitehouse.dm.net/library/1995/4185.TXT, accessed 24 Aug. 1998 at 14:32.

Worth, R. (1998) Clinton's Warriors. *World Policy Journal* 15, 43–8.

Part IV

Cultural Capacities: The American Dream?

Chapter 16

Occult Hollywood: Unfolding the Americanization of World Cinema

Marcus A. Doel

> The history of the cinema is the history of the 20th century.
> (Nowell-Smith, 1990, p. 160)

While moving pictures have a long history, cinema belongs to the twentieth century. Any event worthy of the name has found itself on the screen, that placeless place where all possible worlds coexist, and that timeless time where past, present and future coalesce. Yet when one thinks of almost any other medium – say photography, writing, or drawing – one tends to think of a practice. With cinema, however, one is hard pressed not to think of a place: Hollywood. Whilst there are many *forms* of photography, writing, and drawing; there seem to be many *places* of cinema: Europe, Africa, China, Russia . . . (cf. Everett, 1996; Leong, 1991; Malkmus and Armes, 1991; Petrie, 1992). With cinema, it is not so much form following function; as form following placement (cf. Clarke, 1997).

In this chapter I want to begin to open up Hollywood. Specifically, I want to show that American film, exemplified in Hollywood, is not an entirely 'American' affair, and that its 'Americanization' depended upon the occlusion – shutting out – of numerous 'foreign' practices. What the chapter pursues, then, are some of those foreign bodies that have been shut up and kept hidden within the diasporic space of Hollywood. For just as the squeezing out of warm air between two bodies of colder air belongs to the structure of the occluded front, so too does the squeezing out of 'other' cinemas from American and Hollywood cinema belong to their structure of occlusion. In the conclusion to the chapter, I will suggest that by revealing such acts of dissimulation and disavowal, the 'dominant' or 'major' forms of American cinema become subject to dissimilation. What was once shut up within the occluded structure of Hollywood can now be seen to differentiate and disperse that structure. In this way, cinema remains in the making and it is always traversed by 'minor' variations.

For many, Hollywood threatens to overwhelm world cinema. This is usually dramatized in terms of a geopolitical, geocultural and geoeconomic clash of national cinemas, with Americanization sweeping away 'minority' cinemas. The 'filmic fingerprints' of Hollywood are all over world cinema, to borrow a felicitous phrase from Everett (1996, p. 5). Yet such Americanization is only a fraction of the historical geography of cinema in the twentieth century. Indeed, the hyperbole that surrounds the tightening hegemony of Hollywood conveniently forgets that it is but one force in world cinema. Others include Bollywood, the pornographic film industry, and the world-wide explosion of 'home' movies.

In many respects, world cinema, like the world system, is haunted by a paradox. While it is virtually omnipresent and transparent, it remains enigmatic and resists 'cognitive mapping'. Cinema, like the world-system, seems to be both self-evident *and* indiscernible (cf. Jameson, 1992). In this chapter I want to begin opening up some of the processes that have enabled world cinema to be shut up around American cinema. I want to spread out some of the folds that have sustained the occlusion of world cinema through the occultation of Hollywood. In particular, I want to rework and decentre the familiar narrative of the ongoing Americanization of world cinema.

The Mass Machine

Without a trace of irony, Schrader (1996, p. 201) claims that there are 'two linear, chronological lines running from the beginning to the end of recorded history to the present. One is technology, the other democracy. They are progressive, not cyclical, and are the yardsticks by which art, religion and social conduct can be measured.' For Schrader, American cinema is where these two great movements interlace for the first time, where the free will of the atomistic masses conjugates most forcefully with the dynamism of the machine age. The rapid spread of parlours containing 'peep-show' machines for single viewers throughout American, and to a lesser extent European, cities in the 1890s exemplifies this fusion of spectacle, individualism and economy. The cinema as a collective experience is often traced back to 1902, when a Los Angeles parlour projected the slot-machine image onto a screen for an audience to view; although films had been shown to collective audiences in fairgrounds well before 1902 (Christie, 1996). Just as the mass uptake of the bicycle 'foreshadowed the automobile in providing immediate, democratic access to speed' (Thrift, 1994, p. 199), so many believe that photography and early cinema foreshadowed television in the democratization of art and technology. Such is the myth of American cinema. However, this story of manifest destiny is plagued by contingency. For the conjugation of technology and democracy arose from a strange twist of fate and an accident of geography. 'The upstart film medium dovetailed democracy and art', such that 'movies became the voice of the people', only because its desire to emulate was 'Scorned by Culture' (Schrader, 1996, p. 202). So, American cinema turned to the American masses for legitimation, which it found in the nickelodeon boom of the 1900s. 'In the ultimate insult, [US film makers] fled the East Coast with its stultifying patent laws and cultural prejudices and started anew, lock, stock and lens, in Southern California' (Schrader, 1996, p. 203).

On the eve of the next millennium one could be forgiven for thinking that world cinema has always been an American affair. However, in the first few decades of the century world cinema was dominated not by a patchwork of heterogeneous cinemas, nor by an embryonic American cinema, but by the French company Pathé Frères. Mattelart (1994, p. 16) reminds us that 'it was with the film industry that the first major internationalization of nascent mass culture began'. At the start of the twentieth century, while the French-based projection system of the Lumière brothers competed with that of the US-based Edison system, Mattelart notes how their rivalry was felt not only in Europe and the United States, but around the world, and especially in Latin America. For example, Pathé began opening 'foreign exchanges' from 1903, and soon had a chain of such exchanges encircling the world. A 1908 poster shows the Pathé brothers striding forward under the slogan 'Towards the Conquest of the World'. 'This was no idle boast', suggests Christie (1994, p. 94). 'Within two years, [they] would control nearly a quarter of the world's film trade.' What is striking about the French control of production and distribution is how, in the space of a few years, it was overturned and effectively eliminated in America. So, before focusing on the Americanization of world cinema one should consider what Abel (1995) paradoxically calls the 'Americanization of American cinema'. Moreover, given that accounts of the twentieth century are overwhelmingly dominated by various versions of technological determinism, with all manner of utopian and dystopian morality tales, it is worth recalling that the contingency of the historical geography of cinema extends to the existence of the medium as such.

In the latter part of the nineteenth century the entertainment industry passed through an enormous number of 'attractions', most of which disappeared rapidly. This is the context into which film-making emerged. Given that most early films were non-fiction actuality shots of ordinary life and distant places, many were pessimistic about its potential. Despite the plasticity of cinematic space and time, such as the 'reversing' of action sequences, early films appear to have been less attractive than X-rays, which had both a horrific and erotic quality, since they rendered visible an eerie form of nakedness. Indeed, Bottomore (1996) notes evidence that in the first few years of the twentieth century films slipped down the order of attractions, and disappeared entirely in some places.

The lanternists maintained that the cinema was a sensation of the moment, comparing it unfavourably to the quality and solid tradition of the lantern; many in the variety professions suggested that the movies were but another sensational act, which could only last for a brief time on their bills; early actuality producers wondered whether, once all the world's sights had been recorded on film, there would be anything left for the cameras to do; finally, representatives of the legitimate stage maintained that spectators would never prefer mere celluloid entertainment to the presence of live performers.

(Bottomore, 1996, p. 146)

What appears to have revived cinema in the 1910s, especially in America, was not the medium *per se*, least of all the conjugation of technology and democracy, but the unexpected turn to 'made-up subjects' and the re-staging of newsworthy events that proved impossible to film in actuality, such as battles, crimes and train crashes.

The contingency of the longevity of cinema is matched in its form. According to recent scholarship, the 'cinema of attractions' was not simply the precursor to the 'narrative cinema' that 'supplanted' it (Bowser, 1990; Brewster and Jacobs, 1997; Robinson, 1996). For example, Gunning (1996, p. 71) describes how he 'began to envision early film as less a seed bed for later styles than a place of rupture, a period that showed more dissimilarity than continuity with later film style'. It was not inevitable that films would come to express space-time continuity or realize 'their destiny' of 'telling stories'. One need only compare the conventions of classic Hollywood cinema with those of the European avant-garde or the Soviet constructivists to glean a sense of the specificity of certain forms of American cinema. Consequently, before the American nickelodeon boom of 1905 to 1909, films were *not* groping towards narration; they were attractions. It was less representational than presentational; less optical than haptical (cf. Lant, 1995). One should always remember the specificity of narrative cinema. Narrative solicits the interest of the viewer through the posing of an enigma that lends itself to the dialectical form of suspense and resolution, which in turn tends to evoke a stable and coherent, character-centred diegesis within which the enigma is to be played out. Hence the growing love-affair of the American film industry with the uncanny and enigmatic urban land-scapes of *film noir* (Naremore, 1995–6). By contrast, the haptical cinema of attractions 'invokes an exhibitionist rather than a voyeuristic regime. . . . Rather than a desire for an (almost) endless delayed fulfilment and a cognitive involvement in pursuing an enigma, early cinema, therefore, *attracts* in a different manner. It arouses a curiosity that is satisfied by surprise rather than narrative suspense' (Gunning, 1996, p. 75). Running alongside narrative cinema, then, there is an affective cinema that operates directly on the body. This other cinema is exemplified in slapstick comedy, pornography and action films. As such, it is fully imbricated with the narrative cinema of classic Hollywood films (Shaviro, 1993).

The Occlusion of World Cinema

Having noted the contingency and specificity of cinema I want to return to the Americanization of American cinema in the first decade or so of the twentieth century. With its Paris-based vertically integrated factory system of mass production and mass distribution well established before 1906, Pathé was able to position itself at the core of the massive expansion of cinema in both Europe and America and dominate it. Oddly, while Britain was a leading producer and exporter of films in the period 1896 to 1907, it was not part of the massive world-wide growth in production in the early 1910s, partly because of the ambivalence, and often distaste, of cultural intermediaries such as journalists and scholars. By 1914, 'only 2 per cent of the million feet of film sold in London for exhibition each week was home-produced' in Britain (Christie, 1994, p. 135). In America, the expansion of film production in the late 1900s was centred on Chicago, and America rapidly became Pathé's principal market. In 1907, Pathé 'unilaterally decided to replace film sales by rental, forcing the whole industry to follow suit' (Christie, 1994, p. 94). At that time, while American companies were releasing a film every couple of weeks or so, Pathé was releasing between three and six films each week (and eight to twelve films from its six factories by 1909). Moreover, while the American companies tended to produce a limited range of lengthy 'headliner' films, Pathé produced more varied lengths and attempted to cover all of the genres. Accordingly, 'By 1907 it was estimated that two-thirds of the films released in the United States came from Europe, mainly France' (Robinson, 1996, p. 134).

Following the cinematic explosion of the late 1990s, the stage was set for a geocultural clash between so-called 'American' and 'foreign' interests. Abel (1995) notes that by the Autumn of 1907, Pathé was selling nearly twice as much film stock on the American market than *all* of the American companies combined. Pathé planned to bring production to America to enhance its dominance of the market, yet by the close of 1908, Pathé's dominance of America had been significantly curtailed and then side-lined by a series of machinations instigated by the litigious Edison. This 'film war' was mediated through an overtly *moral* discourse which mobilized an array of powerful interests in the attempt to exclude 'foreign' films from America. Despite an initial turn by exhibitors to 'literary' and 'cultural' imports from Europe, Pathé's films became increasingly associated with immorality and bad taste. Such was the shift from the positive to the negative associations of Frenchness: from technical and aesthetic sophistication to deviancy and degeneration. This was paralleled by moves to replace '*grand guignol* melodrama' with the 'bright, happy denouements' of 'ethical melodrama' (Abel, 1995).

The turn against Pathé in the mid-1900s was overdetermined by more general concerns about early cinema's 'American' audience and who was being constructed as 'American'. As immigration into America peaked in 1909, at nearly a million, many became concerned that the Americanization of immigrants would become unrealizable. Since cinema-going had become a truly mass phenomenon embracing all strata of society it is not surprising that this fear should have been so forcefully focused on films. By 1910, perhaps half of the American population were being 'educated' by the cinema, which attracted in the region of 26 million people each week (Kuhn, 1985a, p. 4). In such a context, Pathé became positioned as part of the unassimilable and therefore undesirable 'foreign Other' that threatened the cohesiveness of the American people, and against which American cinema would become constituted as cinema.

After 1907, there were increasing restrictions on foreign subjects, and the formation of the American National Board of Censorship in 1909 helped entrench them. There was also a strong demand for true 'American subjects.' By 1910, one in five films were Westerns, which enjoyed enormous popular appeal both inside and outside America, especially in Europe. Even Pathé succumbed to the drive for 'American subjects' by producing Westerns; although its focus on 'Indians' and women as much as on cowboys and men put it out of step with the fixation on hardened white masculinities. As the genre waned, and the 'independent' producers gained ascendancy in the American film industry, Pathé found that its 'foreign' films had been all but excluded from the American market and its 'American' films left forlorn.

The Rise of Hollywood

The Americanization of American cinema was achieved through the systematic exclusion of European cinema, although the shutting out of foreign films was not something specific to America (for example, Germany initiated controls on foreign films in 1916–17). By the mid-1910s, that occlusion had been screened off. The turn-of-the-century domination of world cinema by European film-makers was all but forgotten, and by 1911–12, the export of American films – mostly to Europe – surpassed the flow of European films into America. It was at this time that the relocation of much of the American film industry to Hollywood began in earnest. What remained of the European competition collapsed during the First World War (Thompson, 1985). The formation in 1914 of the first national network for American film distribution, Paramount Pictures Corporation, soon gave it a stranglehold over the industry. To counter this, in 1917, 27 of America's largest cinemas formed their own 'independent' network of distribution, First National Exhibitors circuit. The now infamous Hollywood studio system crystallized out of the struggle that ensued for control over the American film industry. One of the casualties of this struggle was the creation of highly individualistic films, which had no place in the rationalized and standardized factory-like production system of Hollywood (Bordwell et al., 1985; Gomery, 1991; Staiger, 1995). A similar concentration of industrial and financial capital arose in the German film industry, and it had been a feature of the French film industry since the turn of the century.

In the 1930s and early 1940s, the tightening grip of Hollywood on world cinema was mitigated against by the introduction of sound, which enabled many non-English-speaking countries to create or to consolidate their national pro-duction (Mattelart, 1994). Although by 1919 about 80 per cent of all films were being made in California, 'the enormous popularity of synchronized sound from 1927 . . . redefined cinema as a cultural and political entity', suggests McQuire (1998, p. 201). 'The new language barriers imposed by sound films split the international market into a number of unequal national markets, enabling Hollywood, which operated in the largest domestic market, to consolidate its dominance of the international scene. . . . By the 1930s, movies had become vir-tually synonymous with Hollywood' (McQuire, 1998, pp. 201–2). Indeed, by 1930, the America film industry exported four times more than the European film industries.

In America, production, distribution, and exhibition became vertically integrated and centred on an organizational structure initially dominated by Paramount, Universal, Fox Film Corporation, First National, and Loews-Metro-Goldwyn-Mayer. This structure was constituted in the period 1912 to 1928, and was only made possible because of the breaking-up of Edison's monopolistic 'Trust' which had sought to stifle the flow of films. However, this structure rapidly reduced the scope for other companies to flourish. By 1930, an oligopoly of five vertically integrated 'majors' had emerged: 20th Century Fox, Loews-MGM, Paramount, Warner Bros (which had acquired First National, and, with the assistance of AT&T, rigged it for sound) and Radio-Keith-Orpheum (established by AT&T's principal rival, Radio Corporation of America). In addition, there were three significant 'minors': Columbia, United Artists, and Universal, which depended on the independent cinemas and those of the big five for exhibition. "Of the 23,000 theatres operating in 1930 the majors controlled only 3,000 but these accounted for almost three-quarters of the annual box-office takings in the US. The majors produced only 50% of the total output of the industry, but this figure represented 80% of the "A" films exhibited in the first-run theatres' (Kuhn, 1985a, p. 7). This left some scope for 'poverty-row studios' to make 'B' movies.

Although the Depression hit the film industry hard – with the bankruptcy of Paramount and the receivership of RKO and Universal – the vertically integrated structure of oligopolistic majors and constrained minors persisted. Meanwhile, a renewed concern for censorship swept through the industry, and by the mid-1930s a powerful regulatory authority had insinuated itself into the entire production process, from scripting to editing. The shift in Warner Bros' gangster films, from focusing on the gangsters to concentrating on the G-men who thwarted them, illustrates this new-found morality. The tight grip of the censor on American cinema lasted until the early 1950s, not least through the ramifications of the anti-Communist purges and the Cold War. This grip slackened when the First Amendment was extended to the film industry; and weakened still further in 1968, when the new production code finally acknowledged the importance of artistic freedom and initiated a rating system for content. Similarly, the increasing importance of the youth market and 'counter-cultural' values helped to refashion both the form and content of Hollywood films.

The Second World War sealed Hollywood's global dominance. 'Hegemonic at the end of the war, as much in America as in Europe, Hollywood – aided by its hatred of protectionism – assured itself of a strong position through the purchase of cinemas, the control of distribution, and the organization of local production' (Mattelart, 1994, pp. 64–5). Similarly, Wasser (1995, p. 428) reports that '1946 was the best year ever for Hollywood', with record-breaking American audiences; although attendance dropped markedly in the following years. While a backlog of Hollywood films swept into post-war Europe, the capital released from the onset of vertical disintegration helped fuel a huge expansion of overseas production in the 1950s and early 1960s. Furthermore,

the rise in the 1970s of the 'pre-selling' of the distribution rights to future pro-
ductions, along with the insatiable demand of the television networks for films,
'allowed foreign film producers to come to Hollywood and make "Hollywood"
pictures independent of the American companies and American financing'
(Wasser, 1995, p. 431). Nevertheless, 'While the Vietnam war showed that the
USA could be defeated militarily, no nation has yet devised a strategy to defeat
the USA cinematically' (McQuire, 1998, p. 206; cf. Easthope, 1988).

Hollywood as a Work of Art

Few would deny that 'since the end of the First World War, Hollywood has
operated both an economic and aesthetic hegemony over world cinema'
(McArthur, 1997, p. 33). Accordingly, McQuire (1998, p. 202) claims that out
of the world's encounter with Hollywood films 'was born . . . a fascination with
"America" – its cities, its cars, its speech, its fashions, its way of life – which
has proved a durable yardstick against which diverse cultures have sought to
measure themselves this century, whether in approbation or opposition.' While
many have mistaken the *enigma* of Hollywood films for an *image* with which
audiences all over the world either identity or dis-identify, one cannot assume,
as McArthur (1997, p. 33) does, that 'since the point of utterance of such representations is American, they reflect an American perception of the world'.

According to Wasser (1995), one consequence of the heightened internationalization of the Hollywood film industry has been its dissociation from the particular needs of the American audience, resulting in what he calls a 'deracinated
transnational media'. Since 'The global audience is too infinite to be knowable'
claims Wasser (1995, pp. 434–5), it should not be surprising to find Hollywood
countering such uncertainty by appealing to 'crude' and 'overly reductive'
codes and styles, such as slapstick humour, outlandish special effects and excessive violence. Given the plurality and incommensurability of 'local' narrative
forms, many globally oriented film products have reinvested in the affectivity
of a cinema of attractions. Consequently, Medved (1992, p. 3) declares that
'millions of Americans now see the entertainment industry as an all-powerful
enemy, an alien force that assaults our most cherished values and corrupts our
children. The dream factory has become the poison factory' (cf. Clover and
Rogin, 1990, which points to the reactionary film-making of Hollywood in the
1980s). While it is doubtful that the American audience is any more knowable
than the global one, and that the cinematic America has ever faithfully represented the so-called 'real' America, it is interesting that Wasser should fear that
American cinema may have fallen victim to the capriciousness of its object of
desire: 'In my conversations with American film executives, it was obvious that
they perceive the world wide market as desiring a certain image of America to
be featured in the movies. . . . film producers set for themselves the task of portraying an "America" that is a dreamscape for "universal" desires rather than
a historic reality' (Wasser, 1995, p. 435). So, Wasser bemoans the flood of
foreign money into Hollywood, which he sees as the fulcrum for globally
oriented, rather than American-oriented, film production. Here, it is worth

recalling that the early 'American' film industry also depended on an enormous amount of European creativity, technology and capital; especially from Germany, France and Italy.

Whatever the connection between the globally oriented product of the Hollywood-based film industry and the American market, one should be wary of accepting too readily the assumed global dominance of Hollywood films. For example, Saunders' (1994) study of the reception of American cinema in Weimar Germany suggests that they had varying degrees of success, were often found wanting – especially in terms of historical sense, narrative coherence, and ethical sophistication – and with the arrival of sound all but disappeared from German cinemas. Other reception studies bear such sentiments out (e.g. Vincendeau, 1992). The dominance of production, distribution and exhibition does not guarantee control over audience reception, which is always bound to cultural specificity and contextual contingency. Reception is always a matter of translation. Similarly, despite the continuing importance of what Bordwell, Staiger and Thompson (1985) characterize as 'the classical Hollywood film style' – which was hegemonic from about 1917 to 1960, and which was articulated around the aesthetic principles of 'decorum, proportion, formal harmony, respect for tradition, mimesis, self-effacing craftsmanship and cool control of the reader's response' (Bordwell et al., 1985, p. 4) – one should acknowledge the heterogeneity of American cinema, and even of Hollywood cinema itself: 'the presence within Hollywood cinema of a number of minority cinemas, some as large as a whole genre or cycle (film noir), others as small as one film' (Butler, 1992, p. 416; see also Horak, 1996; Leong, 1991; Medovoi, 1998; Naremore, 1995/6; Powers et al., 1996; Stevenson, 1997). Yet for all of its heterogeneity, Hollywood films are particularly associated with mass-produced generic products: stock forms; stock narratives; stock motifs; and stock stars. These owe much to the Hollywood labour process. Take Hollywood music:

> the organizattion of production is fragmented into a conveyor-belt routine. Each studio has its music department, whose director parcels out duties of composing and arranging for current productions. Further specialization – some musicians compose nothing but military music, some do nothing but Wild West music . . . – alienates composers from their labour, which is boring and repetitious, and allows them no vision of the whole.
>
> (Gorbman, 1991, p. 275)

The studio system's standardized output was differentiated principally by genre: Westerns; comedies; urban exposés; musicals; gangster movies; *film noir*; horror; thrillers; melodrama; sci-fi; action–adventure; war films; cartoons; etc. But since the familial is given over to repetition it is necessarily haunted by the spectre of tedium. As well as cinematic returns that enable efficient mass production and predictable mass consumption, financial returns are pivotal to the Hollywood film industry. Indeed, the former is largely the product of the latter. Yet as Perkins (1992, p. 195) reminds us: 'Movies may be commercially successful without being widely or greatly enjoyed. A film that fails does so in the terms set by the industry.'

While featured actors and actresses had been central to the theatre, early film was effectively anonymous, partly because of the uncertain nature of the new medium. Not only did this help keep labour costs down, it also ensured that the medium itself remained the star. Although the domination of the Americanized film industry until 1915 by the Motion Picture Patents Company – Edison's 'Trust' – did not weaken this anonymity, the independents eventually did. The costly move to longer feature films helped reinforce the development of a star system; and the subsequent mass production of films by the studios lent itself to differentiation by genre *and* star. Both are sign values whose currency is always already in circulation (Dyer, 1990; deCordova, 1990; Gamson, 1992). 'Publicity, advertising, and "exploitation" crews – organized together like newspaper city rooms and with 60 to 100 employees at their height – would actively create and manipulate the player's image' to create 'an image that did not *appear* to emanate from the studios', reports Gamson (1992, p. 6). 'The appetite for films, film stars, and their movie and private lives had by the 1920s become voracious. By the 1930s, Hollywood was the third largest news source' in America.

The Specificity of Hollywood

I have already noted the contingency of the historical geography of the film industry. Such contingency applies to both the technologies that underpin the medium and the narrative codes through which films are produced and consumed. What I want to emphasize here is how the realism of classic Hollywood cinema was not a necessary development of film-making, and certainly should not be seen as its culmination. Indeed, the shadowy world of early film-making was firmly embedded in the wider *fin-de-siècle* flight from realism. Additionally, the fixation on Hollywood films often blinds us to the specificity of other forms of cinema, especially the extraordinary creativity that came out of the earliest experimentation with film-making. Indeed, much has been *lost* in the changing practices that make up the historical geography of film-making in the twentieth century. For example, the advent of recorded sound in the late 1920s brought to an abrupt end silent cinema's experimentation with mobile framing and reinforced the need for rehearsals. The silent cinema's fondness for close-ups and dramatic cutting gave way to static group shots with much less cutting. However, what emerged in the 1920s was not sound *per se*, but a different articulation of sound and image, since 'silent cinema was never actually silent – or hardly ever' (King, 1996, p. 31). In silent cinema, 'a filmmaker's choice of music . . . could determine the "tonality" of a film's image' (Abel, 1996, p. 5; cf. Gorbman, 1991). Indeed, 'from the beginning cinema aspired to have a "voice" as well as an image' (King, 1996, p. 32).

Just as 'silent cinema' was rarely silent, so too was 'black and white' cinema rarely black and white. Usai (1996, p. 22) reminds us that 'eighty to ninety percent of all films produced in the United States towards the end of the 1910s were entirely or partially colored.' However, post-production colouring became increasingly problematic in the late 1920s and early 1930s, since it adversely affected the quality of the soundtrack recorded onto the film. With the phasing out of post-production colouring the studios had to decide whether to film on the more expensive colour stock. The Depression saw many of the major's withdraw from the added expense of colour, although Walt Disney secured exclusive rights for the production of colour cartoons. Despite an upsurge of interest in colour film in the late 1940s and early 1950s, cinema's competition with black-and-white television was to ensure that the re-emergence of colour cinematography would have to wait until colour television.

In the 1940s, Hollywood managed to achieve a significant articulation of technological innovation and aesthetic style. Early lenses gave a relatively narrow angle of vision and depth of field, which lent themselves to tightly-cropped images and close-ups. Hence 'one of cinema's eeriest early properties – giantism of the moving face' (Lant, 1995, p. 59). This had the dual effect of foregrounding the linking of shots, with all of the attendant implications for experimenting with space and time. In addition, early films were generally layered, rather than cast in perspective; a quality that persists in most cartoons (cf. George, 1990). This 'depthlessness' was exacerbated by early cinema's evenness of lighting, the fixity of the camera at right angles to the profilmic scene, painted backdrops, and fondness for tableaux. Despite the apparent self-evidence of Hollywood's realism, from which other cinematic forms would seem to struggle to depart, it took considerable effort to effect 'the shift from a cinema of frontal presentation to a cinema of articulated depth' (Lant, 1995, p. 53; cf. Salt, 1996). For example, the introduction of unusually wide-angled lenses in the early 1940s had the effect of giving the framed image a greater angle of vision and a deeper focus, drawing the spectator into the image. While the usual 35 mm motion-picture cameras gave an inhuman vision, these lenses were much closer to that of the human eye. Accordingly, the new deep-focus cinematography reduced the need for close-ups, reaction shots and a mobile camera. By naturalizing the image it helped to usher in a certain form of realism. Kuhn (1985b, p. 208) reminds us how 'this "zero point of cinematic style" enjoyed its apotheosis in the Hollywood cinema of the 1930s and 1940s, the era of the "classic" narrative system.' In short, 'the installation of perspective in the cinema – the fulfillment of its "three-dimensional vocation"– was not immediate and obvious, despite the physics of the photographic lens, but was striven for and awkward, producing discontinuity in the spatial worlds early cinema offered spectators' (Lant, 1995, p. 69). Once lost, such discontinuity has been hard for Hollywood to rediscover.

While a certain articulation of sound, colour and deep-focus contributed to the realist thrust of Hollywood cinema, lighting may arguably have worked in another register. Until the 1910s, films were invariably lit by diffuse daylight. Thereafter, artificial lighting came to dominate, which was used to highlight, add atmosphere and to captivate. As with the advent of night-life and the enchantment of commodities in the brightly-lit department stores, artificial light was to hyperrealize film: it lent things an auric *excess* of reality (cf. Schivelbush, 1988). However, such hyperrealism was subordinated to the dominant

Hollywood aesthetic: enough lighting to enchant the real; but not enough to de-realize it. More specifically, this hyperrealism was wedded to the enchantment of the everyday. By combining the aesthetics of commodity fetishism and hyper-real naturalism, Hollywood films became exemplary expressions of the con-suming passions of what Taylor (1996) has aptly dubbed 'ordinary modernity'. However, whether hyperrealist or not, moving pictures have *always* been asso-ciated with something more than realism, not least because of the illusory and uncanny nature of the medium itself. Indeed, film-makers pursued trick pho-tography more or less from the start, and they developed an extensive reper-toire of techniques during the 1890s (Chanan, 1996).

Whilst considering the specificity of Hollywood's brand of cinematic realism it is worth noting that the spectator's credulity cannot be taken for granted. Even in the case of the earliest films there is little evidence to support *the* found-ing myth of cinema: that film-goers were terrified by 'suddenly' encountering 'the absolute novelty' of moving images; most famously the Lumière brothers' (1895) *L'entrée d'un train en gare de la Ciotat*, whose train 'darts like an arrow' out of the screen, and which could be called 'the first static vehicle', to borrow a fine phrase from Virilio (1989, p. 110). Most film-goers were well versed in how to 'suspend their disbelief' in readiness for such dark attractions: 'these screams of terror and delight were well prepared for by both showmen and audi-ence' (Gunning, 1995, p. 129; cf. Brewster and Jacobs, 1997). Indeed, the speci-ficity of the earliest films was probably not so much realism and the audience's panic-stricken credulity as the *uncanniness* of a shadowy, dreamy and in-substantial spectacle drained of all vitality. Yet such irrealism – and proto-surrealism – was not born with the cinema. Many apparatuses had long been associated with the black arts, such as the macabre reputation of the Magic Lantern, the Phantasmagoria of the 1790s, the 'spirit photography' of the 1870s, automata, and the waxwork Chamber of Horrors. Likewise, *camera obscurae* have been dramatizing and de-realizing actuality for centuries. Cinema is but one expression of the ceaseless occultation of actuality, of the interminable folding of this world and its doubles.

All in all, then, recent studies of early film-making have challenged the teleology that treats it as a precursor of Hollywood's 'classic' form: 'so-called "primitive" cinema should be considered a fully realised mode of filmic com-munication, constructed around alternative, not inferior, conceptions of story-telling and spatio-temporal continuity' (Pearson, 1996, p. 154; cf. Brewster and Jacobs, 1997).

The 'New' Hollywood

Owing to their virtual monopoly of first-run cinemas, the Hollywood majors were able to frustrate the entry of other film-makers into the market. However, Hollywood suffered considerable problems in the late 1940s and early 1950s, not just from the 1948 Supreme Court ruling against the Big Five's production–distribution–exhibition monopoly that compelled them to divest themselves of their cinema chains, but also from the two-thirds fall in audiences, which was exacerbated by the growth of television. While weekly cinema attendance in America hit a high of 95 million in 1929, and 90 million in the years immediately after the Second World War, it fell to around 20 million in the late 1960s and has hovered around that figure ever since: much of this audience being young. However, it was not until the late 1960s and early 1970s that new independent production companies began to emerge as the power of the studios weakened, the liberalized production code came into effect, and the demand for made-for-television films grew rapidly. In this new situation that extends to the present, the once autocratic studios became incorporated into huge corporations. They increasingly moved to the short-term contracting of talent (such as writers, directors, and stars), subcontracted significant parts of the labour process, and abandoned others. In the process, studios increasingly relinquished control of their stars' image-maintenance, and delegated it to the stars themselves and a legion of public-image professionals (Gamson, 1992). The once marginal talent agents became recast as key players in the new Hollywood. They assemble 'packages' of scripts, stars and directors that serve as the principal currency in the restructured Hollywood film industry (Powers et al., 1996; Balio, 1985). While the major studios remain dominant they increasingly need to collaborate with both talent agencies and other media, particularly television (cf. Anderson, 1994). Meanwhile, the rise of home video in the 1980s has enabled Hollywood, and especially the pornography industry, to create a new market for films. By the mid-1980s, the profits from video rentals and sales were surpassing those from box-office takings.

It is a moot point whether the 'new' Hollywood is characterized by 'industrial dualism, in which the smaller independent firms would act as shock-absorbers for the "majors" (Aksoy and Robins, 1992, p. 4), or whether it truly is a 'poststudio structure'. For example, Christopherson and Storper have argued that the new Hollywood is an exemplary form of post-Fordism. They maintain that the post-1948 restructuring of the Hollywood studio system saw it shift from oligopolistic mass production through vertically integrated large-firm structures to the flexible specialization of vertically disintegrated agglomerations of small-and-medium-sized enterprises (Christopherson and Storper, 1986; Storper and Christopherson, 1987). 'Post-Fordist' Hollywood is supposedly characterized by dense transactional networks between firms that are bound together through reciprocity, trust, collective risk-taking and creative collaboration; rather than through competition, division and rivalry. Moreover, the heterogeneity and often unpredictability of the various cinematic elements have mitigated against their geographical disaggregation. In this way, the processes that have resulted in 'the material dispersal or "fragmentation" of Hollywood cinema' (Hay, 1997, p. 223) have surprisingly reinforced its spatial agglomeration. Accordingly, Belton (1994, p. 80) maintains that 'each time a new film is made today, a "studio" must be assembled from scratch to produce it'. Production 'has left the factory and returned to a kind of preindustrial workshop in which the final product is custom-made'.

So, in the wake of classic Hollywood's vertical disintegration came the horizontal agglomeration of the new poststudio 'entertainment industrial complex' (cf. Hillier, 1992). However, once one recognizes the continuing dominance of the major Hollywood conglomerates; and the structural asymmetries with regard to reciprocity, trust and risk-taking; the idealized post-Fordist screenplay for the 'new' Hollywood becomes implausible (cf. Aksoy and Robins, 1992; Powers et al., 1996; Prindle, 1993). While more production has been under-taken by the independents, it remains largely financed and distributed by the majors, who themselves have increasingly focused on more costly and more prestigious production. Whilst giving a semblance of diversity and competition, the independents increasingly shoulder the greatest risks in the film industry. 'The major Hollywood companies are being turned into image empires with ten-tacles reaching down, not only to movies and TV programmes, but also to books, records, and even hardware', write Aksoy and Robins (1992, p. 17). 'The feature film business no longer exists in its own right, but is increasingly becoming part of an integrated, global image business, central to the broader media strategies of entertainment companies and conglomerates." In the 1990s, Hollywood is increasingly providing not merely block-buster movies, but a 'total' cinematic experience, which ranges from tie-in merchandising campaigns to theme-park rides (Wasko et al., 1993; cf. Staiger, 1990).

It is also worth noting that the 1980s saw further conglomeration of the studios' parent corporations; along with the onset of vertical reintegration, occa-sioned by a 1984 Supreme Court ruling that parent companies of the studios may own cinemas, effectively reversing the 1948 Anti-Trust ruling. The future of Hollywood will partly depend on whether the production of films will be geared around the specificity and singularity of 'talent', or whether such talent will once again become yoked to an extremely integrated, assembly-line method of production. Meanwhile, in the 1990s, the growth of block-buster films has made the new Hollywood more susceptible to boom and bust cycles. "Since the system which guaranteed a profit has disappeared, these guarantees must be built into the films themselves in the form of bankable stars, sequelization, and presales to cable television" (Belton, 1994, p. 80).

Conclusion

In this chapter I have sought to unfold some of the occlusions that surround and sustain Hollywood: such as the occlusion of 'foreign' films and subjects at the start of the century; the occlusion of early cinematic forms of 'haptical attraction'; the occlusion of 'minority' cinemas; and the occlusion of the 'new' Hollywood. The come-uppance of this unfolding is simply that Hollywood should not be seen as *the* most significant force in world cinema. Hollywood is not a privileged form, but merely an articulation of several modes of cinematic representation amongst others (Bordwell, 1997; Gaines, 1992; Gomery, 1991; Mast and Kawin, 1992). It is a commercially successful, ideologically powerful, and popular articulation to be sure; and one with a tremendous power to affect and incite bodies; but it is *an* articulation none the less. As such, Hollywood films may be folded, unfolded and refolded in many ways; as reception studies clearly demonstrate. The future of world cinema remains in the making. In other words, while Hollywood films remain a 'major' form of world cinema, this does not alter the fact that world cinema – including Hollywood cinema – continues to be traversed by 'minor' variations, such as the hybridization of genres, the splicing of places of enunciation, or the recovery of 'haptical cinema'. Whether the output, profitability and affectivity of the 'new' Hollywood will be over-taken by that of, say, pornography, home movies or Bollywood is a matter that turns on 'majorities'. Running alongside this are all of the cinematic departures that are forever in the making, deviation that owe as much to production and distribution as they do to the context-bound specificity of consumption and recycling. In this sense, perhaps cinema in the next millennium will prove to be an increasingly minority affair.

References

Abel, R., 1995, 'The perils of Pathé, or the Americanization of early American cinema', in *Cinema and the Invention of Modern Life*, eds L. Charney, V. R. Schwartz (University of California Press, Los Angeles), pp. 183–223.

Abel, R. (ed.), 1996 *Silent Film* (Athlone, London).

Aksoy A., Robins K., 1992, 'Hollywood for the 21st century: global competition for critical mass in image markets', *Cambridge Journal of Economics*, 16, 1, pp. 1–22.

Anderson, C., 1994, *Hollywood TV: The Studio System in the Fifties* (University of Texas Press, Austin).

Balio, T. (ed.), 1985, *The American Film Industry* (University of Wisconsin Press, Madison).

Belton J, 1994, *American Cinema/American Culture* (McGraw-Hill, New York).

Bordwell D, 1997, *On the History of Film Style* (Harvard University Press, Cambridge).

Bordwell, D., Staiger, J., Thompson, K., 1985, *The Classical Hollywood Cinema: Film Style & Mode of Production to 1960* (Routledge, London).

Bottomore, S., 1996, ' "Nine days' wonder": early cinema and its sceptics', in *Cinema: The Beginnings and the Future: Essays Marking the Centenary of the First Film Show Projected to a Paying Audience in Britain*, ed. C Williams (Routledge, London), pp. 135–49.

Bowser, E., 1990, *The Transformation of Cinema, 1907–1915* (University of California Press, Berkeley).

Brewster, B., Jacobs, L., 1997, *Theatre to Cinema: Stage Pictorialism and the Early Feature Film* (Oxford University Press, Oxford).

Butler, A., 1992, 'New film histories and the politics of location', *Screen*, 33, 4, pp. 413–26.

Chanan, M., 1996, 'The treats of trickery', in *Cinema: The Beginnings and the Future: Essays Marking the Centenary of the First Film Show Projected to a Paying Audience in Britain*, ed. C. Williams (Routledge, London), pp. 117–22.

Christie, L., 1996, *The Last Machine: Early Cinema and the Birth of the Modern World* (British Film Institute, London).

Christopherson, S., Storper, M., 1986, 'The city as studio; the world as back lot: the impact of vertical disintegration on the location of the motion picture industry', *Environment and Planning D: Society and Space*, 4, 3, pp. 305–20.

Clarke, D. B. (ed.), 1997, *The Cinematic City* (Routledge, London).

Clover, C. J., Rogin, M. (eds), 1990, 'Special forum: entertaining history: American cinema and popular culture', *Representations*, 29, pp. 1–123.

deCordova, R., 1990, *Picture Personalities: The Emergence of the Star System in America* (University of Illinois Press, Urbana).

Dyer, R., 1990, *Stars* (British Film Institute, London).

Easthope, A., 1988, 'Hollywood and Vietnam', in *Tell Me Lies About Vietnam: Cultural Battles for the Meaning of the War*, eds A. Louvre, J. Walsh (Open University Press, Milton Keynes).

Everett, W. (ed.), 1996, *European Identity in Cinema* (Intellect, Exeter).

Gaines, J. (ed.), 1992, *Classical Hollywood Narrative: The Paradigm Wars* (Duke University Press, Durham).

Gamson, J., 1992, 'The assembly line of greatness: celebrity in twentieth-century America', *Critical Studies in Mass Communication*, 9, 1, pp. 1–24.

George, R., 1990, 'Some spatial characteristics of the Hollywood cartoon', *Screen*, 31, 3, pp. 296–321.

Gomery, D., 1991, *Movie History: A Survey* (Wadsworth, Belmont).

Gorbman, C., 1991, 'Hanns Eisler in Hollywood', *Screen*, 32, 3, pp. 272–85.

Gunning, T., 1995, 'An aesthetic of astonishment: early film and the (in)credulous spectator', in *Viewing Positions: Ways of Seeing Film*, ed. L. Williams (Rutgers University Press, New Brunswick), pp. 114–33.

Gunning, T., 1996, ' "Now you see it, now you don't": the temporality of the cinema of attractions', in *Silent Film,* ed. R. Abel (Athlone, London), pp. 71–84.

Hay, J., 1997, 'Piecing together what remains of the cinematic city', in *The Cinematic City,* ed. D. B. Clarke (Routledge, London), pp. 209–29.

Hillier, J., 1992, *The New Hollywood* (Studio Vista, London).

Horak, J.-C., 1996, *Lovers of Cinema: The First American Film Avant-Garde, 1919–1945* (University of Wisconsin Press, Madison).

Jameson, F., 1992, *The Geopolitical Aesthetic: Cinema and Space in the World System* (British Film Institute, London).

King, N., 1996, 'The sound of silents', in *Silent Film*, ed. R. Abel (Athlone, London), pp. 31–44.

Kuhn, A., 1985a, 'History of the cinema', in *The Cinema Book*, ed. P. Cook (British Film Institute, London), pp. 2–56.

Kuhn, A., 1985b, 'The history of narrative codes', in *The Cinema Book*, ed. P. Cook (British Film Institute, London), pp. 207–20.

Lant, A., 1995, 'Haptical cinema', *October*, 74, pp. 45–73.

Leong, R. (ed.), 1991, *Moving the Image: Independent Asian Pacific American Media Arts* (UCLA Asian American Studies Centre/Visual Communications, Los Angeles).

McArthur, C., 1997, 'Chinese boxes and Russian dolls: tracking the elusive cinematic city', in *The Cinematic City*, ed. D. B. Clarke (Routledge, London), pp. 19–45.

McQuire, S., 1998, *Visions of Modernity: Representation, Memory, Time and Space in the Age of the Camera* (Sage, London).

Malkmus, L., Armes, R., 1991, *Arab and African Film Making* (Zed, London).

Mast, G., Kawin, B. F., 1992, *A Short History of the Movies*, fifth edn (Macmillan, New York).

Mattelart, A., 1994, *Mapping World Communication: War, Progress, Culture* (University of Minnesota Press, Minneapolis).

Medovoi, L., 1998, 'Theorizing historicity, or the many meanings of *Blacula*', *Screen*, 39, 1, pp. 1–21.

Medved, M., 1992, *Hollywood vs. America: Popular Culture and the War on Traditional Values* (HarperCollins, New York).

Naremore, J., 1995/6, 'American film noir: the history of an idea', *Film Quarterly*, 49, 2, pp. 12–27.

Nowell-Smith, G., 1990, 'On history and the cinema', *Screen*, 31, 2, pp. 160–71.

Pearson, R. E., 1996, 'The attractions of cinema, or, how I learned to start worrying about loving early film', in *Cinema: The Beginnings and the Future: Essays Marking the Centenary of the First Film Show Projected to a Paying Audience in Britain*, ed. C. Williams (Routledge, London), pp. 150–7.

Perkins, V. F., 1992, 'The Atlantic divide', in *Popular European Cinema*, eds R. Dyer, G. Vincendeau (Routledge, London), pp. 194–205.

Petrie, D. (ed.), 1992, *Screening Europe: Image and Identity in Contemporary European Cinema* (British Film Institute, London).

Powers, Rothman, D. J., Rothman, S., 1996, *Hollywood's America: Social and Political Themes in Motion Pictures* (Westview, Boulder).

Prindle, D., 1993, *Risky Business: The Political Economy of Hollywood* (Westview, Boulder).

Robinson, D., 1996, *From Peep Show to Palace: The Birth of American Film* (Columbia University Press, New York).

Salt, B., 1996, 'Cut and shuffle', in *Cinema: The Beginnings and the Future: Essays Marking the Centenary of the First Film Show Projected to a Paying Audience in Britain*, ed. C. Williams (Routledge, London), pp. 171–83.

Saunders, T., 1994, *Hollywood in Berlin: American Cinema in Weimar Germany* (University of California Press, Berkeley).

Schivelbush, W., 1988, *Disenchanted Night: The Industrialization of Light in the Nineteenth Century* (Berg, Oxford).

Schrader, P., 1996, 'Don't cry for me when I'm gone: motion pictures in the 1990s', in *Cinema: The Beginnings and the Future: Essays Marking the Centenary of the First Film Show Projected to a Paying Audience in Britain*, ed, C. Williams (Routledge, London), pp. 201–6.

Shaviro, S., 1993, *The Cinematic Body* (Minnesota University Press, Minneapolis).

Staiger, J., 1990, 'Announcing wares, winning patrons, voicing ideals: thinking about the history and theory of film advertising', *Cinema Journal*, 29, 3, pp. 3–31.

Staiger, J. (ed.), 1995, *The Studio System* (Rutgers University Press, New Brunswick).

Stevenson, J., 1997, 'From the bedroom to the bijou: a secret history of American gay sex cinema', *Film Quarterly*, 5, 1, pp. 24–31.

Storper, M., Christopherson, S., 1987, 'Flexible specialisation and regional industrial agglomerations: the case of the US motion picture industry', *Annals of the Association of American Geographers*, 77, 1, pp. 104–17.

Taylor, P. J., 1996, 'What's modern about the modern world-system? Introducing ordinary modernity through world hegemony', *Review of International Political Economy*, 3, 2, pp. 260–86.

Thompson, K., 1985, *Exporting Entertainment: America in the World Film Market 1907–34* (British Film Institute, London).

Thrift, N. J., 1994, 'Inhuman geographies: landscapes of speed, light and power', in *Writing the Rural: Five Cultural Geographies*, eds P. Cloke, M. Doel, D. Matless, M. Phillips, N. Thrift (Paul Chapman, London), pp. 191–248.

Usai, P. C., 1996, 'The color of nitrate: some factual observations on tinting and toning manuals for silent films', in *Silent Film*, ed. R. Abel (Athlone, London), pp. 21–30.

Vincendeau, G., 1992, 'France 1945–65 and Hollywood: the *policier* as inter-national text', *Screen*, 33, 1, pp. 50–80.

Wasko, J., Phillips, M., Purdie, C., 1993, 'Hollywood meets Madison Avenue: the commercialization of US films', *Media, Culture and Society*, 15, 2, pp. 271–93.

Wasser, F., 1995, 'Is Hollywood American? The trans-nationalization of the American film industry', *Critical Studies in Mass Communication*, 12, 4, pp. 423–37.

Chapter 17

Global Disney

Alan Bryman

The immense success of Disney symbols abroad, most notably the global influence of its cartoon characters, is well known and well established. This chapter, however, will emphasize the global reach of the Disney theme parks. It will do so in two ways. First, it will examine studies which have addressed the degree to which the non-US Disney theme parks (Tokyo Disneyland and Disneyland Paris) represent a simple transfer of the US parks to a foreign context. Secondly, it will examine how British visitors 'consume' the parks, in other words what they make of the parks' signs and symbols. The latter will draw upon the preliminary results of an exploratory interview study of British visitors to Disney parks.

In recent times, the theme of globalization has been undergoing considerable revision as a result of critical commentary on some of the more immoderate claims of some exponents. One has to do with the question of whether globalization is happening at all in the sense that the term is often employed. In place of an all-encompassing globalization, a number of commentators have suggested that while a globalization impulse can be seen at work, it is brought down to earth by the need to make local accommodations. As a consequence of this recognition, a number of neologisms have been coined to signal the dialectic at work when global and local forces conjoin, such as 'glocalization' (Robertson, 1992), 'creolization' (Hannerz, 1987) and 'hybridization' (Pieterse, 1994), while Barber (1995) refers to the contrasting directions of McWorld and Jihad. The second strand is the claim that globality has to be assessed at the point of reception/consumption (Tomlinson, 1991: 44–6). Ultimately, the two are highly related. One aspect of hybridization (or glocalization) may be the way in which foreigners make sense in their own terms of supposedly global forces. Van Eltern, for example, has suggested that the notion of Americanization does not adequately 'take into account that the European reception of American culture implies selective borrowing and active appropriation among the receivers' (1996: 52). In other words, do foreign audiences passively absorb global products in the way in which cultural critics presume they do on the basis of their examinations of the global propensity of the texts of (usually) American exports?

Exporting Disney

In addition to its films, merchandise and television channel, Disney has been an exporter of its theme park concept. The first non-American Disney park, Tokyo Disneyland, opened in April 1983; the second, Disneyland Paris (formerly referred to as Euro Disneyland) opened in April 1992. The parks differ in two major respects. The Japanese park is owned by a Japanese company which approached Disney about the possibility of building the park. Disney was and still is heavily involved in its creations and receives royalties. At Disneyland Paris, Disney owns 49 per cent of the equity in the company known as Euro Disneyland. The second difference is that while Tokyo Disneyland has been a runaway success, its French counterpart has struggled and at one time was reputed to be on the verge of closure (Bryman, 1995: 75–80). How, then, is the confrontation of the global and the local played out in these two parks?

From the outset, the designers of Tokyo Disneyland sought to create an American version of Disneyland in the heart of Japan in such a way that little concession was made to the local culture. However, this does not represent mindless Americanization on the part of Disney: it was the local owners who tempered the desires of the Disney Imagineers to adapt to local cultural conditions. In the words of an executive with the Japanese owner: 'When we built Tokyo Disneyland, we made the decision to keep the American flavor as much as possible' (cited in Simons, 1990: 43). According to Brannen (1992: 219, 227), this 'keeping the exotic exotic' represents not the assimilation of Japanese culture by American culture but a means of differentiating Japanese cultural identity from that of the West. Yoshimoto (1994) argues that to have sought too much of an accommodation with local customs and traditions would have run into the problem that the Japanese revel in simultaneously asserting their national identity and valorizing American products to the extent that American cars are unpopular in Japan because they are not sufficiently American!

There *are* certain differences between Tokyo Disneyland and the US Disney parks that reflect an accommodation to Japanese traditions (Brannen, 1992; Van Maanen and Laurent, 1992; Yoshimoto, 1994). Main Street USA has been replaced by World Bazaar, not because of the former's nostalgic Americanism, but because of a Japanese preference for up-market shopping opportunities, to which the Bazaar has been designed to cater. In place of Hall of Presidents or Great Moments with Mr Lincoln there is Meet the World which is a 16-minute introduction to Japanese history using film and audio-animatronics. Other differences include: drivers of people movers wear white gloves; cast members' name badges display last rather than first names; there is a picnic area close to the park; and greater ad-libbing by drivers of people movers.

The designers of Disneyland Paris have also sought to accommodate to local customs and traditions but have displayed greater uncertainty than their Japanese counterparts over how far to go. The names and attractions are in French, most of the cast members are European and most of these are French, wine and beer were allowed in certain restaurants a year or so after opening the park, and certain American attractions like the Jungle Cruise were missed out in order not to offend post-colonial sensitivities. However, a year after the park opened, Euro Disney's new chairman (Philippe Bourgignon, a Frenchman appointed in January 1993 to replace the US incumbent of the post) felt that the park was insufficiently American:

> Each time we tried to Europeanise the product we found it didn't work. Europeans want America and they want Disney, whether French intellectuals like it or not.
>
> (Cited in Skapinker and Rawsthorn, 1993: 18)

This view contrasts sharply with that of many commentators at the time who suggested that the company had quite simply failed to harmonize with European culture and habits sufficiently (Langley, 1993). This was certainly the view of Steve Burke, vice-president of Euro Disney, who in an interview in August 1994 said:

> We should never have tried to be as American as we were. It was a mistake. At one point we had four hundred Americans over here . . . Our advertising in the beginning was very American. It assumed that you knew what a Disney park was all about. So it assumed that all you had to do was say 'it's open' and they will come.
>
> (In Lovelace, 1995: 36)

Burke also said:

> Most of the changes we've made to Europeanise the product we've gone back on . . . the merchandise was initially much more upscale than the US – we've come down. The food was much more sit-down – we've gone back.
>
> (In Lovelace, 1995: 49)

Unlike Tokyo Disneyland, the designers of Disneyland Paris appear to have been more uncertain about the degree to which its global product needed to be tempered by a recognition of local conditions and circumstances. In the case of both parks, however, we see glocalization or hybridization in action in the sense that, as a number of writers now acknowledge, apparently culturally imperialist products invariably represent a meeting place for the global and the local.

Receiving Disney

While it is fruitful to reflect upon the process of globalization and its various accommodations, a number of writers have suggested that the phenomenon needs to be explored at the point of reception. Are the supposed carriers of globalization interpreted as such? This line of reasoning borrows from at least two impulses in the field of media and cultural studies. The first is the postmodern distaste for metanarratives and a corresponding preference for a cacophony of competing and equally compelling micronarratives. The second, which is by no means independent of this tendency, is a disdain in many quarters for intellec-tualized readings of texts (and theme parks can be considered texts ripe for inter-pretation) and a parallel pride of place being allocated to meaning at the point of reception. Variously known as 'active audience' and 'audience reception' analysis (see Fenton et al., 1998: 2–8 for a summary), this second related strand of development reflects death of the author thinking, in emphasizing the ways in which audience members engage with a text, frequently challenging or at least undermining its manifest messages and deriving from it their own autonomous pleasures.

When considered in relation to the supposed signs of globalization, this popu-list position raises the question of whether and how far international audiences come up with their own distinctive readings of global texts. In the context of US exports which span the globe, such a view means attending to such issues as whether the American nature of the product is both recognized and signifi-cant to audience members and how far they adopt distinctive readings of the texts in question. Liebes and Katz's (1990) study of the reception of the televi-sion series *Dallas* in different cultures is emblematic of this approach. Specifi-cally in relation to Disney products, like its theme parks, we might wonder what non-Americans make of this prime mover of global commodities, an issue that has attracted surprisingly little attention (for exceptions see Real, 1977; Wasko, 1996).

The rest of this section reports some preliminary findings from a very exploratory study of the responses of British visitors to Disney theme parks. Twenty interviews have been conducted, though in some cases couples were interviewed. The style of questioning was very informal and open ended, with an emphasis upon giving the respondents a free rein. Respondents had been sampled on the basis of word of mouth and/or with initial contacts often nomi-nating others (snowball sampling). This is in no sense a sample that has been selected through a rigorous random method. Of the 20 interviewees, only 4 had not been to Disney World. Three had visited Disneyland and no other Disney park. Six had been to Disneyland Paris, but only one of these had not been to an American Disney park as well. In other words, 19 interviewees had visited one of the US parks on at least one occasion (many were multiple visitors). The interviewees' accounts of their reception of the Disney parks is particularly likely to be instructive because in all but one case, the reception is largely in connec-tion with the American parks and therefore largely unmediated by any adapta-tion to local circumstances.

The following findings are based largely on sections of the interview tran-scripts where interviewees were encouraged to discuss matters such as whether they felt the parks were relevant to them, and what they thought of the repre-sentation of cultural differences in the parks. Specific questions on 'American-ization' were not asked in case they forced lines of thinking down a certain channel.

In considering the findings, it is helpful to introduce a classification of dif-ferent audience readings suggested by Liebes and Katz (1990) and subsequently elaborated by Lutz and Collins (1993) in their study of readers of *National*

Geographic: *ludic* (playful); *aesthetic* (emphasizing attractiveness or ingenuity); *moral* (judgemental about programme characters); and *ideological* (emphasizing covert messages with political connotations). Liebes and Katz found that viewers of *Dallas* varied in their responses in terms of this classification according to their cultural background. Lutz and Collins's study of Americans' reflections on *National Geographic* photographs rarely included an ideological dimension.

The interviews with Disney visitors, however, were overwhelmingly infused with a ludic response to the parks. This was the case even among visitors who claimed they were not enthusiastic about visiting a Disney park prior to their first visit (usually going because other members of the family wanted to go). Only one interviewee was adamant about not liking the Disney parks; the rest described the parks as better than they had expected prior to their first visits. As a result, the transcripts were suffused with enthusiasm for the unbridled fun that was meted out for the interviewees during their visits, which in turn created an overwhelmingly ludic tone to the nature of their responses. This is not to say that critical readings were absent even among the unbridled enthusiasts: some were able to articulate ideological readings when prompted, such as the absence of blacks in the representation of historical themes. However, when drawing attention to such critical nuances, they often observed that their recognition of these features was in many respects an artefact of the interview, rather than something that had concerned them or that they had been aware of during their visits. As one respondent said:

> *Interviewer*: Did you think there were any special emphases [in the representation of different cultures]?
> *Interviewee*: Well thinking about it now, because I hadn't really given this any consideration before you started asking about it, but thinking about it now, it was only really representative of the developed world, you know, Britain, America, Japan, world leaders in technology, and there was nothing of the Third World there.
>
> (M, 64, DW, DP)[1]

Reflecting on this and some other issues that the same interviewee commented on during the interview, he said: 'We were there for fun, so the rest wasn't important, rightly or wrongly.' In this, as in most other cases where it occurred (the chief exception being the person who disliked the parks), a critical, ideological interpretation was secondary to a predominantly ludic reading.

In Walt Disney's original vision of Disneyland three prominent thematic frames can be discerned. First, Disneyland as a place of unbridled fun and fantasy. But this element was if anything treated by him as secondary to two other related motifs in his thinking: Disneyland as a place of education and as a celebration of America's ideals and accomplishments. While the publicity for the parks nowadays tends to emphasize the fun factor, with copious allusions to fun, fantasy, magic and making dreams come true, Walt Disney drew upon an ideology which equated utopia with America: 'Disneyland would be a world of Americans, past and present, seen through the eyes of my imagination';

'Physically, Disneyland would be a small world in itself – it would encompass the essence of the things that were good and true in American life'; 'Disneyland is dedicated to the ideals, the dreams, and the hard facts that have created America'; and 'Disneyland will be the essence of America as we know it, the nostalgia of the past, with exciting glimpses into the future' (Disney, 1994: 29, 30, 34; see also, Bryman, 1995: 140–2). On the face of it, this is not a recipe that would be attractive to non-US visitors, who might be expected to cavil against the prospect of rampant jingoism.

One way in which an ideological reading of the parks surfaced was in terms of an awareness of what was seen as a very American focus of many of the attractions and in terms of a very American style of presentation. The following respondent commented on the Carousel of Progress, an attraction in which the audience revolves around several tableaux from different epochs in modern American history, including the future (Bierman, 1976; Weiner, 1997):

> Funny, but fascinating. I liked that. I couldn't understand why I liked it 'cos it was so American; it was like watching an old black and white American soap opera, but I still liked it and there was a lot you could identify with. Yeh, there was a lot of history, a lot of looking at what America's done and achieved.
>
> (F, 42, DW)

In other words, some interviewees recognized, in common with many commentators on the representation of history at Disney parks (e.g. Bryman, 1995; Fjellman, 1992; Wallace, 1985), that Disney history is very much a panegyric to American progress (and indeed, as previously noted, Walt Disney intended it to have this quality). Similarly, they were also sensitive to the fact that there was a Disney way of doing things that pervaded the parks. Comments on such features as the 'slick' presentation of many attractions, as one interviewee put it, and the friendliness and helpfulness of Disney employees, or 'cast members' to give them their official titles, were common. Such comments are redolent of an aesthetic reading of the parks and their attractions. But the heavily Americanized tone and emphasis was largely disregarded, in part because of the predominantly ludic stance taken by most visitors and in part because it was expected and therefore they were relatively inured to it:

> *Husband*: I remember laughing, it's all the good news of how wonderful America is as a country of liberty and freedom for all, so even the slavery issue was neutered to make it look good for the all American dream. Yes, it was a bit cheesy.
> *Interviewer*: Did that bother you at all?
> *Husband*: No, I expected it, it's a big advert for America, Disney . . .
> *Wife*: You don't take it seriously really, it's not an issue, I don't think.
>
> (M, 38; F, 37: DW; PD)

And again:

> *Interviewer*: Did you think there was anything lacking about the content of the presentation [of the American Adventure attraction] at all, did anything strike you, or did you think there were any particular emphases?

Interviewee: Well I mean it's all very, again as you'd expect, very sort of hand on heart patriotic American, which it was bound to be really, sort of celebrating America, but that's not such a bad thing because going in to the country it's good to learn something about its history and in a way that they would wish to portray it rather than just reading it in a book or something, so I suppose it's bound to be far more of a celebration than perhaps if we did something on America but then that tells you something on its own, doesn't it?

(F, 20, DW)

Interestingly, the first of these two passages derives from an interview with people who had visited one of the American parks and Disneyland Paris and who felt that an American approach to the parks was a feature that French employees failed to carry off with the same aplomb as their American counter-parts. This implies that for these visitors the French park was insufficiently Disneyized and Americanized. This came across in the following passage from an interview with a couple:

Wife: In general I don't think it was done as well [in Paris]. I think Disney is still probably better in America.
Husband: That's where the strengths are, yeh.
Interviewer: Could you pinpoint what was weak about that?
Husband: Well I think the staff . . .
Wife: Attitudes.

(M, 50; F, 42: DW(2); PD)

It would seem, then, that the visitors who were interviewed were patently aware of the heavily American tenor of the parks and their attractions, but were largely unconcerned about this facet. In fact, Disneyland Paris was regarded as disap-pointing to those who had visited the American parks as well, almost because it was insufficiently American. Perhaps in the process some of the Disney magic was lost for these interviewees. The predominantly ludic stance that was taken to the parks and their attractions resulted in a subordination or suppression of concerns about biases in the various presentations or about an American empha-sis. The tendency for some of the interviewees to feel that Disneyland Paris was insufficiently American clashes somewhat with the suggestion that executives at the park felt at one point that it was too American and that they needed to Europeanize. The apparent contradiction may reflect the fact that our inter-viewees were from Britain, which probably has a closer cultural affinity with America than the French, who make up by far the major nation supply-ing visitors to Disneyland Paris.

It is not surprising, therefore, that when questioned about whether the Disney parks were aimed at particular nationalities, the majority of interviewees felt that there was a more or less international appeal. Frequently, they mentioned the great variety of nationalities that were represented among Disney park visi-tors in the USA. The general impression gleaned was that people felt that there was not an excessively American emphasis. To a certain extent, this is surpris-ing. A number of attractions, such as American Adventure in EPCOT Center and the Hall of Presidents in the Magic Kingdom are patently American in tone and emphasis, while the Carousel of Progress has decidedly American overtones, as one of the quoted interviewees recognized. There are a number of possible reasons for the fact that our interviewees seemed by and large to be unfazed by the Americanness of the parks. These reasons are by no means mutually exclusive. First, it may be that the distinctively American emphases are viewed as largely confined to certain attractions (as with the American Adventure) and these are either ignored by non-Americans or are simply treated as unique attractions which are separable from the rest of the parks. In a sense, the American focus can be felt in many quarters of the parks, such as Frontierland in the Magic Kingdoms, but it is a very Hollywood version of America's past that is provided and as such is one to which many visitors can relate. Secondly, it may be that the predominantly ludic mode of reading the parks' texts inoculates against an excessive concern about these signs of Americanness. A third possibility is that as previously suggested, the British may be particularly culturally receptive to American exports. The fourth possible explanation relates to the theming process itself. As a purveyor of themes with considerable global reach, Disney is able to create a sense of universal appeal. Many of its images are instantly recognizable to a large portion of the globe and can be related to across a wide spectrum of nationalities. This capacity may offset feelings of cultural ennui among many visitors.

With one exception among our interviewees, Disney's global offerings in the form of its theme parks were received and consumed in a predominantly ludic frame of reference, doubtless of exactly the kind that the Walt Disney Corporation seeks to cultivate. In its advertising material the company repeatedly emphasizes the fun, fantasy and magic that await visitors to the parks. These are its master themes. While many of the interviewees were sensitive to the ideological dimensions that have suffused the commentaries by critical intellectuals over the years, a game that shows few signs of abating (e.g. Weiner, 1997; Hiassen, 1998), these concerns tended to be subsidiary (with the one exception), and in certain respects only seemed to surface as a result of our questioning.

The findings and analysis reported here go against the grain of recent theorizing about globalization to a certain extent, in that they suggest that the predominantly ludic frame of reference with which the parks are apprehended by most visitors off-sets concerns about biases and Americanism. As we have seen, some visitors mention these factors, suggesting that some people are able to recognize 'the many silences of the Disney parks' (Bryman, 1995: 188) and to reflect upon the American ideology that undergirds them. However, for only one of our interviewees was this a paramount concern:

> *Interviewer*: Did you think it [the presentation of history] did have any particular emphases, then?
> *Interviewee*: Yeh, yeh. It was kind of the winners' view of history, really, it was the white, middle-class kind of America, the sort of apple pie type thing, you know, what you would think of, the Brady Bunch type America and I just thought it's very, very, it's not multi-faceted at all, it was just very exclusively from the aspiring sort of, but basically it wasn't multi-cultural, it didn't seem to represent people

who were in there. An awful lot of them were American, there were black families in there, there were Asian families in there, Chinese families in there, all obviously American, but I thought well where is the actual representation of this melting pot? There's not one at all, it's a Brady Bunch America.

(F, 27, DW)

This comment came from the one highly jaundiced visitor who was interviewed. It would be a mistake to view such negative views as solely the preserve of British visitors, since Americans who are not cultural critics can also be prone to similarly adverse reactions (see Bryman, 1995: 186 for a summary of research on this aspect). The chief point to be gleaned, however, is that resistance in the form of ideological critique was by no means a common response among those interviewed. But more important than this is that Disney is only able to create the kind of environment in which a ludic reading of the parks takes place because of the association of the company with a certain kind of entertainment and with signs that betoken a recognizable context for having fun (Mickey and the other characters, the castle, etc.). These features are in fact indicative of the company's global reach: they are recognizable across the globe. By and large, those people (whether Americans or non-Americans) who recognize but dislike all that Disney stands for simply do not go to the Disney parks, unless they are culture critics intending to write the latest exposé of what Disney World *really* means!

Indeed, even among those interviewees who were reluctant visitors (usually because they had been invited to join others but had not planned to visit a Disney park) there was a recognition, which was sometimes grudging, that the park had been much better than they had expected and their critical faculties were left in suspension. For such visitors an aesthetic reading of the parks frequently surfaced in tandem with a ludic one. Following a question on the representation of history in Main Street USA, one such interviewee commented:

I didn't, it didn't even occur to me that it was. I mean I just kind of, I didn't know anything about Disneyland before I went and all I saw was like the detail of it, I mean I suppose yes, I suppose that looking back the buildings, the way that the buildings were, the architecture of it, I can see now what you're saying, but at the time it didn't occur to me at all.

(F, 42, DL)

Commenting on the question of whether the parks were aimed at particular nationalities, the same interviewee felt that it was not and commented:

I mean if people go to America, they go to see what's American about things, so I mean in a way it's like, it, you know, it's very American, it's what you expect as being a typically American type of entertainment, um, so no, I don't know.

Griswold (1987) has proposed that certain literary texts possess 'cultural power' which affects the extent to which a variety of audience interpretations arise out of them. Cultural power is affected by two factors: a fair amount of consensus concerning the book's main themes and topics (otherwise it is regarded as abstruse) and a capacity for some variation in interpretation (otherwise it is regarded as formulaic). Some writers are critical of the notion of texts 'possessing' certain qualities, such as cultural power (e.g. DeVault, 1990). However, other writers have emphasized the capacity of powerful organizations in the culture industry like Disney to employ globally recognizable symbols of fun and fantasy, such as its stable of cartoon characters, in order to cultivate an atmosphere of unbridled play, while simultaneously promoting certain values which incorporate a veneration of America in the process of creating silences in its texts (e.g. Bryman, 1995). The foregoing examination of visitors' interpretations of the Disney parks suggests that they provide texts with some cultural power. Visitors typically agree upon the notion of the parks as arenas for unbridled fun, but sometimes are able to pick up more critical interpretations which can be regarded as indicators of resistance (Abercrombie and Longhurst, 1998). However, the dominant frame within which the parks are read is very much in tune with that which Disney seeks to promote. Moreover, while occasionally irked by the strident Americanism of certain attractions, in the vast majority of cases it is absorbed by visitors in their stride and does not interfere with the main item on the agenda – having fun. However, it is difficult to agree with Willis's (1995: 1) assessment that such reactions on the part of visitors are 'uninteresting': on the one hand, as she recognizes, they reflect an acceptance of Disney ideology, which is itself interesting; on the other hand, as the present analysis suggests, the interplay between the focus on fun by visitors and more critical assessments is worthy of attention.

Conclusion

This chapter has sought to examine the Disney theme parks in the context of two current areas of debate among writers on globalization: the idea that hybridization or one of its synonyms better describes what is going on in relation to supposedly global exports, and the notion that globalization needs to be explored at the point of reception in order to determine whether global exports are redefined. An examination of Tokyo Disneyland and Disneyland Paris suggested that the hybridization idea is very relevant to the design of the parks. However, one might want to query the relative significance of such local accommodations for the broader picture of homogenization that is supposed to be a feature of globalization (Ritzer, 1998: 86). The reception argument fares less well. On the one hand it is clear that the park designers have been forced to respond to the specific requirements of Europeans; on the other hand, at the level of interpretation, the British visitors interviewed about their visits to Disney theme parks seemed to have imbibed the Disney emphasis on fun and showed little evidence of being disturbed by its Americanism. This latter finding unmistakably identifies the cultural power of Disney's global offerings.[2]

Notes

1 This simple notation identifies the gender of the interviewee(s), age at time of interview, and which park or parks were visited (with the number of times in brackets

when more than once). DW refers to Walt Disney World in Orlando, Florida; DL is Disneyland in Anaheim, California; and DP is Disneyland Paris.

2 I should like to thank Robin Allan and Alan Beardsworth for his helpful comments on an earlier draft of this chapter and Alyson Grove for her fieldwork assistance.

References

Abercrombie, N. and Longhust, B. (1998) *Audiences*, London: Sage.

Barber, B. R. (1995) *Jihad vs. McWorld*, New York: Times Books.

Bierman, J. H. (1976) 'The Walt Disney robot dramas', *Yale Review*, 66: 223–36.

Brannen, M. Y. (1992) ' "Bwana Mickey": constructing cultural consumption at Tokyo Disneyland', in J. J. Tobin (ed.), *Re-Made in Japan: Everyday Life and Consumer Taste in a Changing Society*, New Haven: Yale University Press, pp. 216–34.

Bryman, A. (1995) *Disney and his Worlds*, London: Routledge.

DeVault, M. L. (1990) 'Novel readings: the social organization of interpretation', *American Journal of Sociology*, 95: 887–921.

Disney, W. (1994) *Famous Quotes*, Walt Disney Company.

Fenton, N., Bryman, A., and Deacon, D. (1998) *Mediating Social Science*, London: Sage.

Fjellman, S. M. (1992) *Vinyl Leaves: Walt Disney World and America*, Boulder, CO: Westview.

Griswold, W. (1987) 'The fabrication of meaning: literary interpretation in the United States, Great Britain, and the West Indies', *American Journal of Sociology*, 92: 1077–117.

Hannerz, U. (1987) 'The world in creolization', *Africa*, 57: 546–59.

Hiassen, C. (1998) *Team Rodent: How Disney Devours the World*, New York: Ballantine.

Langley, W. (1993) 'Euro-dismal', *Sunday Times*, 22 Aug. (section 1): 9.

Liebes, T. and Katz, E. (1990) *The Export of Meaning: Cross-Cultural Readings of Dallas*, Oxford: Oxford University Press.

Lovelace, B. (1995) 'Mickey Mouse in Europe: Euro Disney as a fantastic site of cultural contestation', B.Soc.Sc. Dissertation, University of Birmingham.

Lutz, C. A. and Collins, J. L. (1993) *Reading National Geographic*, Chicago: University of Chicago Press.

Pieterse, J. N. (1994) 'Globalisation as hybridisation', *International Sociology*, 9: 161–84.

Real, M. (1977) *Mass-Mediated Culture*, Englewood Cliffs, NJ: Prentice-Hall.

Ritzer, G. (1998) *The McDonaldization Thesis: Explorations and Extensions*, London: Sage.

Robertson, R. (1992) *Globalization: Social Theory and Global Culture*, London: Sage.

Simons, C. (1990) 'Business and leisure', *Landscape Architecture*, 80: 42–5.

Skapinker, M. and Rawsthorn, A. (1993) 'An older, wiser Mickey Mouse', *Financial Times*, 10 April: 18.

Tomlinson, J. (1991) *Cultural Imperialism*, London: Pinter.

van Eltern, M. (1996) 'Conceptualizing the impact of US popular culture globally', *Journal of Popular Culture*, 30: 47–89.

Van Maanen, J. and Laurent, A. (1992) 'The flow of culture: some notes on globalization and the multinational corporation', in S. Ghoshal and D. E. Westney (eds),

Organization Theory and the Multinational Corporation, New York: St. Martin's Press, pp. 275–312.

Wallace, M. (1985) 'Mickey Mouse history: portraying the past at Disney World', *Radical History Review*, 32: 33–57.

Wasko, J. (1996) 'Understanding the Disney universe', in J. Curran and M. Gurevitch (eds), *Mass Media and Society*, 2nd edn, London: Arnold.

Weiner, L. Y. (1997) ' "There's a great big beautiful tomorrow": historic memory and gender in Walt Disney's "Carousel of Progress" ', *Journal of American Culture*, 20: 111–16.

Willis, S. (1995) 'The problem with pleasure', in The Project on Disney, *Work and Play at Disney World*, Durham, NC: Duke University Press, pp. 1–11.

Yoshimoto, M. (1994) 'Images of empire: Tokyo Disneyland and Japanese cultural imperialism', in E. Smoodin (ed.), *Disney Discourse*, New York: Routledge, pp. 181–99.

Chapter 18

'The Kind of Beat Which is Currently Popular'[1]: American Popular Music in Britain

Tim Cresswell and Brian Hoskin

I have in mind rather the kind of milk-bar – there is one in almost every northern town with more than, say, fifteen thousand inhabitants – which has become the regular evening rendezvous of some of the young men. Girls go to some, but most of the customers are boys aged between fifteen and twenty, with drape-suits, picture ties, and an American slouch. Most of them cannot afford a succession of milk-shakes, and make cups of tea serve for an hour or two whilst – and this is their main reason for coming – they put copper after copper into the mechanical record-player. About a dozen records are available at any time a numbered button is pressed for the one wanted, which is selected from a key of titles. The records seem to be changed about once a fortnight by the hiring firm; almost all are American; almost all are 'vocals' and styles of singing much advanced beyond what is normally heard on the Light Programme of the BBC.

(Hoggart, 1958: 203–4)

In perhaps the best-known invective against 'Americanization' Richard Hoggart paints a picture of shiny and shallow world of imported mass culture threatening the warm and authentic working-class culture of northern England. Kids with their 'American slouch' hang out in their chrome plated paradise listening to American music with the 'kind of beat which is currently popular'. Down the road working-class culture continues in the pub and in the living rooms which, compared to the milk-bar, 'speak of a tradition as balanced and civilised as an eighteenth-century town house' (Hoggart, 1958: 203). The lads in the milk-bar enjoying a kind of 'spiritual dry-rot amid the odour of boiled milk' were, he argued, living in a myth-world 'compounded of a few simple elements which they take to be those of American life' (Hoggart, 1958: 204).

In this chapter we focus on the interaction between music thought to be American and music thought to be British. Our intention in doing this is to undermine the conception of any kind of music as 'American' in any easy sense. Hoggart's is but one reaction against the perceived threat of Americanness associated with popular music. Below we trace some other reactions and interactions which point to the irrelevance of 'Americanization' in popular music at the end of the century. First we discuss largely academic reactions to the perceived hegemony of American music. The most prominent of these is the cultural imperialism thesis.

Cultural Imperialism and Beyond

Hoggart's account is an early version of what later became known as the cultural imperialism thesis. In his account 'America' represents a shallow, showy, repetitive and standardized culture which is spoiling a warm, local, authentic and rooted culture. Music, as the quotation reveals, plays an important role in this perceived process. Americanization has been a term denoting commercialization, standardization and banalization since the early part of the century when American industry positioned itself at the heart of popular culture production in music, film and television. Europeans, in particular, reacted against this perceived threat by invoking any number of alternative folk and elite cultural forms. Keith Negus has suggested that such a phobia was behind cultural policy in both the Stalinist Soviet Union and Nazi Germany (Negus, 1996). Critics such as Hoggart, Adorno and Leavis were united by their fear of Americanization. New products in the cultural sphere were increasingly likely to travel across the Atlantic from the United States. By the 1970s a seemingly one-way flow of music and other forms of popular culture had been firmly established. Morley and Robins have recently argued that European cultural policy still identifies America as a threat to be countered by a genuinely European culture (Morley and Robins, 1995).

The history of popular music, in recorded and transmitted form, certainly has its roots in the United States and Europe. Since the inception of popular music at the turn of the century the industry has been extraordinarily centralized around fewer than ten major recording companies which account for over 70 per cent of legally distributed music. In the last twenty years centralization has increased to the point where 80 to 85 per cent of the world's recordings are produced and distributed by six major companies – Sony, EMI, MCA, Polygram, BMG and Warner – only one of which (Warner) is 'American'. Oliver Boyd-Barrett has referred to 'media imperialism' as the 'the process whereby the ownership, structure, distribution or content of the media in any one country is subject to substantial external pressures from the media interests of any other country or countries – without proportionate reciprocation of influence by the country so affected' (cited in Negus, 1996: 168). The argument is that the ownership of media forms leads to these forms exhibiting basic characteristics derived from the country of origin – particularly the United States but also the United Kingdom, Continental Europe and Japan. The cultural imperialism thesis is not just about the content to the music but, more importantly, about the structure of the industry, the choice of communications vehicle (radio, CD, etc.), the industrial arrangements that produce music (mainly capitalist multinational companies), and the general characteristics of the product (mainly English language, predominantly standard types of instruments, etc.).

The idea of cultural imperialism in the music business and more generally has been subject to a number of linked criticisms. Clearly the idea has little respect for the audience who can and do use and interpret cultural products in many ways. The proponents of active audience research are quick to point out that audiences across the world are not simply the dupes of imperial power. As Hebdige has argued:

> American popular culture – Hollywood films, advertising images, packaging, clothes and music – offers a rich iconography, a set of symbols, objects and artefacts which can be assembled and reassembled by different groups in a literally limitless number of combinations. And the meaning of each select one is transformed as individual objects – jeans, rock records, Tony Curtis hairstyles, bobby socks etc., are taken out of their original historical and cultural contexts and juxtaposed against other signs from other sources.
>
> (Hebdige, 1988: 74)

There may, in fact, be little correspondence between ownership of the music industry and its mediation through processes of consumption. This is connected to the observation that the ownership of the music industry may have little to do with what it produces. Japan, for instance, plays a leading role in the modern popular music industry through Sony's ownership of CBS/Columbia. Few would argue, however, that Japanese music is taking over the world. Rather it is the case that Sony are quite happy to produce and distribute Bob Dylan and any number of other 'American' music icons. The cultural imperialism thesis also tends to assume that a core of fairly corrupt and artificial cultures are spoiling and polluting simple, innocent marginal cultures in the same way that Hoggart thought British working-class culture was being made shallow by Americanization. Dualisms as simple as this rarely hold much water.

A less conspiratorial view on the relations between American music and the rest of the world can be summed up by the word 'Globalization' – a term which has been applied to general processes of transculturation which produce transnational music. Unlike the cultural imperialism thesis there is not an explicit form of domination posited here. The idea is that there are a number of global interconnections not directed by a central imperial power such as the United States. Under the globalization thesis American music is no longer seen as essentially American but rather the product of series of interconnections across time and space. Thus, the argument goes on, Michael Jackson and Madonna are popular across the world because of inherently international appeal – their music is already, to quote the labels of clothing everywhere, the product of more than one country. As Negus has pointed out, such ideas virtually ignore any analysis of the systems of production and distribution that create 'popularity' and are thus subject to the criticism that they have gone too far in rejecting arguments of brute economic power. What systems of production and distribution made it possible for Michael Jackson to be supposedly universal? 'There may be many cultures around the world that have produced cultural forms that could be universally enjoyed. But some are more likely to get to a position to be enjoyed than others' (Negus, 1996: 178). Central to these arguments about music, however, is the term 'American' and what it might mean in relation to music.

What Does 'American' Music Mean?

Hoggart seemed fairly confident in his labelling of music and milk-bars as 'American'. His musings also seem to infer that 'American' means both place of origin and a particular type of style which he interprets as shallow. But what can 'American' mean in the world today? Cultural imperialism arguments suggest that the place of production and distribution is the most significant factor. While it is certainly true that popular music in some senses originated from processes of recording and distribution first fully developed in the United States, it is no longer the case that American companies dominate the popular music industry. While Warner is still an American company, Sony is Japanese and MCA (Music Company of America) is actually owned by the Canadian company Seagrams. EMI is British. To give an indication of the complexity of geography of production in the modern music industry consider the CD *Time Out of Mind* by Bob Dylan. This album is, of course, recorded by an artist who is about as American as American can be. The music is heavily influenced by the blues and the subject matter of the lyrics revolves around dirt roads and LA streets. The CD labels reads 'Columbia' – a company that was once American, Columbia has its origins in the late 1880s as the Columbia Graphophone Company. It was the result of patents awarded to scientist Charles Sumner Tainter and Alexander Graham Bell's cousin Chichester A. Bell for a recording disc and cylinder. The purchase of patents from Tainter and Bell led to the formation of the North American Phonograph Company, who produced office dictating machines. The rights for the production of these machines were sold to the Columbia Phonograph Company which operated in the Baltimore/ Washington market. Columbia expanded into entertainment by recording military marches, popular songs and speeches. In 1904 Columbia developed the double-sided 78 rpm disc, and by 1913 it had ceased producing cylinders and changed its name to the Columbia Graphaphone Company. In 1926 the first of many mergers took place when Columbia took over Okeh (the Otto Heinemann Phonograph Corporation), whose catalogue included Louis Armstrong and Lonnie Johnson. This was followed in 1934 by the purchase of Columbia and Okeh by ARC–BRC (American Record Company–Brunswick Record Company) which was in turn bought by Columbia Broadcasting System (CBS) in 1938. Ten years later Columbia developed the $33^{1}/_{3}$ rpm LP which became the accepted form of musical recording for the next forty years. In 1968, CBS joined forces with the Sony Corporation in order to market CBS products in the Asian markets. In 1982 Sony developed the compact disc and CBS were instrumental in introducing it in western markets. Six years later the Sony Corporation bought the CBS Records Group. In 1994, to make matters more complicated, Sony reorganized into four music labels including the Columbia records Group – the label that appears on Dylan's CD.[2]

So how do we locate cultural products in the age of the multinational? Dylan's CD is on an historic American label who have been centrally involved in innovations in popular music. The company is now owned by a Japanese multinational. Further inspection of the label reveals that the CD is actually manufactured in Austria. Despite all this it seems likely to us that most people would think of a Dylan CD as American.

If we put aside the idea that American music is music produced in America then there is another problem with identifying some music as American. The United States is, after all, not an homogenous thing. Many people have attempted to argue that there are superorganic and essential characteristics of Americaness. Wilber Zelinsky (1973) famously reduced American culture to four themes: individualism, mobility, messianic perfectionism and a mechanical view of the world. John Kouwenhoven (1961) has made similar comments. In his essay 'What's American about America', Kouwenhoven argues that the defining feature of Americanness is 'process'. It is the idea of process, he argues, which links such obviously American things as the New York skyline, chewing gum, urban grid patterns and jazz. The link is the open-endedness of simple and infinitely repeatable units. Jazz is widely held to be an American cultural product (and the source of much anxiety in inter-war Britain). Kouwenhoven argues that jazz is differentiated from European music by its distinctive rhythm – 'an alternation of syncopations and notes played on the beat' (Andre Hodeir cited in Kouwenhoven, 1961: 51). He goes on to suggest that jazz is the product of individuals playing in such a way that a unity is produced almost spontaneously, just as the different buildings in the New York skyline, although not planned together, produce a delightful unstructured harmony. 'What it adds up to' he goes on, is the 'first art form to give full expression to Emerson's ideal of a union which is perfect only when "all members are isolated"' (Kouwenhoven, 1961: 53). Although Kouwenhoven's arguments have immediate appeal, it is hard to think of them as little more than a well-argued repetition of American ideology of progress and individualism mapped on to a musical form. Music produced by poor Puerto Rican kids in New York is not likely to be the same as that produced by affluent white people in Chicago's suburbs or African-American people from the 'deep south'. Still, we have the problem of working out what 'American' music means. This problem, however, has not weighed heavily on the minds of observers in Britain who have rallied against the perceived threat of American music. In the following section we look at some moments in the history of American music in Britain in order to disturb the easy assumption that music can be mapped onto nations.

American Music/British Music

Hoggart wrote out against what he considered to be the new forms of (American) mass entertainment of the 1950s, fearful that they were threatening the traditional working-class culture of his 1930s childhood. The new American music to be heard on mechanical record-players he infers to be inferior to that which could normally be heard on the Light Programme of the BBC.

Hoggart's concern over the pervasiveness of imported American music was, however, not a new one. Almost 30 years earlier, in 1929, the 'popular' music magazine *The Melody Maker* expressed similar concerns, although they had no worries over the quality or type of the American music, just over the fact that it was American:

> If there is a reason for the failure of British songs to compete with American numbers, it is only because they are lacking in merit. Music lovers want to hear and buy music for no other reason than they like it; they do not care who wrote it, who published it, or whether it has a pretty picture on the cover.
>
> British song writers who can produce songs as good as those by the Americans will inevitably succeed.
>
> (1929: 289)

As for the output of the Light Programme of the BBC, the same edition of *The Melody Maker* (p. 277) also features a review of a Columbia Graphophone Co. Ltd gramophone record by 'Jack Payne and His BBC Concert Orchestra', of *Mississippi (A Tone Journey), Parts 1 and 2*. This composition was by American dance musician, composer and arranger Ferdi Grofé, formally of the Paul Whiteman Band. Paul Whiteman and his band provided jazz music that was acceptable to the white American public. Their sound was more restrained, more polished than the great African-American jazz bands of the era, whose aesthetic ideals led them to play with a rawer, raunchier, dirty sound. The Paul Whiteman Band's recordings would have sold in greater numbers than African-American bands, being targeted at more affluent white audiences of both America and Britain; and it was bands like Whiteman's and Jack Hylton's that provided the inspiration for the BBC's resident dance-band leaders Jack Payne and Henry Hall (Frith, 1983). Indeed, the British music industry was keen to maintain a distinction between respectable dance music and other, apparently threatening, forms of music such as jazz. Music that was recognizably jazz was excluded from the BBC and only found a welcome in clubs with names like The Manhattan.

The advent of the phonograph and the end of the First World War had seen the increased popularity of ragtime and jazz records which became associated with the comparatively liberated dance techniques of the Tango (from Argentina), the Foxtrot and the Bunny Hug (both from North America). Bands with names like the Versatile Rag Pickers, the American Five and the Original Dixieland Jazz Band played at the Hammersmith Palais and the Astoria (Chambers, 1986). The success of light jazz music was followed by other popular forms from across the sea. The likes of Bing Crosby (who was at one time vocalist with the Paul Whiteman Band) and Frank Sinatra were followed by the tunes of Broadway musicals and the works of Cole Porter, George Gershwin and others.

Iain Chambers has written of the ways in which each successive wave of American music in Britain was gradually tamed to appeal to the apparently fragile British sensibilities.

The lack of propriety in bodily movements that the undisciplined steps and wild sounds of 'jazz' apparently unleashed were gradually eradicated by a growing army of dance instructors led by Victor Sylvester, and the increasingly predictable metronome rhythms of white dance bands. What jazz might have been about was flattened beneath the dancing master's shoes and the band leader's baton to an acceptable dance hall decorum.

(Chambers, 1986: 136)

The BBC, white dance musicians and commentators were all busy controlling the threat of American, particularly African-American, music long before the 1950s and the advent of rock 'n' roll. The perceived domination of American song and dance styles in Britain was clearly not resisted by Hoggart alone. Americanization always went hand in hand with commercialization in the minds of many. In response some looked towards British authenticity to be found in apparently rooted 'folk music'. British popular music has also retreated into nationalistic moments with strong anti-American sentiments. Blur most famously have been tremendously successful in the UK but virtually ignored in the US. As one critic put it:

Their own particular collage of our pop heritage, a mix of music hall whimsy and sweeping ballads, punky anthems and jokey references, borrows from everywhere and brings to mind everyone, not just The Beatles, The Who and The Kinks, but The Jam, The Buzzcocks and Madness as well. . . . But here, as elsewhere, the heightened Englishness which makes *Parklife* such a wonderful record is likely to be their undoing. Blur don't like America, and Americans don't much like those who don't like them.

(David Runciman cited in Mitchell, 1996: 20)

In this respect Blur are the latest in a long line of British quasi-nationalist popular music including the Jam, most of the punk movement, the Smiths, the Pet Shop Boys and Blur's rivals, Oasis.

The simplicity of musical nationalism is perhaps appealing but deeply misleading. The slightest peep into the history of British folk and popular music quickly becomes an insight into incredibly complex sets of connections and flows between the two nations – flows which have never been one-way. The following is just an example.

Connecting the Copper Family

The Copper Family can trace their roots back to the late 1500s. The collection of their songs inspired the foundation of the English Folk Song Society in 1898. Their songs have been passed down through generations of the family; in 1950 Jim and John Copper and their sons Ron and Bob began singing on BBC Radio's *Country Magazine*, in 1997 Bob toured America with the current generation of Coppers; John, Jill and Jon. The Copper Family appear to be the antithesis of the, supposedly, shallow American music which Hoggart despised. They seem to embody the kind of local, 'rooted' culture he held dear. A recent interview with them (Anderson, 1998), however, begins to reveal the interconnections, which act to complicate and undermine those notions which seek to attach music to place.

Bob Copper, while generally only performing traditional, unaccompanied songs, has recently recorded some blues including the Sleepy John Estes song 'Going Down to Brownsville'. He has loved the blues, which he first heard on imported 78s, since the 1930s and, when asked what he listens to most, replies 'Son House' (Anderson, 1998: 27). Bob Copper was not the first singer to take inspiration from the songs of Sleepy John Estes. In 1930 Estes recorded 'Milk Cow Blues', which became a hit when covered by Kokomo Arnold in 1934. It was Arnold's version that Elvis Presley copied in 1954 as 'Milk Cow Blues Boogie', one of his first sides for Sun Records. Several of Presley's early recordings borrowed from African-American blues or R'n'B, 'Hound Dog' (1956) was based on a Big Mama Thornton's R'n'B hit, 'That's All Right, Mama', was a Big Boy Crudup song, and his million-selling 'Teddy Bear' (1957) was derivative of Blind Lemon Jefferson's 1927 'Teddy Bear Blues'. This copying of African-American songs by famous white American artists was common. Bill Haley's 'Shake Rattle and Roll' was a bowdlerized cover of a song which had been a big hit amongst black audiences for Joe Turner; Jerry Lee Lewis and Carl Perkins both recorded Blind Lemon Jefferson's 'Matchbox Blues' and Eddie Cochran also recorded Estes' 'Milk Cow Blues'.

Just as British cultural critics feared the moral degeneration, 'levelling-down', and homogenization they saw in Americanization, many white Americans feared the effect black music would have on the nation's youth. Just as Paul Whiteman was more acceptable than black jazz, so Bill Haley and Elvis Presley were more acceptable than black R'n'B.

However, it was not just through early white rock 'n' roll recordings that African-American music arrived in Europe. Black American musicians had long been touring Europe, and some, like Sidney Bechet, made their homes there. European jazz and blues enthusiasts bought recordings imported from the US and learned to play from them. In 1956 Lonnie Donegan[3] was remaking African-American music in Britain, where he had a hit with 'Rock Island Line', which he learned from an import by black American artist Leadbelly. Donegan repeated his success in the States; in April 1956 he reached number 5 in the *Variety* chart, and then followed this with a successful tour (Groom, 1971).

As Hoggart was concerned over American music invading British milk-bars, Donegan was taking his version of American music to the coffee bars and colleges of the US. Donegan's record sold more than a million copies worldwide and started a craze for skiffle in Europe and the US. One band that started during this craze was Liverpool's Quarry Men, which featured John Lennon. Lennon, like the rest of the Beatles, heard American music on the radio from Radio Luxembourg and the show Voice of America, and could buy imported records from American seamen. During this period the BBC were still acting as cultural/moral gate-keepers and refusing to play rock 'n' roll (let alone R'n'B). The Beatles were greatly influenced by American (and particularly African-

American) music, and during the late fifties were covering songs by Chuck Berry and Little Richard.

The Beatles' interest in African-American music was evident on their first visit to the US. An American reporter asked them what they hoped to see whilst there, and they said 'Muddy Waters and Bo Diddley', to which the reporter responded 'Muddy Waters? Where's That?' (Cook, 1995). One of the greatest figures of African-American music was better known to young Liverpudlian musicians than to the American press.

Muddy Waters was a tractor driver in Mississippi, where he learned to play acoustic guitar influenced by (Bob Copper's favourite) Son House. Following a field-recording session for the Library of Congress in 1942, he moved to Chicago to make a career out of music (Lomax, 1993). He started playing electric guitar and by 1948 recorded for the label that was to become Chess Records. In 1955 he told a young hopeful, Chuck Berry, who had come to see him play, that he too should audition for Chess (Groom, 1971). In 1957 he toured Britain with Chris Barber's Jazz Band (which Lonnie Donegan had left two years earlier) (Gillett, 1983). Amongst his audience in London was Alexis Korner, who then abandoned skiffle for blues and went on to encourage the careers of Charlie Watts, Jack Bruce, Ginger Baker, Mick Jagger and Paul Jones, who went on to play in bands like Manfred Mann, Cream and the Rolling Stones. The Rolling Stones, who took their name from the Muddy Waters song 'Rollin' Stone Blues', were, among others, largely responsible for spreading the popularity of American musical forms (like blues and R'n'B) back across America and around the world. Interest in British blues-playing bands, like the Stones, Cream, the Groundhogs and Fleetwood Mac, was instrumental in regenerating interest in the music of African-American artists like Robert Johnson, Muddy Waters and John Lee Hooker.

The blues might be considered the quintessential American music although (or, perhaps, because) it is a product of creolization: made up of musical and lyrical elements, drawn together in America, from cultural forms from Africa, America and Europe. Now this African-American musical form has completed another cultural circuit, as it can be heard as an influence in the music of African musicians. For example, the Malian musician Ali Farka Toure has assimilated a blues guitar style he learned from imported John Lee Hooker albums into his renditions of traditional Malian songs.

Coda

Our intention in this short chapter has been to examine the idea of American popular music through the lens of British observers and musicians. Arguments around notions of Americanization and cultural imperialism need to be understood in relation to the music itself, which is not easily situated in an interconnected world. If an argument is made which suggests that music from (in a loose sense) one particular nation-state has been involved in processes of imperialism, then it is important that we unpack that argument to see what it might mean.

Clearly arguments about Americanization or cultural imperialism cannot rest on the place of ownership of production. The giants of the music industry are located (if that word is still appropriate) in Japan, Britain, Germany and Canada as well as the United States. Music thought to be 'American' is produced by all of these companies. The CDs are materially manufactured in still more countries. If the geography of the production process is not the most important then the other option is the form that the music and lyrics take. What we have done in this paper is to point towards the impossibility of locating music in this way through a few snapshots of musical flows between the United States and Hoggart's Britain.

The concerns of cultural critics, like Richard Hoggart, that imports of mass culture's 'shiny barbarism' from America would lead to homogenization, 'levelling-down' and the end of a traditional culture have proved to be unfounded. American music, which is itself the product of cultural hybridity, developed from interconnected elements drawn from multiple sources from within and beyond America's borders, and has had an immense influence on much British popular music during this century. However, British music and musicians, whether playing skiffle, blues, punk or whatever, have also had an immense influence over American music. From Lonnie Donegan, the Beatles and Stones, through to the present day, a large number of nominally British musicians have had great commercial success and artistic influence in America.

As Hebdige (1988) has pointed out, the great diversity of musical styles that have found some degree of popularity during this century is demonstrative that worries over homogenization were unfounded. As for the notion that an influx of American culture would bury an existing rich traditional English culture, the fact that the Copper Family are still going strong after several centuries would seem to refute this. They recently returned from a tour of America, where their songs varied from traditional ones Bob Copper had learned from his grandfather, to the traditional African-American song 'John the Revelator', which he learned from the singing of blues giant, Son House.

Whatever country, or countries, the musicians or the record companies might superficially appear to originate from, to situate the music within a specific country is practically impossible. We hope to have demonstrated in this chapter that the popular music which has dominated this century is best viewed as a complex collage of interconnections and influences with little or no central orchestration. Perhaps the best that can be said about the adjective 'American' in popular music at the end of the century is that it is irrelevant.

Notes

1   Richard Hoggart (1958) page 203.
2   This history is to be found at: <http://www.sonymusic.com/world/aboutus/history.htm>.
3   Donegan's actual name was Tony Donnegan, but he changed it in homage to the black blues/jazz guitarist Lonnie Johnson.

References

Anderson, I. (1998). Family Business. *Folk Roots*, 22–7.

Chambers, I. (1986). *Popular Culture: The Metropolitan Experience*, London: Methuen.

Cook, B. (1995). *Listen to the Blues*, New York: Da Capo.

Frith, S. (1983). 'The Pleasures of the Hearth', *Formations of Pleasure*, London: Routledge and Kegan Paul.

Gillett, C. (1983). *The Sound of the City*, London: Souvenir Press.

Groom, B. (1971). *The Blues Revival*, London: Studio Vista.

Hebdige, D. (1988). *Hiding in the Light*, London: Routledge.

Hoggart, R. (1958). *The Uses of Literacy*, London: Pelican.

Kouwenhoven, J. (1961). *The Beer-Can by the Highway*, Baltimore: Johns Hopkins University Press.

Lomax, A. (1993). *The Land Where the Blues Began*, London: Methuen.

Melody Maker. (1929). British Songs Can Succeed. IV, 289.

Mitchell, T. (1996). *Popular Music and Local Identity: Rock, Pop and Rap in Europe and Oceania*, London: Leicester University Press.

Morley, D. and Robins, K. (1995). *Spaces of Identity: Global Media, Electronic Landscapes and Cultural Boundaries*, London: Routledge.

Negus, K. (1996). *Popular Music in Theory: An Introduction*, Cambridge: Polity Press.

Zelinsky, W. (1973). *The Cultural Geography of the United States*, Englewood Cliffs, NJ: Prentice-Hall.

Chapter 19

American Philanthropy as Cultural Power

Morag Bell

Since the opening years of this century, American philanthropy has been closely associated with the country's industrial growth. As grant-dispensing foundations, the philanthropic trusts have developed significant cultural power through their ability to define, produce, translate and circulate 'useful' knowledge. They also represent a particular projection of American influence overseas as, during the course of the century, several of the major foundations have developed a global reach. Whilst frequently characterized as agents of American foreign policy, these institutions are widely varied in their origins, objectives and identity. In this chapter I highlight some of this philanthropic variety as a basis for a more nuanced interpretation of their influence and actions. By reference to concepts of space, place, time and knowledge, I provide evidence, through the foundations, of the complex, dynamic and at times contradictory nature of American cultural power. The chapter is divided into three sections. First, it briefly outlines the roots of the American foundations and some critiques of their institutional authority. Second, it focuses on the trusts of Andrew Carnegie and their hybrid character as products of his Scottish roots filtered through American industrial capital. The third sections concentrates on the Carnegie Corporation of New York and two Inquiries into poverty in South Africa which it funded during the late 1920s/early 1930s and in the 1980s respectively.

The Foundations, Science and Circuits of Knowledge

A distinctive feature of twentieth-century American cultural power has been the birth and growth of philanthropic foundations. As private grant-making institutions, among the largest in the world are located in the United States. These include, the Ford, Rockefeller, W. K. Kellogg Foundations, and the Carnegie Corporation of New York. During the nineteenth century only five general foundations were established in the United States, by the First World War there were just under 30 and in 1930 nearly 200 (Whitaker 1974). Since the 1940s, the rate of increase has accelerated dramatically, stimulated initially by the higher taxation levels resulting from the Second World War and the Korean War. Whilst within the North the practice of charity has been variously associated with the values of organized religion, the Enlightenment, liberal, feminist and socialist reformers, during this century modern industrial empires have provided a major source of wealth for philanthropic trusts. Many of the largest foundations and arguably the most famous within the North, owe their origins to the expansion of American industry. The first of the tycoon-philanthropists was Andrew Carnegie whose fortune from iron and steel was used to establish a suite of trusts at the turn of the century. The Rockefeller Foundation, based on wealth from the oil industry, was established in 1913, whilst the growth of the automobile industry of Henry Ford led to the formation of the Ford Foundation in 1936.

Unlike charities which depend upon public appeal for revenue funds, as non-profit-making bodies the foundations are endowed with assets which enable them to operate independently of public opinion. Their legal and fiscal status, including exemption from taxation, is nevertheless dependent upon an avoidance of activities which are interpreted as seeking to influence or undermine American legislation. Within this somewhat ambiguous legal framework, the character of the American foundations differs widely in size, objectives and in the functions that they conceive for themselves. As products of individual wealth and personal visions of human 'improvement', they also vary in their spatial reach. Much American philanthropy is confined within national or state borders. However, in offering an alternative to more inward-looking responses to modernity, during this century a minority of grant-dispensing foundations have developed a broader geographical vision and a spatial reach extending beyond national boundaries. The majority of overseas activities are funded by the Ford, Rockefeller, Carnegie, Commonwealth, Johnson, Kellogg and Guggenheim Foundations. As a context for the 'globalization' of philanthropic ventures, a response to American-perceived needs associated with the First and Second World Wars, the Cold War, the independence of Europe's colonial empires, and the dynamics of development practice, have been profoundly important.

Unsurprisingly, new critiques have emerged as, following the Second World War in particular, foundation activities have expanded overseas, often in support of more 'radical' projects than would be possible at home. In their ability to evade many of the controls imposed at national level, the international operations of the foundations have been likened to those of multinational corporations. As supposed agents of foreign policy, their role in American imperial expansion has also been highlighted. As if to confirm this claim, for many years Rockefeller and Ford divided the world, the former concentrating its activities in Latin America and the Far East, the latter specializing in Africa, the Middle East and the Indian Subcontinent. Granted, over twenty years ago Ben Whitaker (1974, p. 179) noted that, in response to criticisms of what appeared to be their imperialist practices, 'many of the most intelligent foundation people' were concerned that their international programmes should not 'degenerate into condescending paternalism or into smug demonstrations of exported national habits'.

Notwithstanding these sensitivities, the extent to which the foundations, like other non-governmental organizations have the ability and/or the willingness to transcend the instrumental approach to international interactions implicit in political realism, has remained a source of tensions between North and South (Edwards and Hulme 1997). Debate persists over their role in the elaboration and extension of a world-view commensurate with the economic, military and political hegemony of the United States.

Within this context, the association between philanthropy and scientific endeavour has a particular significance. As grant-dispensing institutions, the foundations have enormous power to define, produce, translate and circulate 'useful' knowledge.[1] The notion that science associated with philanthropy is different from other forms of institutionalized knowledge has some persuasive power. When specifically associated with advancing human welfare and social justice, scientific study funded by the foundations, as with other non-governmental organizations, acquires a distinctive moral authority. During this century, however, critics have frequently attacked the concept of philanthropy itself as one which undermines the significance of rights by diverting attention from the need for structural changes both at national and international levels (Prochaska 1988). In the specific case of scientific investigations with reformist goals, not only has the supposedly stultifying moralism with which they are associated been interpreted as obstructing theoretical endeavour (Abrams 1968); as components in the geopolitics of knowledge their links with the cultural hegemony implicit in ethnocentric values and identities has also been emphasized. Berman (1983) argues that whilst the American foundations are not engaged in overt forms of economic and military imperialism, they have perfected complementary activities in the cultural sphere through their educational and scientific programmes which produce and disseminate ideas congruent with their own and United States foreign policy objectives.

Notwithstanding broad connections between scientific philanthropy and American overseas influence, these views are open to debate.[2] In interpreting science as neither a neutral source of objective knowledge about how the world works nor simply a socially determined vehicle of ideology and culture, closer analysis of the nature of this form of scientific practice in specific contexts is worthy of investigation. Attention could usefully be given to how scientific studies have been justified in particular times and places and to the role of institutional values and networks in shaping their form and outcome. The Carnegie Corporation and its funding of two Inquiries into poverty in South Africa during this century, provide illustrations. The Inquiries draw together three twentieth-century episodes in the Corporation's grant-making. The first focuses on the turn of the twentieth century when the ideals of Andrew Carnegie were formulated and his vision for a 'better' world became enshrined in the terms of his charitable trusts. Following his legacy, the Inquiries provide two subsequent historical moments, one in the late 1920s/early 1930s, the other in the early 1980s. In view of the importance of Carnegie's Scottish roots in shaping his vision of improvement, the Inquiries highlight the dynamics of not merely a dual, but a triangular, set of associations between the United States, Britain and South Africa in the operation of the Corporation's philanthropy.

Andrew Carnegie, Scientific Philanthropy and his 'Mission to Better the World'[3]

The Carnegie Corporation is the biggest and was the last of Andrew Carnegie's trusts to be established in the opening years to this century. Founded in 1911 with a self-imposed mandate to promote 'the advancement and diffusion of knowledge and understanding among the peoples of the United States', since its foundation it has developed a franchise to govern in important indirect ways. Whilst it is one of the largest trusts in the world and has its headquarters in New York, it is also a hybrid institution with roots in the Calvinist principles of its Scottish founder filtered through his American industrial power. Indeed the Corporation's various programmes and priorities combine the interests of metropolitan east-coast America with a commitment to Britain's former colonies and dominions, supported through a small proportion of the income from the trust set aside annually. This institutional hybridity and the web of cultural and scientific connections across the globe which it has forged during this century, displace the nationalist framework within which much empire writing has been set. In providing a commentary on the shifting relations between Britain and two members of its former empire, the United States and South Africa, the roots of both Inquiries lie in the opening years of this century. They are products of one person's movement from social and spatial marginality in Britain to success in a rival global economic centre.

Born in 1835 in Dunfermline, Scotland, an emigrant to America in 1848, Andrew Carnegie became a highly successful product of nineteenth-century industrial capitalism and, by the year of his retirement in 1901, was among the wealthiest men in the United States. For Carnegie, the coincidence of a new century with new times was of profound symbolic and material importance. Retirement brought to an end the Carnegie Company Ltd, an immense steel empire under the watchful paternalism of its owner and a major contributor to one of the defining ideals of American multicultural society, the 'technological sublime' (Nye 1994). It was replaced by an equally powerful cultural force, the United States Steel Corporation, one of the vast super-corporations which have come to control the twentieth-century global economy. For Carnegie it also marked the end of an 'era of accumulation' and its replacement by a 'century of redistribution' (1920, p. 255). Notwithstanding the ambiguous relationship between business ethics and philanthropy, for Carnegie the two were interrelated. Designed to achieve his 'lofty' aspirations – 'his distribution of wealth, his passion for world peace and his love of mankind' (Carnegie 1920, editor's note), the task of redistribution itself became big business (Wall 1970).

The opening years of the new century provided a context for, and a validation of, new philanthropic practices. The numerous trusts established by

Carnegie represented a conscious attempt to break with the past. Each one enshrined specific objectives. Collectively they were intended to achieve his commitment to twentieth-century scientific philanthropy, a concept which included the promotion of scientific expertise with a view to shaping 'reformist' public policy. In an era in which the principles of scientific management and efficiency were widely interpreted as consistent with the nation's needs (Taylor 1911), Carnegie liked to think of himself as a pioneer in the philanthropic field. The roots of his scientific philanthropy in America and beyond can be traced to his Scottish origins, his British national identity, his role as paternalist entrepreneur in America's industrial heartland and to the citizenship he acquired in his adopted home. This biography profoundly shaped his approach and contribution to a prevailing political, intellectual and social ethos on both sides of the Atlantic which linked science with social progress, and the public consumption of institutionalized knowledge with the ideals of citizenship. Whilst the vast majority of his philanthropic resources are dispensed within the United States, Carnegie's experience as both outsider and insider in Britain and America, shaped the form and location of his initial endowments.[4] He records that his first gift, a public library, was to his 'native town', Dunfermline (Carnegie 1920, p. 260). The Carnegie-Dunfermline Trust was established in 1903 for the 'betterment of social conditions in the town', followed in 1908 by the Carnegie Hero Fund Trust, and in 1913 by the Carnegie United Kingdom Trust for the 'improvement of the well-being of people of Great Britain and Ireland'.

Coupled with this allocation to his place and country of birth, Carnegie's philanthropy was, and remains, supportive of projects which analyse and communicate knowledge of 'other' societies and environments. Two trusts are particularly significant in this respect; the Endowment for International Peace established in 1910 and the Carnegie Corporation of New York. At a time when American foreign policy was struggling to come to terms with the country's emergence as a world power, the Endowment was intended to strengthen the 'weak beginnings' of American internationalism in favour of peaceful diplomacy as against military might (Fabian 1985, p. 1). Its first president was Elihu Root, a member of Theodore Roosevelt's cabinet for many years and Nobel Peace Prize winner in 1912. The work of the Endowment, including its support for 'scientific' analyses of conflict, reflected Carnegie's belief that harmony between nations depended, at least in part, on knowledge rather than ignorance (Lester 1933).[5] Similar principles apply in the activities of the Carnegie Corporation. With a strong emphasis on the financing of civic institutions of public instruction, notably, schools, universities and libraries, coupled with support for scientific research and 'useful publications', the Charter of the Corporation extends beyond the United States to include the former British dominions and colonies. Whilst the portion of the income from the trust that could be used for activities outside America was fixed at under 10 per cent, projects supported by this source have played a crucial role in the founder's 'global' reach.[6] It is within the context of these distinctive priorities that the Inquiries into poverty in Southern Africa can be placed.

The Carnegie Corporation and South Africa:
The Poor White Question

'Africa has been overlooked in world movements'; a Carnegie programme, 'although necessarily modest, could be of immediate practical service' (Carnegie Corporation 1958, p. 1). So wrote the President and Secretary of the Corporation following a visit to the continent in 1927. One year later the Carnegie trustees voted the first sizeable grants to Africa and the first made beyond Canada with a view to fulfilling the founder's mandate regarding the British dominions and colonies. In inaugurating what was to become the African Program, funds were included for a *Commission of Investigation into the Poor White Problem*, the Poor White Study, as it became popularly known. Initiated in the decade after the First World War, the Poor White Study was launched at a time when European optimism over the meaning of progress had been shattered by the carnage of war. Woodrow Wilson's inclusionary vision of international citizenship based on a partnership of peoples not merely governments, provided a persuasive political and cultural context for the Study and also a validation of its goals. It chimed effectively with a commitment within America to the 'international mind', a concept actively supported both by the Corporation and the Peace Endowment. Based on the principle that nationalism and internationalism could be mutually reinforcing identities, it represented a particular variant of the liberal developmentalist ideology which, from the late-nineteenth century to the mid-twentieth century, couched American imperial interests in terms of altruism, evolution and world progress (Rosenberg 1982).

The focus of the South African Study on poverty gave it further public credibility. During the inter-war years social and humanitarian concerns for world health and poverty were first conceptualized by politicians on an international scale in terms of development, underdevelopment and basic needs (Rimmer 1981). These found expression in international studies of disease and nutrition funded by the League of Nations and secular charities. Based on the 'objective' measurement of human needs, underpinning these studies was a faith in the munificence of new scientific knowledge through its practical application in public health policies and in international nutrition campaigns. That the Corporation could assist this process in Africa, and in the British territories in particular, resonated with the trustees. The inter-war years marked a new phase of colonial occupation in Britain's African domain in which scientific survey became integrated with colonial administration (Worthington 1938). Whilst only limited public resources were committed by Britain to social welfare in its overseas territories during the 1930s, the Poor White Study in a former British territory preceded many scientific investigations supported by charitable foundations, including the Corporation in British colonial Africa, prior to the Second World War.

In the formulation of the research and in its implementation, the strangeness of an unfamiliar continent was mitigated for the Corporation by connections with South Africa. An expanding institutional system comprising 'a handful of

pioneer experts . . . from American universities' provided the agencies through
which the Corporation could work (Fleisch 1995, p. 352). According to Lester
(1937, p. 174) these bonds of familiarity, coupled with the use of 'local quasi-
trustees' and 'advisers', assisted 'the recipients of Corporation subsidies to
secure the most advantageous returns for their efforts in educational and social
progress'. One such was Dr C. T. Loram. A powerful figure in the Union Government, he had studied at Columbia University, New York, and had spent
several months in the American South studying 'negro' educational institutions,
as they were then described. Among South Africa's principal advocates of segregation, his approach to the 'native problem' strongly reflected his experience
in America's southern states. Appointed as one of the local Carnegie trustees
responsible for advising on the Poor White Study, he proffered advice to those
within South Africa who formally approached the Corporation for funds.[7]
Through these professional networks supposedly common problems were identified and shared solutions sought through the deployment of 'universally-
applicable' modern science. That the first Carnegie Inquiry should address the
Poor White Question in South Africa, a topic which had deep resonances within
Afrikanerdom, served not only to highlight social divisions within the population of European origin; it also sought out the 'familiar'. It found echoes in a
deep unease within sections of American society over what, from the opening
years of the century, had been perceived as widespread poverty across racial
groups, wasteful natural resource use and the rapacious values of citizenship,
and which had become the focus of 'expert' attention.[8] More particularly, it
reinforced eugenist fears held by groups within both countries that white 'civilization' could decline. Thus, the inauguration of the Poor White Study reflected
on the part of its architects, a particular reading of history and the present. Both
the New York-based funding agency and the instigators of the Inquiry in South
Africa acknowledged, albeit from markedly different perspectives, that such a
study would initiate new times, both optimistic and challenging. For the purposes of coordination, the project was located within a single institution, the
Africaans-speaking University of Stellenbosch with which the Corporation also
had connections.

The five-volume report which emerged from the Poor White Study in 1932
had dual objectives: to shape the policy process in South Africa and to communicate useful knowledge within the United States (Report of the Carnegie
Commission of Investigation 1932). In defining more precisely than hitherto
a particular category of whiteness, it established a deviation from 'normal' South
African citizenship and emphasized the need for ameliorative strategies to ensure
the reintegration of this section of the population into the general body of
European society in South Africa. At a time when segregation between the
'races' remained significant in the American South and similar principles
were being adopted with increasing force in South Africa, the promotion of an
integrationist approach for 'deviant' members of the white population was of
profound symbolic importance. In seeking to prevent the 'ultimate nightmare'
of a class movement which would unify the poor across racial lines, it served
to heighten 'race' as opposed to class difference as the significant social cate-

gory in policy discourse (Dubow 1989; Marx 1997; Freund 1992, p. xvii). The Corporation's Annual Report for 1933 (p. 29) noted that the five-volume study had attracted an extraordinary amount of attention, not only in South Africa but in other parts of the world. Widely circulated among universities and colleges in the United States, it also provoked comparative studies between the two countries. Within South Africa itself, under the direction of H. F. Verwoerd, Professor of Psychology at the University of Stellenbosch and a strong advocate of American approaches to social science, the actions which followed from the Inquiry were profoundly important in tackling poverty among the country's white population. This included the early development of the South African social welfare movement which, drawing upon American experience and research, Verwoerd was instrumental in shaping, and which provided him with a prominent position from which to launch a political career after 1936 and, in his role ultimately as Prime Minister, to pursue the ideology of apartheid (Miller 1993).

The Second Carnegie Inquiry into Poverty in Southern Africa

By the early 1980s, a further 'scientific' analysis of poverty in South Africa had, once again, a powerful resonance. Conducted at the conclusion of a decade in which the United Nations had placed a priority on meeting basic human needs, the Inquiry can be interpreted as one of a series of country studies conducted at the time with a view to promoting (once again) this internationalist humanitarian ideology. However, for its architects, the Inquiry enshrined a more specific reading of history and the present and a profoundly different form of internationalism from the Poor White Study. Following the implementation of apartheid, the Corporation, like most American and British philanthropies, chose to redirect its resources to the new nations of the Commonwealth (Murphy 1976, p. 52). However, by the late 1970s, the strategy of the Corporation had shifted. Notwithstanding the muted condemnation of apartheid by the American administration at this time (Ó Tuathail 1992), Alan Pifer, the Corporation's President, reopened a selective programme in the country with a view to supporting 'the first glimmerings of a prospect for peaceful change' (Carnegie Corporation 1980).[9] A Second Inquiry, the impetus for which came from prominent public intellectuals in South Africa, confirmed the Corporation's active agenda in the country under the new director of the international programme, David Hood. By drawing on particular interpretations of space, time and place, it secured an intellectual and moral authority which apartheid sympathizers could not easily discredit.

For both the Corporation, and for those working to end apartheid within South Africa, this Second Inquiry had strategic importance as a legitimate expression of North–South collaboration in the interests of democratic values and racial equality. Far from an 'unknown' territory, by the late 1970s following the Soweto riots and the death of Steve Biko, South Africa had become a major focus of international media attention. As part of a discourse of social justice and transformation, a Second Inquiry offered the opportunity to under-line the failure of successive apartheid regimes to confront the evidence of inequality within South African society as a whole, as opposed to a small section of it. It also chimed with a particular concern of Alan Pifer who, on becoming President in 1967, and in response to urban riots and the civil rights movement in America, had sought to make poverty a major focus of the Corporation's programme. However, as part of a strategic phase in the Corporation's grant-making, for its President the form of cooperation across two continents was also crucial in freeing it from criticisms of ethnocentrism. In order to achieve this, emphasis was placed not on common ground but on difference. In acknow-ledging the climate of political uncertainty which accompanied P. W. Botha's early years in office, and the Corporation's own lack of detailed knowledge and experience of the country, the trustees affirmed that it was within South Africa rather than outside it that the necessary expertise lay. Alan Pifer's presidential report to the Corporation in 1980 had already hinted at this. Whilst American experience had underpinned the first study, the value of such an approach was, by this stage, severely questioned. In emphasizing the need for humility and sensitivity in seeking to affect change in another country, Pifer stressed that America 'cannot be held up as an unquestioned model for South Africa, however useful certain pieces of our experience may be' (Carnegie Corporation 1980, p. 14).

The institutional arrangement finally reached and the cultural technologies used in the research process also sought to satisfy competing interests within South Africa itself, notably the need to establish a kind of scientific inquiry which would genuinely look forward. The outcome was a study of hybrid character: on one level it was conformist, on another it was transgressive and transformative. Responsibility for designing, stimulating and coordinating the Inquiry was assigned to Professor Francis Wilson and the Southern Africa Labour and Development Research Unit (SALDRU) in the University of Cape Town. One of South Africa's oldest and most prestigious institutions of higher education for the white population, it was also among its most outspoken critics of apartheid (Jansen 1991). In a background briefing on the Second Inquiry, Wilson noted that the funds made available for the study 'came with no strings whatever' (undated p. 2).[10] Whilst reflecting on the part of the Corporation a far more flexible approach than the earlier Investigation, within the public domain the principles of historical continuity were used specifically to validate a follow-up study. At a time of acute political tension in South Africa when radical voices, including participants in the Second Inquiry, sought to break with the past, two features of the earlier study were profoundly important in ensuring that a second could take place: time was used as a landmark in the numerical recording of social change and a similar institutional language was employed. For the purposes of national publicity, a gap of fifty years since the Poor White Study, funded once again by a 'respected' Northern foundation, offered a legitimate reason for a follow-up study across the population as a whole but with similar objectives. Coordinated once again by an academic insti-tution and retaining the label of a formal Inquiry, this arrangement gave it, like its predecessor, the status of an official investigation and public legitimacy through its independence from official manipulation.

This deliberate deployment of continuity for the purposes of external consumption allowed the Second Inquiry to offer a profoundly different kind of study from its predecessor. Thus, an orthodox image of the university as the intellectual space for politically-neutral, objective science was used to full effect. Much was made in the study of its role in uncovering the 'facts' about poverty and of the university as a public forum for debate about these facts. But in their production, this same setting provided the protected space, the laboratory, within which to experiment; to recast the nature of scientific practice and the concept of expertise. Following a gap of a half-century and within an intellectual environment openly hostile to epistemologies rooted in the North, the second study offered a less prescriptive and narrowly focused approach than its predecessor. Moreover, in contrast to the prevailing technocratic policy discourse within South Africa at the time which sought to depoliticize contentious issues by perpetuating the notion of 'development' as an 'objective' science (Tapscott 1995), in the process of knowledge production, notation and dissemination, the imperatives were social critique, plurality and participation. Not only was the relevance of disciplinary traditions questioned but, more particularly, the boundaries of useful knowledge were redefined and democratized as the study encouraged participation beyond the spheres of academe and professional practitioners to include also those closest to the experience of poverty such as social workers and community leaders.

The Inquiry drew upon the experience and recommendations of over 200 people of different racial groups from across the country. Unlike the Poor White Study which depended on established intellectual networks for its implementation, these contributors created diverse circuits of knowledge as part of the research process. These circuits were consolidated at a public conference hosted by the University of Cape Town in April 1994 with a view to encouraging the discussion and dissemination of knowledge accumulated in the course of the Inquiry and also as a basis for practical action. For the Corporation's representatives at the conference, there was a strong emphasis on the Inquiry as a learning experience. The address by Alan Pifer's successor as President, David Hamburg, on the final day of the conference ranged widely over twentieth-century North–South relations and the role of the Corporation in these. In highlighting the limited vision of society implicit in the First Inquiry, he noted the constructive role of time in shaping the Corporation's subsequent activities. Drawing on American experience, he emphasized how within less than a decade, and in support of a more inclusive notion of social justice, Gunnar Myrdal had been commissioned by the Corporation to examine the relations between 'race', poverty and inequality in the United States, the published output of which, *The American Dilemma* (1944), had exposed for the first time the extent of racism in American society (Carnoy 1994). Notwithstanding the country's democratic ideals and constitution, it highlighted the massive 'flaw in the fabric of American democracy' which enabled both individual and collective white prejudice to perpetuate racial discrimination (Carnegie Corporation 1981, p. 5). As

President of one of the most powerful American philanthropic institutions, Hamburg also challenged the comfortable assumption that poverty happens 'elsewhere' by referring specifically to the importance of the Second Carnegie Inquiry to the multiple dimensions of poverty within the United States.

For the architects of the Inquiry in South Africa, few would argue that it made any significant contribution to ending apartheid beyond further discrediting the system (Wilson 1986; Wilson and Ramphele 1989). But as useful knowledge, of greater importance at the time was the public attention drawn to the nature and extent of poverty and to the research process itself in giving a voice to many of those to whom it had been denied, in anticipation of a post-apartheid future. As an exercise in North–South cooperation, for the Corporation, the study enhanced its reputation. Much publicity was given to the results of the Inquiry in the national and international press, an important achievement in itself at the time. The public credibility of the Corporation was also enhanced by a major dissemination exercise which extended well beyond South Africa into Europe and North America during the 1980s. But these public benefits obscure the internal debates extending over some ten years over the position of the New York-based Corporation as funding body and supporter of radical science advocates. Here was an institution which, through its venture into the 'unknown', became quite effectively harnessed by public intellectuals in the South. In the valedictory report of its President given to the trustees in 1996, David Hamburg openly admitted that the 'extraordinary experience' of the Inquiry had been one of the major features of his presidency.

Conclusions

Whilst during this century American philanthropy has contributed to the projection of American cultural power overseas, the foundations are complex and varied institutions. Reference to the origins and overseas activities of the Carnegie Corporation highlights the need to move beyond a narrowly nationalist framework in analysing their ideals and 'global' reach. In seeking to retain, over time, a degree of national and international credibility, the foundations have been required to respond to shifting North/South relations, to changing interpretations of citizenship and the boundaries of 'useful' knowledge. Nor have they been slavish mimics of American foreign policy. As agents in the production and diffusion of knowledge, scrutiny of specific investigations highlights the significance of time and place in shaping the form and nature of their scientific endeavours across national boundaries and the extent to which they have operated independently of, or in opposition to, particular administrations both at home and overseas. Notwithstanding their willingness to engage in politically challenging activities, the Carnegie Inquiries in South Africa demonstrate how they have also sought to fulfil their own objectives. In doing so, however, they have not necessarily overlooked the interests of their funding recipient. Scrutiny of the research process itself is significant in this respect. In the design and promotion of particular investigations, concepts of political and cultural similarity and difference have served to justify their implementation and their mutually beneficial nature. The cultural power of the foundations can not, therefore, be assigned merely to their position within a broad structure of American global domination. Nor can it be interpreted merely as power over its funding recipients. Such an inflexible formulation fails to acknowledge the complex and dynamic nature of the research process, the dialogues involved, and the shifting power relations within recipient countries as an outcome of particular investigations.

Notes

Source materials for this chapter were obtained from the Carnegie Corporation of New York archives, Columbia University Rare Book and Manuscript Library. I am grateful to Professor Francis Wilson, University of Cape Town, for allowing me access to material associated with the Second Carnegie Inquiry.

1 Ellen Lagemann (1989) discusses this process within the United States with reference to the activities of the Carnegie Corporation during this century. The influence of the Rockefeller Foundation on the work of the California Institute of Technology is discussed by Lily Kay (1993).

2 Notwithstanding a revisionist tendency to characterize the overseas activities of the foundations as imperialistic, recent research on the Rockefeller Foundation (Desowitz 1997) counters this by emphasizing its role in combating yellow fever and hookworm in the United States and Mexico during the early part of this century when the civilian US government was doing very little. Equally, in a comparative study of the United Nations Development Programme, the World Bank and the Ford Foundation in their approach to gender issues in development since the 1970s, Kardam (1990) found that the latter had been most responsive. This reflected the Foundation's particular value system which emphasized equity and anti-discrimination, the priorities it attached to the production of new knowledge in seeking solutions to social problems, and a decentralized style of management which encouraged creativity and permitted flexibility among staff members.

3 Carnegie Corporation 1992, p. 6.

4 Carnegie's support for American democracy, republicanism and commitment to the reunion of Britain and America, are discussed in Carnegie (1886; 1897; 1900; 1920); Hendrick (1993); Lagemann (1989); Wall (1970).

5 The organization funded a monumental 152-volume study of the economic and social history of the First World War.

6 During 1909–11, Carnegie Hero Funds in Europe were established in France, Germany, Norway, Switzerland, the Netherlands, Sweden, Denmark, Belgium and Italy. The fund in Germany is no longer active. Carnegies's notable personal gifts include his contribution to the construction of the Peace Palace at The Hague, the Pan American Union Building, now called the Organization of American States Building, in Washington DC, and the Central American Court of Justice in San José, Costa Rica.

7 In 1929 Loram became the first Chairman of the South African Institute of Race Relations. In 1931 he took up the appointment of Sterling Professor of Education at Yale University and, two years later, became Chairman and Director of Studies in the Department of Culture Contacts and Race Relations.

8   The South Africa study anticipated Roosevelt's New Deal for heartland America which, infused with a spirit of modernity, and based on comprehensive scientific survey, promised sweeping welfare reforms.
9   The Corporation's involvement in the country nevertheless came under attack from opponents of apartheid in America at this time who argued that such involvement lent legitimacy and respectability to the regime (Carnegie Corporation 1980, p. 12).
10 Carnegie Corporation Grant files Series 2, Box 461.

References

Abrams, P. 1968, *The origins of British sociology, 1834–1914*. University of California Press, Berkeley.

Berman, E. 1983, *The influence of Carnegie, Ford and Rockefeller Foundations on American foreign policy: the ideology of philanthropy*. State University of New York Press, New York.

Carnegie, A. 1886, *Triumphant democracy or fifty years march of the Republic*. London.

Carnegie, A. 1897, The reunion of Britain and America. A look ahead. *Contemporary Review*, Nov.

Garnegie, A. 1900, *The gospel of wealth and other timely essays*. The Century Company, New York.

Carnegie, A. 1901, British pessimism. In *The gospel of wealth and other timely essays*. The Century Company, New York, pp. 309–30. (Reprinted in *Nineteenth Century and After*, June 1901.)

Carnegie, A. 1920, *Autobiography*. Constable and Co., London.

Carnegie Corporation. 1958, Africa. *Carnegie Corporation of New York Quarterly*, 6, p. 1.

Carnegie Corporation. *Annual reports*. Carnegie Corporation of New York.

Carnoy, M. 1994, *Faded dreams. The politics and economics of race in America*. Cambridge University Press, Cambridge.

Desowitz, R. 1997, *Tropical diseases*. Harper Collins, London.

Dubow, S. 1989, *Racial segregation and the origins of apartheid in South Africa c.1919–1936*. Cambridge University Press, Cambridge.

Edwards, M. and Hulme, D. (eds) 1997, *NGOs, states and donors: too close for comfort?* Macmillan, London.

Fabian, L. 1985, *Andrew Carnegie's peace endowment. The tycoon, the President and the bargain of 1910*. Carnegie Endowment for International Peace, Washington DC.

Fleisch, B. 1995, Social scientists as policy makers: E. G. Malherbe and the National Bureau for educational and social research, 1929–1943. *Journal of Southern African Studies*, 21, pp. 349–72.

Freund, B. 1992, The poor whites: a social force and a social problem in South Africa. In Morrell, R. (ed.), *White but poor. Essays in the history of poor whites in Southern Africa, 1880–1940*. University of South Africa Press, pp. xiii–xxiii.

Gross, P., Levitt, N. and Lewis, M. (eds) 1997, *The flight from science and reason*. Johns Hopkins Press, Baltimore.

Hendrick, B. 1993, *The life of Andrew Carnegie. 2 Volumes*. Doubleday, Doran and Co., New York.

Jansen, J. (ed.) 1991, *Knowledge and power in South Africa*. Skotaville, Johannesburg.

Kardam, N. 1990, *Bringing women in: the women's issue in international development*. Lynne Rienner.

Lagemann, E. 1989, *The politics of knowledge. The Carnegie Corporation, philanthropy and public policy*. Wesleyan University Press.

Lester, R. 1933, Review of grants to Carnegie Endowment for International Peace 1911–35. Carnegie Corporation of New York *Review Series* vol. 1, no. 13 (printed for the information of the Trustees, 1 Oct. 1933).

Lester, R. 1937, Summary of grants from the British Dominions and Colonies Fund 1911–1935. Carnegie Corporation of New York *Review Series* vol. 2, no. 20 (printed for the information of the Trustees, 20 Sept. 1935).

Marx, A. 1997, *Making race and nation: a comparison of the United states, South Africa and Brazil*. Cambridge University Press, Cambridge.

Miller, R. 1993, Science and society in the early career of H. F. Verwoerd. *Journal of Southern African Studies*, 19, pp. 634–61.

Murdoch, J. 1997, Inhuman/nonhuman/human: actor-network theory and the prospects for a nondualistic and symmetrical perspective on nature and society. *Environment and Planning D: Society and Space*, 15, pp. 731–56.

Murphy, E. Jefferson, 1976, *Creative philanthropy. Carnegie Corporation and Africa, 1953–73*. Teachers College Press, New York.

Myrdal, G., Sterner, R. and Rose, A. 1944, *An American Dilemma: the negro problem and modern democracy*. Harper and Brothers.

Nye, D. 1994, *American technological sublime*. MIT Press, Cambridge, MA.

Ó Tuathail, G. 1992, Foreign policy and the hyperreal. The Reagan administration and the scripting of South Africa. In Barnes, T. and Duncan, J. (eds), *Writing worlds*, pp. 155–75.

Prochaska, F. 1988, *The voluntary impulse. Philanthropy in modern Britain*. Faber and Faber, London.

Report of the Carnegie Commission of Investigation on the poor white question in South Africa. 1932, *Part I. Economic report: Rural impoverishment and rural exodus. Part II. Psychological report: The poor white. Part III. Educational report: Education and the poor white. Part IV. Health report: Health factors in the poor white problem. Part V. Sociological report: (a) The poor white in society. (b) The mother and daughter in the poor white family*. Pro Ecclesia-Drukkery, Stellenbosch.

Rimmer, D. 1981, 'Basic needs' and the origins of the development ethos. *The Journal of Developing Areas*, 15, pp. 215–38.

Rosenberg, E. 1982, *Spreading the American dream: American economic and cultural expansion 1890–1945*. New York.

Tapscott, C. 1995, Changing discourses of development in South Africa. In Crush, J. (ed.), *Power of development*. Routledge, London, pp. 176–91.

Taylor, F. 1911, *The principles of scientific management*. New York. (Sections reprinted in Pursell, C. 1969, *Readings in technology and American life*. Oxford University Press, Oxford.)

Wall, J. 1970, *Andrew Carnegie*. Oxford University Press, Oxford.

Whitaker, B. 1974, *The foundations. An anatomy of philanthropy and society*. Methuen, London.

Wilson, F. 1986, *South Africa. The cordoned heart*. Gallery Press, Cape Town.

Wilson, F. and Ramphele, M. 1989, *Uprooting poverty: the South African challenge*. Norton, New York.

Worthington, E. B. 1938, *Science in Africa. A review of scientific research relating to tropical and southern Africa*. Oxford University Press, Oxford.

Chapter 20

Between North and South: Travelling Feminisms and Homeless Women

Claudia de Lima Costa

Introduction: Travelling Theories, Travelling Theorists

As pointed out by Clifford (1989), it is an indisputable reality these days that theories are travelling as never before through the fulcrums opened up by a process of globalization that shrinks distances between different localities and brackets the meanings of place, home, and community. Displaced (from its 'home' in the West), and appropriated, resisted and transformed by increasingly larger transnational academic communities, theories are now migrating through ever more complex itineraries. Radhakrishnan (1996) reminds us that in light of the disappearance of clearly demarcated routes characterizing post-colonial landscapes, it is often the case that the travel may go wrong, take sudden detours, or encounter unexpected pitfalls. As a result, the theorist, especially when speaking and writing across different contexts, temporalities and narratives, must be self-reflexive about her location (understood as more than just a geopolitical situation, but – after Mohanty (1987) – also in the metaphorical senses of a position of debate and of an imagined, political, cultural and psychic location). Moreover, the critic in her travels should not lose sight of the fact that any place is irredeemably punctuated and fractured by differences and multiple tensions, by circuits and borders that exceed binary formulation of the relations of power.

As Bhabha (1994) has already pointed out, location and locution are two inextricably intertwined terms. Depending on one's location in the structures of privilege, one may be empowered or disempowered to speak. Indeed, in light of the multiple fractures within the condition of Woman, followed by her continual decentrings in poststructuralist times, it has become increasingly difficult to theorize in feminism. Feminism's once foundational concepts are rapidly yielding their explanatory power and being forced into invisibility in the critical vocabulary of the 'post'.

In my attempt to address the travels of theories across the North/South divide, I will map out – situating myself between Brazil and the US – the poststructuralist debates taking place in the North and their repercussion, or lack thereof, in the South (as well as their differential appropriations), trying to stake out some of the itineraries of feminist theories in their movement 'between cultures, languages, and complex configurations of meaning and power' (Mohanty, 1987: 42). To accomplish that, I will draw on my experiences of dislocation between these two countries so as to reflect upon the mediations demanded from the feminist critic to speak (and write) within different geopolitical contexts. On a more personal note, I want to tackle the issue of how travels to and from different contexts of knowledge production and reception may supply one with certain types of analytical baggage that can alter one's perceptions of subalternity, privilege, intellectual work and feminism. I voluntarily left Brazil in my early twenties to pursue my undergraduate and graduate training in US academic institutions, where I was introduced to feminist theories of the eighties and nineties. After having finished my studies, I travelled back to Brazil with a brief detour through Venezuela, where I taught for one year. Back home I spoke at many conferences and workshops about feminist poststructuralist theories and had to continually respond to challenges from local audiences about the relevance of such theories to Brazilian women's concerns. In response to these exchanges, I began the painstaking task of critically re-assessing the very theories that until then had guided my intellectual pursuits. That process of cultural translation was furthered by my fieldwork: collecting life histories of a group of women from the Sem Teto (homeless) movement (in Florianópolis) to understand how their narratives constructed their identities and reconstructed their experiences. It was at this point – at the intersection of US postmodern/poststructuralist feminist theories, my return 'home', and my encounters with the lived experiences of Brazilian Sem Teto women – that I discovered uneasy silences. While I didn't want to discard my hard-won theoretical baggage, I embarked on the difficult path of mediation between these two geopolitical and theoretical contexts.

This chapter – which is divided into three sections that are roughly demarcated by three geographical locations: the US, Latin America and, more specifically, Brazil – represents an attempt to understand what such mediations require. More particularly, I am interested in gauging the uneven flows of feminist theories between North and South. Emphasizing one specific and most recent trend within US feminism – the debates around historical versus cultural materialism – and drawing on fieldwork examples, I want to point out that unequal theoretical exchanges have pushed 'third world' women out of the realms of postmodern/poststructuralist ludic politics (accessible only to 'first world' women) and how these marginalized subjects, by claiming their right to the pleasures of ludic politics concomitantly with assertion of rights to citizenship, are pushing themselves back into the texts of 'first world' feminist theories. My contention is that US feminist discourse and its hegemony in the region at the century's end has not gone unchallenged by Latin American and Brazilian women's movements.[1] As my local story illuminates, women – academic and non-academic and positioned at the other side of the border – are deploying in subtle and creative ways feminist postmodern/poststructuralist discourses issuing from the North to open up a space for the assertion of their (once erased) subjectivities, identities and politics toward the construction of what has been termed 'transnational' feminist practices and alliances.

Feminist Debates in the US

Succinctly put, poststructuralist theories, covering a wide range of works in diverse disciplinary quarters, and theorizing 'from within the horizon of the linguistic turn' (Fraser, 1995: 157), mounted an overwhelming critique of knowledge and representation that configured a 'fundamental historical shift away from objectivist and rationalist cognitive paradigms and towards a social and material understanding of the situatedness of what we have till now called reason' (Ryan, 1989: 1). That is, instead of appealing to universally grounded propositions or overarching narratives, this body of often mutually incompatible theories – or better yet, critical practices – offers a more local, plural and immanent ground for social criticism. In contesting Western logocentric traditions (that is, the view that ideas transcend their own representation), questioning representation, decentring identity, undermining foundationalism and essentialism, and in insisting on self-reflection, poststructuralist theories have unmasked the coupling of knowledge and power, and the theorist's very complicity with that power – signalling the end of an innocence that had in fact never been innocent. As a consequence of poststructuralism's focus on language and discourse, we have a shift away 'from social science's preoccupation with things and towards a more cultural sensibility to the salience of words' (Barrett, 1992: 205). In the (now false) things/words dichotomy, words are seen as also pertaining to the 'materiality' of things. That is, matter does not exist outside the boundaries of forms of representation; rather, it is a constitutive effect of regulatory practices, linguistic tropes, discourses and norms.

Western feminism, in the span of twenty years (from the 1970s to the 1990s) – and in all its varying taxonomies (e.g., liberal, socialist, Marxist, radical, cultural and, more recently, postmodern, poststructuralist) – has gone through several paradigm shifts that very early on exposed, on the one hand, the contingent, perspectival, contradictory, and historically situated nature of its theoretical activity and, on the other, its political commitment to the struggle against domination. However, in attempting to explain the cause of women's oppression across diverse geographical landscapes, feminists in the 1970s – as Fraser and Nicholson (1988) put it – formulated 'quasi-metanarratives' of women's oppression. For Elizabeth Weed (1989), a profound contradiction or tension began to permeate feminism from the late 1970s through the 1980s. Positioned at the conflicting intersection of 'identity politics' and the poststructuralist critique of liberal humanism, feminism, while espousing the notion of a universal female identity anchored in Enlightenment values such as rights, equality and freedom, was concomitantly engaged in the deconstruction of the humanist discourse of modern theory.

Given some affinities between the feminist and the deconstructionist projects, their alliance has enabled in significant ways a more differentiated approach to issues of subjectivity, selfhood and agency (Best and Kellner, 1991). For Hutcheon (1989), the '"de-doxifying" work on the construction of the individual bourgeois subject' in postmodernism (and poststructuralism, I would add) 'has had to make room for the consideration of the construction of the *gendered* subject' (142), along with de-naturalization of the private and the public, the personal and the political.

However, despite the positive aspects of the exchange between these two political-cultural currents, their partnership has also been met not infrequently with both fierce opposition from some feminist quarters and cautious reticence from others. There is, moreover, a generalized consensus that the two cannot and must not be conflated if feminism is to resist being subdued by postmodernism.[2] Because the quandaries are so numerous, taking so many major and minor twists, my discussion here will be of necessity symptomatic, not exhaustive. I will highlight what I believe are the most pertinent to the central concerns of this chapter. Schematically arranged, these are:

First, there is the claim that the dispersion of the subject, history and philosophy into language games carries with it the abjuration of subjectivity, the ideals of autonomy, reflexivity and accountability – all requisites for any emancipatory project. Benhabib (1995) argues that in so far as subjectivity, identity and agency are interpreted as the sole effects of discursive practices – following a certain linguistic foundationalism – postmodernism and feminism remain incompatible projects. For a more politically oriented view on these crucial issues, Benhabib continues, we need to look beyond the myopic hermeneutical circle of narratological strategies and meaning-constitution to explore broader social and structural processes.

Second, and in the same critical vein as Benhabib, other authors (Bordo, 1990, 1992; Flax, 1990; hooks, 1990), while earnestly engaged with postmodernism, and although acutely aware of the dangers of essentialist definitions of identity, find themselves none the less hesitant to embrace wholesale the fragmentation of the subject along with the undermining of the authority of her experience. They maintain that the decentring of the subject appears to attain a positive aura only in theory; in practice and in the context of everyday struggles for the (re)construction of a sense of self, location and agency by subaltern subjects, such a theoretical and methodological creed risks eviscerating feminism of its affirmative politics.

Third, the emphasis on questions of perspective or on the inescapably situated nature of one's gaze upon the world brought about by postmodernism and feminist theories alike have turned into a powerful political and epistemological platform from which feminists launched piercing attacks on a number of postulates concerning the relationship between the knower and the known. After having overcome the Enlightenment pretension concerning the objectivity and neutrality of knowledge and reason – i.e., the 'view from nowhere' (Bordo, 1990: 139; Hartsock, 1990: 18), some garden-varieties of poststructuralist feminism, in their excessive attentiveness to and espousal of plurality, heterogeneity and the endless play of difference, wound up replacing the inauspicious 'view from nowhere' with an equally unpropitious and short-sighted 'view from every-where' (Bordo, 1990).

In reaction to the above cross-currents in feminism, materialist feminists, claiming that 'the struggle over meanings, over language, over cultural practices and politics of representation must be grounded in the priority of economic struggles' (Ebert, 1996: 123), call emphatically for a return to the Marxist all-determining – or last instance – contradiction between capital and labour to account for women's oppression. According to a welcome message on an Internet materialist feminist discussion list,

> Materialist feminist work is distinguished by the claim that the critical perspective of historical materialism is historically necessary and empowering for feminism's oppositional project. Materialist feminism calls for a consideration of the ways class, divisions of labor, state power, as well as gendered, racial, national and sexual subjectivities, bodies, and knowledges are all crucial to social production. While materialist feminists have made use of postmodern critiques of empiricism to develop analyses of the role of ideology in women's oppression, they have also insisted that ideology is one facet of social life. This *systemic view* – the argument that the materiality of the social consists of class, divisions of labor, state power, and ideology – is one of the distinguishing features of materialist feminist analysis.
>
> ('Welcome to Matfem' at MatFem@csf.colorado.edu; May 1997;
> emphasis added)

More recently, however – and in order to bring attention to the political economy of capitalist patriarchy and to countervail the spread of post-Marxism in feminism – materialist feminists such as Rosemary Hennessy have delved even deeper into the troubled waters of economic reductionism. Turning away from what they regarded as indeterminate types of 'systemic analysis', and stressing the priority of causal analysis, some materialist feminists state that to make sense of women's oppression one needs to abandon appeals to 'abstract, ahistorical, or merely cultural categories like desire, matter, or performativity' (Hennessy and Ingraham, 1997: 2) in favour of a renewed focus on the direct effects of capitalism's class system in women's lives. In materialist feminism's hierarchy of causal determination, poor third world women occupy a unique – and not enviable – place. Because of utter exploitation and oppression, they cannot afford the privilege and pleasures of post-class ludic politics aimed at semiotic freedom. These impoverished third world women are given the arduous task of (nothing less than) connecting 'cultural politics to the revolutionary struggle to fundamentally change the conditions of production, appropriation of surplus labor, and unequal distribution of wealth' (Ebert, 1996: 124).

Notwithstanding both Ebert's and Hennessy's important contribution in arresting the infernal slipping and sliding of much of contemporary Western academic feminist theorizing, I none the less think that one should attempt to evade the binary trap that has framed the debates between postmodern/post-structuralist feminism versus historical materialist feminism, libidinal liberation versus revolution. To resist the easy temptation of making cultural practices (again) into epiphenomena of social relations of production, it is crucial that feminists conceive culture as an articulated ensemble, that is, 'as the point of intersection and negotiation of radically different kinds of vectors of determination – including material, affective, libidinal, semiotic, semantic, etc.' (Grossberg, 1997: 19). To expand the textual limits of meaning (so as to encompass its larger historical con/text), feminists need to unravel the ways cultural practices shape and are also shaped by other social forces, such as imperialism, capitalism, patriarchy, racism, homophobia and – the latest newcomer to the list – globalization.

Drawing on my fieldwork observations of the weekly Sem Teto movement (homeless movement) meetings in Florianópolis, Brazil (attended mostly by women), I will argue that an exclusive and narrow focus on material needs by movement leaders – to the neglect of other aspects of people's social existence – contributed in significant ways to alienate members from this local social movement. My fieldwork example is intended as one illustration of the limits that both materialist and poststructuralist-informed feminist theories encounter when carried to other geopolitical contexts. As I travelled South, homeward bound, such contradictions and difficulties became even more visible, thus forcing deeper reflections regarding not only questions about displacement of theories but of theorists as well.

Travelling Feminisms and the Latin American Difference

To re-enter Latin America, both physically and intellectually, after a period of nearly fifteen years of voluntary exile (from 1976 to 1991), represents a daunting task for any person. Upon my return from this long absence, my Brazilian 'home' (in the broader sense of the term) appeared unrecognizable. Hyperinflation, government corruption followed by presidential impeachment, dwindling job opportunities in the formal sector, high criminality, widespread fear in the middle and upper-middle class sectors of the urban population leading them to voluntary confinement in huge condominiums that have become 'cities of walls' (Caldeira, forthcoming), and rampant narcotraffic that turned the *favelas* (Shantytowns) into war zones controlled by rival gangs – along with their *balas perdidas* (stray bullets) – had all become an integral part of the (invisible) city horizon in the postcard pictures on display in newsstands.[3] As I was greeted by this scenario, I felt 'homeless' in several respects: I not only no longer recognized what lay ahead of me in the streets, buildings, shopping malls, parking lots and hillsides, but was also struck by a sense of my own 'foreignness' in terms of my 'Brazilianness' (or Latin Americanness), so to speak. With my gaze transformed by 'the immigrant experience' and my immersion in critical discourses in the US, I realized that the 'unrecognizable' landscape now before me in fact bore a strong resemblance to the nefarious socio-political scenario engendered by the military dictatorship. But at the time I embarked on my Northbound travel, I had not ventured outside of the narrow confines of my 'Teflon existence' (West, 1988: 277) within the rigid structures of Brazilian class and racial privilege.

Therefore, holding onto the experience of standing in this unfamiliar familiar space (in both the physical and theoretical senses of the word), what I will
seek to accomplish in this section is to briefly sketch out what is distinct about
Latin America that makes a difference *vis-à-vis* the feminist polemic about the
postmodern turn in social and cultural theory as presented above.

It is sufficient to set foot on Latin American soil to understand immediately
that Western postmodern *grand récit* about the ends of Man (*sic*), of representation, of ideology, of History, finds little echo.[4] This lack of resonance is not
due to the fact that, as some writers have defended, that part of the world has
yet to attain 'modernity', or that the subject inhabiting that location still
appears, nostalgically, undivided and centered.[5] Rather, it is pointed out that the
postmodern narrative becomes partially *déjà-vu* because, for the vast majority
of those living in Latin America, the subject has never been allowed accession
to Western subjecthood – imagined or otherwise – in the first place. In fact,
Beverley argues that 'the decentered subject of poststructuralist theory is more
often than not "cheap labour" these days' (1993: 108). According to Mouat
(1993), although many parts of Latin America are still very distant from becoming 'modern', the discourse about the postmodern in the region is

> rendered necessary not because Lyotard and Habermas, or deconstructionists and
> Marxists, have gone at it in Europe but because the material conditions and social
> attitudes that underlie it – consumerism, technocracy, state-of-the art technology,
> the culture of the image, etc. – have themselves carried over into selected sectors
> of Latin American social and political life.
>
> (155)

It is generally contended in the literature that the variegated range of competing and asymmetrical realities (or particularisms) that make up the Latin
American scene have constructed for it a specific sort of postmodernism *avant
la lettre*, whereby aesthetic manifestations have gone hand in hand with social
struggles for the enfranchisement of all citizens. It is in that sense that Yúdice
(1989) argues, defending the idea of a 'postmodernism of resistance', that in
Latin America this cultural-political movement speaks for the 'multiple local
aesthetic-ideological responses/proposals in the face of, against, and within
transnationalization' (108). Yúdice's definition of Latin American postmodernism emphasizes two important interrelated points. First, the heterogeneity of
cultural formations that coexist (in conflict and dialogue) in the South of the
Americas cannot be reduced to a homogenous modernity (something that Nelly
Richard (1995) also continually stresses). This is to say, in Latin America there
are several modernities and the postmodern condition does not mean a rupture
with the latter, but 'the recognition that these are multiple modernities and that
no one model nor subject will determine the course of history' (Yúdice, 1989:
118). Second, these formations (a mosaic of different ethnic and communal identities with its own forms of knowledge and subjectivities) represent the possibility of democratic intervention in, and restructuring of, social formations,
particularly in the face of successive economic crises ravaging most Latin
American countries. Postmodernism, for Yúdice, is 'a discussion about the pos-

sibilities of a democratic culture' (110) without, however, assuming that the experience of marginality *per se* automatically impels the subject into transgression. It is at this point in the discussion that I would like to return to the question of feminist (postmodern) theorization across the North/South divide.

In an edifying discussion on feminism, experience, and representation, Chilean cultural critic Nelly Richard (1996) points out that, in the global division of labour, the traffic in theory to and from metropolitan centres and peripheries remains tied to an unequal exchange: while the academic centre theorizes, the periphery is expected to supply the former with case studies. In other words, it is reduced to the practice side of theory (or, in another perverse binary opposition, to the concrete body as opposed to the abstract mind of metropolitan feminism). Trinh Minh-ha vividly captures the inscription of the experiences of third world women in the Western feminist repertoire when, reflecting upon her own status as an immigrant woman of colour in the US, she states that

> Now I am not only given the permission to open up and talk, I am also encouraged to express my difference. My audience expects and demands it; otherwise people would feel as if they have been cheated: We did not come here to hear a Third World member speak about the First (?) World. We came to listen to that voice of difference likely to bring us *what we can't have*, and to divert us from the monotony of sameness.
>
> (in Bulbeck, 1998: 207)

In the context of the South of the Americas, Richard further asserts that some Latin American feminists are guilty of falling into this hierarchical binary trap when they turn practice (understood as the body of pre-discursive experiences) into the truth and emblem of Latin American feminism. However, I would argue that if, on the one hand, the recent and justifiable suspicion of universalism in feminist discourse has largely contributed to the construction of the third world woman as a repository of unadulterated difference (oftentimes translated as authenticity); on the other, Latin American feminists are rightfully wary of the lure of postmodern/poststructuralist discourses ever transforming the materiality of women's experience of oppression and resistance into redemptive tropes of Western feminist discourse.

To occupy the middle ground *vis-à-vis* these two discursive positions seems to me a more productive intellectual and political alternative out of the impasse. If, on the one hand, materialist feminists in the first world have a tendency to reduce poor third world women to a long list of *carências* (material needs) at the same time that they make them the bearers of utopian revolutionary practice (with no pause for the pleasures of cyborg politics); on the other, poststructuralist-informed (post-Marxist) feminists feel inclined to take the easy way out by dissolving material needs into matters of language. One way of avoiding the above trap and to dismantle the reductive language and reality binarism requires reconceiving reality as a structure of multiple planes of effects deployed by practices which traverse, intersect and disrupt one another (Grossberg, 1992). Practices – having no inherent essence – are defined by their effects, which are never guaranteed but always real and effective. For

Grossberg, inasmuch as economic practices can have effects other than economic ones, cultural practices may also have effects that exceed or even bypass
meaning effects. I want to contend, drawing on my fieldwork, that articulation
– and not apriorisms such as the ones advanced by either materialist – is what
allows one to build linkages between practices and effects, needs and affects, as
well as to connect them to the larger context.

Homeward Bound: The Brazilian Context

I will begin my discussion on the relatively recent encounter between postmodernism and feminist theorizing in Brazil on a personal note. In 1992, one
year after my return to Brazil (after an absence of fifteen years), I applied for a
fellowship, sponsored by a major Brazilian research institution, to do my fieldwork among women living in three poor working-class neighbourhoods in
Florianópolis, and who were also participants of the local *Movimento dos Sem
Teto* (Homeless Movement).

As a requirement for the grant, I was to attend a seminar along with the
other grantees, on which occasion I presented my research project. Since at that
time I had still no field data to show the audience – made up mostly of social
scientists from disciplines such as sociology, anthropology and political science
– I decided to centre my presentation on the theoretical framework informing
my project: the study of the constitution of the subject through life-narratives.
After a rather brief but none the less elaborate mapping of the epistemological
issues guiding my research concerns, one of the first questions directed at me
by a juror of the grant competition was 'how will it be possible for you to reconcile postmodernism with women living in the *favelas* (shantytowns)?' In my
view, this very pertinent question raises two related issues. First and foremost,
it nicely captures the following paradox confronting feminist theories south of
the US border: while, on the one hand, Latin American identities and cultural-
aesthetic practices and discourses are often cited in the literature as being postmodern *avant la lettre* (or as setting an example of how to think of identity and
difference as politically grounded); on the other hand, in the realm of Latin
American feminisms – claimed by Richard (1989) and Raquel Olea (1995) as
the most radical expression of Latin American postmodernism – poststructuralist theories have been kept at bay for quite some time. Second, in the above
question posed to me there was a (not so veiled, albeit understandable) in-
sinuation that I was importing first world theories to third world contexts.[6]

Conceding materialist feminists some points, and not wanting to deny the
reality and materiality of the oppressive conditions circumscribing the lives of
underprivileged third world women, I would like to suggest that one cannot –
in a reverse move – simply reduce these women's very lives to the terrain
of Marxist class struggle. On the contrary, the task at hand is to untangle the
complex ways through which people construct their identities and loyalties,
interpret their needs and struggles, and understand their place in the structures
of inequalities so as to begin the formidable work of changing them. I want to

argue that attempts to unravel the multifarious dimensions of people's lives require that we expand the domains of analysis from emphases on economics and formal political analysis to a preoccupation with cultural politics and the dynamics of identity formation, thus making the insights of poststructuralist theories indispensable.

According to Cynthia Sarti (1996), until recently, and under the influence of an economistic orthodoxy in the social sciences, the poor had only been an aggregate of faceless statistics. Representations of the poor, especially emerging in the 1960s and 1970s, followed a logic of negativity: they were what they lacked (Tereza Caldeira, 1984: 154). As Sarti puts it, the net result of such a standpoint is that no attention was given to the positive dimension of the social and symbolic life of the poor:

> The tendency to think of the poor from the perspective of the relations of production revealed a conception of the human being as *homo economicus*, one typical of a Marxist-inclined sociology. In this view, the poor – identified as destitute of material means . . . – were only considered in their position of dominated. Embodying this material need, they were mechanically preempted of any symbolic resource.
>
> In other words, the poor were theorized as if their social identity were exclusively constructed from their class position, or, from another angle, as if their actions were or ought to be motivated by their interest in satisfying material needs. . . . The class determination of the poor population living in the cities, although structuring their position in the society as a whole, does not constitute the only reference from which they operate and construct both their explanations of the world and their place on it.
>
> (20–1; my translation)

It is with the waning of a Marxist production-centred paradigm and the turn to culture in the 1980s in Latin America (coinciding with US materialist feminists' revival of the production paradigm against all odds) that another, less instrumental, understanding of the poor came into view which did not privilege so strongly the category of work, therefore opening up some space for explorations of the symbolic and representational aspects of the lived experience of the popular classes. However, as Sarti rightly remarks, anthropological studies of the poor often fell prey to a narrow culturalist approach that neglected, on the one hand, the political dimension of the symbolic domain and, on the other, its incorporation into the larger dynamic of society so as to avoid framing this social group as a marginally autonomous, and internally homogenous, cultural formation.

The writings of a new batch of Brazilian social scientists in the 1980s and 1990s – among which, besides Sarti and Caldeira, one can also cite Ana Maria Doimo, Claudia Fonseca, Vera Telles, Lícia Valladares, Eunice Durham and Alba Zaluar – were influential in breaking social analysis free from a simplistic 'us' versus 'them' (the poor) binary frame. These studies, examining complex aspects of the social and cultural formations of the poor – ranging from analysis of the circulation of children among the poor in the building of neighbours' and kin's support networks, the construction of a place of residence in the periphery as an imagined community around which to assert a collective identity, to the mediations between traditional rural values and modern ones acquired upon insertion in complex urban *milieux* – revealed the multiple differences and hierarchies cross-cutting the popular classes in the constitution of heterogeneous and fluid social identities in asymmetrical configurations of power.[7]

In what follows I will be drawing on my discussion – based on life history interviews and ethnographic observations – of the life narratives of five homeless women living in a squatter settlement in Florianópolis, Brazil, and who were participants in the local Sem Teto (homeless) movement. In my larger study (de Lima Costa, 1998), and working from within the realms of subjectivity, identity and politics, I attempted to assess the extent to which theories about the subject and agency in feminism and poststructuralism speak to the material reality of the women interviewed. By juxtaposing epistemological debates and autobiographical reflections, I tried to interrogate both the limits of theory and lived experience in feminism. Among other concerns, I strove to explore the possibilities and limits of identity politics and its rhetorical strategies as they were deployed by the leaders of the local Sem Teto movement.

In the present chapter, and against the backdrop of the feminist debates around historical versus cultural materialism, I will be relying on fieldwork interviews with homeless women to counter some of the claims put forth by materialist feminists. In doing that, I want to underscore the importance that the poetic and ludic dimensions of politics may have even for the most disadvantaged sectors of the population.

The Poetics and Politics of (E)motion

To get to Florianópolis by car or bus from the metropolitan centres of Rio de Janeiro, São Paulo and/or Curitiba, one has to exit highway BR-101 by turning right into Via Expressa at the road sign to the city. This expressway ends at the bridge connecting the continent to the Island of Santa Catarina, where downtown Florianópolis is located. The fast, monotonous and endless flux of automobiles, trucks and buses in and out of the island is only interrupted by eventual traffic jams indicating an accident. Unfortunately, accidents happen with a certain tragic regularity given that, throughout the muddy edges of the expressway, situated dangerously near the fast lanes, a growing number of the homeless population have set up residence in makeshift dwellings built out of discarded construction materials and cardboard. Further away from the road, occupying a vast extension of land, lie the less precarious shelters of a squatter settlement, the result of organized (as opposed to disorganized) occupations of idle government land. What differentiates the former, more provisory, settlement on the edges of the road from the latter, situated farther away from the expressway, is that its layout follows a certain order: the land plots are of equal size, although quite small, with unpaved streets criss-crossing them, and ample room for future recreational, day care and health centre areas. Those who live there are in the process of obtaining their land entitlement due to the mediation of the Movimento dos Sem Teto/MST (urban homeless movement).[8] The MST, after having carried out land seizures, also provided political and strategic support to the communities so that they could demand from local government goods, services and (citizenship) rights.

The weekly assemblies with the leaders of the MST were oriented towards the construction of an identity of homeless (Sem Teto) and the shaping of needs discourses of its constituency in order to provoke debates on rights – from the need for housing, running water, sewage, electricity to the need for justice. There was, on the part of the movement leaders, a concerted effort to construct a consensual 'ethical-political (discursive) field' on needs around the Sem Teto identity, hence building what Fraser denominates as 'ramified chains of 'in-order-to' relations' (1989: 163). As Doimo explains in another context, when issue-specific movements are articulated to others, there is not only an increase in the number of demands, but also a redefinition/reinterpretation of these demands and needs:

> From the struggle for day care one moved to confront the problem of the cost of living which, in turn, suggested the struggle for health care and which, moreover, was connected to the problem of sewage. And in the place of residence, one discovered readily the absence of regularized or legalized land parcels or lots which, to be sure, unveiled the unequal distribution of lands and the scarcity of housing. But, what of the bus to take one to work? And how many can even find work?
> (Doimo, qtd. in Alvarez, 1998: 92)

However, according to Fraser, when connecting one network of 'in-order-to' relations to another, some other chains are necessarily neglected or denied political importance. As an example of such neglect, I could notice that the exclusive focus on material needs (framed by discourses in which war tropes – such as battle, struggle, strife, combat, fight – were recurrent) and organizational order, at the expense of other equally important realms of social and political existence (e.g., sexuality, marriage, community life and so forth), contributed – coupled with conjunctural determinants – to put the MST to dormancy up to the present moment.

Given that social movements are also symbolic struggles for the legitimization of identitarian categories and the construction of a sense of community, there was an identity issue at stake in the assemblies of the MST. To construct a collective subject of rights, the MST's rhetorical strategy consisted in producing maps of meaning that could be articulated to the social (and political) identity of homeless through processes of symbolic inversion. Accordingly, it deployed in its discourses a particular subject-position – now re-signified in empowering terms – around which excluded groups, which so far had counted but as faceless statistics, were able to coalesce.[9] Being a Sem Teto provided its constituency not only with a political incentive to pursue the fulfillment of material needs, but also with new and positive ways of self-construction in terms of a collective (and valorized) identity of a subject of rights, thus articulating demands for shelter with struggles for access to full citizenship and cultural capital or, in Franco's words, 'interpretive power' (1989: xi). None the less, while land tenure was being granted, houses built, sewage implanted and roads paved, the identity of Sem Teto acquired less positive overtones. It began to feel disempowering, that is, as something to which movement participants could not relate, let alone develop any sense of belonging.

By then it had become evident that a major motivation for the interviewees' participation in the MST, besides the acquisition of material goods (land and shelter), was the building of networks of friendship and affective ties. Unfortunately, however, this latter incentive was not echoed in the discourses of the militant left and some of the movement leaders. These discourses – encoded in the tropology of strife and sacrifice – in the end produced a disidentification of the participants in relation to the collective subject that this rhetoric sought to construct.[10] In the words of a women interviewed, people were feeling the need to engage in a different kind of militancy, one that 'speaks to the poetics of life but which doesn't cease to be political'. That is, a 'lighter, more romantic way' (uma forma mais leve e mais romântica) of doing politics; a more ludic manner of engaging in popular participation which heralds a new time in social movement activism: 'I have perceived that there is something different in the Movement today. I think that what is different . . . is a search for a poetics of work' (Joana).

The opposition between 'need for justice' (carência de justiça) and 'need for affect' (carência de afetividade), drawn by the women of the MST in describing their reasons for participation in the movement, remits one back to Fraser's (1989) well-known debate about the juxtaposition and asymmetries between discourses about rights and discourses about needs interpretation. This author argues that it is often the case that existing (and publicly legitimate) political discourses about needs interpretation are inadequate in that they tend to mirror the interests and languages of dominant social groups, whose 'structures of feeling' are usually out of synchrony with the 'structures of deprivation' (Steedman 1987: 17) of those situated at the margins and outside the law. That is to say, radical leftist discourse displays an insensitive blindness to the particular experiential (subjective and emotional) circumstances of those who, though living in the exile of capitalism, may harbour wants and desires that do not necessarily coincide with those of the collective subject prescribed by social movements theory.

Based on my fieldwork interviews, I would say that there is a pressing need for more engaged theorizations about the place of (e)motions and desires (affect) in recruiting people to social movements. One may become tired – as the women I interviewed clearly did – of living in a crystallized collective subject-position, and having one's self anchored in a fixed identity of Sem Teto (homeless) in spite of one's changing material circumstances and experiences. Once such identities become closed and immutable, they no longer respond to the specificities and contingencies of one's experience, thereby hindering the process of building identification with the subject-positions offered in movement discourses of strife and struggle. There is the exigency, as implicit in most of the women's life narratives, of thinking of identities as in process, that is, as being continually (re)invented along the way and in consonance with one's location in the structures of feeling and deprivation. All of the women I interviewed were searching for a different way of doing politics, one with more (e)motion. One, I would add, which also admits of playfulness – something that is allowed only to the privileged circle of US poststructuralist academic feminists.

Going back to the debates in feminism between historical materialism versus postmodernism/poststructuralism, I believe that the Sem Teto women in Florianópolis spoke louder. From the standpoint of their concrete experiences they claimed their right to have material needs fulfilled as well as a right to ludic politics – that is, to other pleasures as well.

Conclusion: The Politics of (Dis)location

Lata Mani (1989) has written very sharply on the multiple mediations demanded from the critic when she attempts to speak (theorize) within different historical moments and in distinct contexts. One way to make such mediations while being sensitive to context and to a feminist politics is to develop attentiveness to one's geopolitical and historical location(s). As argued by Mohanty (1991), the 'imagined community' of third world feminism – a result of political articulations and oppositional struggles in constructing transpersonal contiguities – does not rest on any fixed notion of colour, sex or even culture, but on the political links that are forged among and between specific, sometimes divergent, groups in different social locations. As Mohanty explains,

> The idea of imagined community is useful because it leads us away from essentialist notions of third world feminist struggles, suggesting political rather than biological or cultural bases for alliance. Thus, it is not color or sex that construct the ground for these struggles. Rather, it is the *way* we think about (our experiences of) race, class, and gender – the political links we choose to make among and between experiences and struggles. However, clearly our relation to and centrality in particular struggles depend on our different, often conflictual locations and histories.
>
> (1991: 4)

I would like to conclude by recounting my first field encounter with a group of women from the SMT in Florianópolis. I remember having showed up for the first time at the squatter settlement of Novo Horizonte (the result of land-seizure by the movement), on a Sunday morning, with two university students to help out clear an area on which the future community centre would be erected. This was my opportunity to explain to them my intention to be involved in the homeless movement while also pursuing my project of collecting their life histories.

After being introduced to those present, without any further delays, each of us, along with several women who lived there, grabbed a shovel and started working to remove a huge mound consisting of a combination of red dirt and garbage. After I had dug into that pile for about fifteen minutes, some women jokingly pointed at my feet. I then realized that my white sneakers, as well as my white socks, white jeans and white T-shirt were all stained with the red mud. I was sweating (under the inclement sun) and had a fast-developing blister on my right hand. The women present began mocking me given my inappropriate attire for the occasion. From that early moment on, I gave up any attempts to 'go native' under any guise, including my summer leisure outfit. I put the shovel aside (we had come to the conclusion that the mound was literally non-removable except by a tractor), sat under the shade of one of the few trees and chatted with them, at the same time as I blamed myself for not having done my 'homework'.

As a result, my first experience 'in the field' was marked by a deep sense of failing, of which my white leisure outfit stood as the most telling allegory. The inadequacy of my clothing made me acutely aware of the manifold levels of privilege structuring Brazilian society and my own location in it. Despite my voluntary and multiple displacements through geographic, economic, political, cultural, sexual and theoretical landscapes that further destabilized my subject-position *vis-à-vis* dominant configurations of power, coming back 'home' re-inserted me within impermeable and relatively fixed class and race systems of value. My transnational border crossings did not alter my white, upper-middle class privilege (it may well have even strengthened it). I was not feeling unfit simply because I had been removed for many years from a context of glaring social inequalities and poverty. Even if I had lived next to this squatter settlement for my whole existence, mine and my interviewees' respective life trajectories would be still worlds apart, disassociated by rigid class and racial divisions that articulate different 'structures of feeling' and experiences of inclusion and exclusion. From the specificity of my place, I still would have remained clueless about what to wear.[11] What I can do now is to try to make sense of both the life histories of the women I interviewed and my own perception of impropriety and difference (my sense of dislocation) as stuff out of which to develop less complacent political and cultural analyses. Such analyses become especially critical in light of the increasing demands of 'transnational' feminist alliances toward 'horizontal discursive travels' (Thayer, 1998) in redressing the power imbalances of feminist theories in the Americas.

Notes

1 For a study of Latin American feminisms at the turn of the century, see Sonia E. Alvarez (1998).
2 In this respect, Hutcheon writes that '[f]eminisms will continue to resist incorporation into postmodernism, largely because of their revolutionary force as political movements working for real social change. . . . Postmodernism has not theorized agency; it has no strategies of resistance that would correspond to the feminist ones. Postmodernism manipulates, but does not transform signification; it disperses but does not (re)construct the structures of subjectivity. . . . Feminism must' (1989: 168).

3 Today the *favela* tours are a high point in the several specialized tourist agencies' brochures, remarkably after Michael Jackson and his entourage's stop over in one (Morro Dona Marta) for videoclip shooting purposes in 1995.

4 Román de La Campa (1996) argues that the postmodern debate in Latin America has taken more cultural and political overtones than in the context of the relatively independent spaces of North American universities.

5 Escobar claims that in Latin America, 'the differentiation and transnationalization of cultural and economic systems presuppose and produce a mixture of pre- or nomodern, modern, postmodern, and even antimodern forms' (1992: 67).

6 For a discussion about the travels of theory between North and Latin America, see Francine Masiello's (1996) instructive essay.

7 Recently, Paulo Lins – a black informer of the anthropologist Alba Zaluar – published a highly acclaimed novel about a poor neighborhood in Rio de Janeiro, *Cidade de Deus* (where he lives), that mixes with piquancy both ethnography and fiction in the description of the multiply layered lives and characters of its inhabitants. See also Norma Chinchilla (1997) for an account of how, in Latin America, the contradictions of capitalism have created a plurality of social subjects and have made class struggle far more complex and multi-dimensional.

8 The MST organized land invasions in Florianópolis with the support of the Church-linked Comunidades Eclesiais de Base/CEBs (Christian Base Communities), progressive sectors of the Catholic Church, militants of the leftist Partido dos Trabalhadores (Worker's Party), labour unions, and engaged university professors and students, among other movement sympathizers and activists.

9 There is a remarkably vast literature on the relationship between collective/cultural identity and social movements that I cannot do justice here. For two notable examples of the range and complexity of debates in this area, see the anthologies edited by Alvarez, Dagnino and Escobar (1998), and Escobar and Alvarez (1992).

10 James Holston makes a similar point in his analysis of auto-construction among the working class. He argues that residential organizations demobilize as soon as they attain their goals, and the political identities constructed dissolve due to a tendency to reduce politics to material interests 'that can be satisfied within the existing system of social relations'. However, he further contends that these organizations give their members valuable experience of mobilization that can be used to other similarly valuable ends.

11 I wish to thank Sonia E. Alvarez (personal communication) for this sharp observation.

References

Alvarez, Sonia E. 1998. 'Latin American Feminisms "Go Global": Trends of the 1990s and Challenges for the New Millennium.' *Cultures of Politics, Politics of Cultures: Re-visioning Latin American Social Movements*, eds Alvarez, Sonia E., Evelina Dagnino and Arturo Escobar. Boulder: Westview Press.

Alvarez, Sonia E., Evelina Dagnino and Arturo Escobar, eds. 1998. *Cultures of Politics, Politics of Cultures: Re-visioning Latin American Social Movements*. Boulder: Westview Press.

Barrett, Michèle. 1992. 'Words and Things: Materialism and Method in Contemporary Feminist Analysis.' In *Destabilizing Theory: Contemporary Feminist Debates*, eds Michèle Barret and Anne Phillips, 201–19. Stanford: Stanford University Press.

Benhabib, Seyla. 1995. 'Subjectivity, Historiography, and Politics: Reflections on the Feminism/Postmodernism Exchange.' In *Feminist Contentions: A Philosophical Exchange*, eds Seyla Benhabib, Judith Butler, Drucilla Cornell and Nancy Fraser, 107–25. New York: Routledge.

Best, Steven and Douglas Kellner. 1991. *Postmodern Theory: Critical Interrogations.* New York: The Guilford Press.

Beverley, John. 1993. *Against Literature.* Minneapolis: University of Minnesota Press.

Bhabha, Homi. 1994. *The Location of Culture.* New York: Routledge.

Bordo, Susan. 1992. 'Postmodern Subjects, Postmodern Bodies.' Review Essay. *Feminist Studies* 18 (1): 159–75.

Bordo, Susan. 1990. 'Feminism, Postmodernism, and Gender-Skepticism.' In *Feminism/Postmodernism*, ed. Linda Nicholson, 133–65. New York: Routledge.

Bulbeck, Chilla, 1998. *Re-orienting Western Feminisms: Women's Diversity in a Postcolonial World.* Cambridge: Cambridge University Press.

Caldeira, Tereza do Rio Pires. *Cities of Walls.* Forthcoming.

Caldeira, Tereza do Rio Pires. 1984. *A Política dos Outros: O Cotidiano dos Moradores da Periferia e o que Pensam do Poder e dos Poderosos.* São Paulo: Editora Brasiliense.

Clifford, James. 1989. 'Traveling Theories/Traveling Theorists.' *Inscription* 5: 177–88.

Chinchilla, Norma. 1997. 'Marxist, Feminism, and the Struggle for Democracy.' In *Materialist Feminism: A Reader in Class, Difference, and Women's Lives*, eds Rosemary Hennessy and Chris Ingraham, 214–26. New York: Routledge.

de La Campa, Román. 1996. 'Latinoamérica y sus Cartógrafos: Discurso Poscolonial, Diásporas Intelectuales y Enunciación Fronteriza.' *Revista Iberoamericana* lxii (176–7): 697–717.

de Lima Costa, Claudia. 1998. 'Off-Center: On the Limits of Theory and Lived Experience.' Unpublished Ph.D. Thesis, Department of Speech Communication, University of Illinois, Urbana-Champaign, Illinois.

Ebert, Teresa. 1996. *Ludic Feminism and After: Postmodernism, Desire, and Labor in Late Capitalism.* Ann Arbor: University of Michigan Press.

Escobar, Arturo, 1992. 'Culture, Economics, and Politics in Latin American Social Movements Theory and Research.' In *The Making of Social Movements in Latin America: Identity, Strategy, and Democracy*, eds Arturo Escobar and Sonia E. Alvarez, 62–85. Boulder: Westview Press.

Escobar, Arturo and Sonia E. Alvarez, eds. 1992. *The Making of Social Movements: Identity, Strategy, and Democracy.* Boulder: Westview Press.

Flax, Jane. 1990. *Thinking Fragments: Psychoanalysis, Feminism, Postmodernism in the Contemporary West.* Berkeley: University of California Press.

Franco, Jean. 1989. *Plotting Women: Gender and Representation in Mexico.* New York: Columbia University Press.

Fraser, Nancy. 1995. 'Pragmatism, Feminism, and the Linguistic Turn.' In *Feminist Contentions: A Philosophical Exchange*, eds Seyla Benhabib, Judith Butler, Drucilla Cornell, and Nancy Fraser, 157–71. New York: Routledge.

Fraser, Nancy. 1989. *Unruly Practices: Power, Discourse and Gender in Contemporary Social Theory.* Minneapolis: University of Minnesota Press.

Fraser, Nancy and Linda Nicholson. 1988. 'Social Criticism without Philosophy: An Encounter between Feminism and Postmodernism.' In *Universal Abandon? The Politics of Postmodernism*, ed. Andrew Ross, 83–104. Minneapolis: University of Minnesota Press.

Grossberg, Lawrence. 1997. *Bringing it All Back Home: Essays on Cultural Studies.* Durham: Duke University Press.

Grossberg, Lawrence. 1992. *We Gotta Get Out of This Place: Popular Conservatism and Postmodern Culture*. New York: Routledge.

Hartsock, Nancy. 1990. 'Postmodernism and Political Change: Issues for Feminist Theory.' *Cultural Critique*, 14: 15–33.

Hennessy, Rosemary and Chris Ingraham. 1997. Introduction. In *Materialist Feminism: A Reader in Class, Difference, and Women's Lives*, eds R. Hennessy and C. Ingraham, 1–16. New York: Routledge.

Holston, James. 1991. 'Autoconstruction in Working-Class Brazil.' *Cultural Anthropology* 6 (4): 447–65.

hooks, bell. 1990. *Yearning: Race, Gender, and Cultural Politics*. Boston: South End Press.

Hutcheon, Linda. 1989. *The Politics of Postmodernism*. New York: Routledge.

Lins, Paulo. 1997. *Cidade de Deus*. São Paulo: Companhia das Letras.

Mani, Lata. 1989. 'Multiple Mediations: Feminist Scholarship in the Age of Multinational Reception.' *Inscription* 5: 1–23.

Masiello, Francine. 1996. 'Tráfico de Identidades: Mujeres, Cultura y Política de Representación en la Era Neoliberal.' *Revista Iberoamericana* LXII (76–7): 745–66.

Mohanty, Chandra T. 1991. 'Under Western Eyes.' In *Third World Women and the Politics of Feminism*, eds Chandra Mohanty, Ann Russo, and Lourdes Torres, 51–80. Bloomington: Indiana University Press.

Mohanty, Chandra T. 1987. 'Feminist Encounters: Locating the Politics of Experience.' *Copyright* 1: 30–44.

Mouat, Ricardo Gutiérez. 1993. 'Postmodernity and Postmodernism in Latin America: Carlos Fuente's *Christopher Unborn*.' In *Cultural Theory, Cultural Politics and Latin American Narratives*, eds Steven M. Bell, Albert Le May and Leonard Orr, 153–79. Notre Dame: University of Notre Dame Press.

Olea, Raquel. 1995. 'Feminism: Modern or Postmodern?' In *The Postmodernism Debate in Latin America*, eds John Beverley, Jose Oviedo and Michael Aronna, 192–200. Durham: Duke University Press.

Radhakrishnan, R. 1996. *Diasporic Mediations: Between Home and Location*. Minnesota: University of Minnesota Press.

Richard, Nelly. 1996. 'The Cultural Periphery and Postmodern Decentring: Latin America's Reconversion of Borders.' In *Rethinking Borders*, ed. John C. Welchman, 71–84. Minneapolis: University of Minnesota Press.

Richard, Nelly. 1995. 'Cultural Peripheries: Latin America and Postmodernist De-Centering.' In *The Postmodern Debate in Latin America*, eds John Beverley, Jose Oviedo and Michael Aronna, 217–22. Durham: Duke University Press.

Richard, Nelly. 1989. *La Estratificacion de los Margenes*. Santiago: Francisco Zegers Editor.

Ryan, Michael. 1989. *Politics and Culture: Working Hypotheses for a Post-Revolutionary Society*. Baltimore: Johns Hopkins University Press.

Sarti, Cynthia. 1996. *A Família como Espelho: Um Estudo sobre a Moral dos Pobres*. São Paulo: Editora Autores Associados.

Steedman, Carolyn K. 1987. *Landscape for a Good Woman: A Story of Two Lives*. New Brunswick: Rutgers University Press.

Thayer, Millie. 1998. 'Traveling Feminisms: From Embodied Women to Gendered Citizenship.' Paper presented at the XXI International Congress of the Latin American Studies Association, Chicago, April 24–6.

Weed, Elizabeth. 1989. Introduction: Terms of Reference. In *Coming to Terms: Feminism, Theory, Politics*, ed. Elizabeth Weed, ix–xxi. New York: Routledge.

West, Cornel, 1988. Interview. In *Universal Abandon? The Politics of Postmodernism,* ed. Andrew Ross, 269–86. Minneapolis: University of Minnesota Press.

Yúdice, George. 1989. 'Puede Hablarse de Postmodernidad en America Latina?' *Revista de Critica Literaria Latinoamericana* 29: 105–28.

Chapter 21

The Americanization of the World

Pablo González Casanova
(translated by John Page)

The Foreigner

'The concepts of Americanization and Americanism brand a danger that henceforth threatens the European spirit. Machinism,' writes Armand Mattelart, 'gregarious democracy, downward leveling, regimentation, materialism, all these accusations seek to outline the stakes of this confrontation with the 'American jungle' and its cult of the power of money.

(Mattelart 1996)

This brief outline of what is understood by Americanization refers to a broad current of opinion that exists not only in Europe, but in other parts of the world as well. It involves the false logic of xenophobia that attributes everything the society considers wrong to a harmful foreigner. The use of this logic thwarts the thought that 'nothing human is foreign to us'. It is found in the antipodes of the terse logic that led Valery to write: 'Without the Persians, without the Semites, without Hellenic culture and the unexpected developments that it received, Europeans would be far inferior in refinements and inventions to the peoples of the East and the Far East.' (Valery 1974) The difference between these two ideas lies not only in that the first refers to criticism of the second, country or culture, and the second to recognition of what one, Europe in this case, owes to another. The difference stems also from the fact that whereas one rationale fails to invite definition and verification in history, letters and the sciences, of the defects to which it refers, the second summarizes prior historical, literary and scientific studies, leaving open the chance to deny, confirm, refine or broaden them.

Thinking about America

The problem posited is both scientific and political. How do stereotypes of towns, nations, ethnic groups and civilizations impede thought based on verifiable knowledge? More concretely, how can one think of America and the world, or the Americanization of the world, without our preconceptions or pre-judices taking over the discourse? In other words, how can we abandon our deeply rooted essentialist tradition, and in what coarse or sophisticated way is it expressed in stereotypes?

Rather than sticking to rules, this study will attempt an unstereotyped analy-sis of America and Americanization, that responds to significant questions.

The Word 'American'

Just as stories begin with 'once upon a time', this study begins with the con-ventional dictionary definition: an American is one born in the United States or who has adopted that nationality.

José Martí spoke of 'Our America', which included Hispanic America and the Caribbean countries threatened by the United States; he also spoke of 'Greater America', including Latin America, the United States and Canada.

The American adjective was not used in the early nineteenth century as it is now, when Bolívar and Hidalgo called themselves Americans. But the dominant application of the term by the denizens of the United States to them-selves and their own came to be accepted over time. It grew with the weight of the nation at the expense of Indians and Mexicans. It then reached global dimensions.

Many writers and politicians, either on principle or because they felt defrauded, spoke of North America and North Americans, rejecting that the entire region be called America or that its inhabitants call themselves Ameri-cans. The linguistic resistance seemed to collapse when NAFTA included Mexico in North America and when Mexicans supposedly became subsumed among North Americans. It continued, however, with the accentuated use of the term *estadounidenses*, applied to 'yankees' (*latu sensu*) or 'gringos', as they are called in Mexico.

These lexical, logical and political observations suffice to point out the uses of the word, and to remind the reader that it has several meanings. In what follows, the term Americans will be used as the Americans use it.

Let 'The Other' Speak

To have a more precise idea as to what 'the other' is like, he must at least be allowed to speak. If you want to know the other let the other speak. The problem is fascinating. A vast study would have to consider different types of individuals and groups or social conglomerates. It would observe the relations between what they think, write, say and do; between the wealth of humanist, scientific and artistic American literature, and the historical and current records of surveys; and research on social movements as cognitive movements; and linguistic studies on word-acts, or cyber-anthropology applied to the American Indians and WASPS, – there is a wealth of sources.

But if we set out only to listen to a few Americans and what they have to say about a few subjects like 'America', humanity, capitalism and democracy, we will strike a sharp blow at our stereotype of ONE America, good or bad; the America Europeans and other peoples of the world believe they criticize, or the one praised by technocrats and pochos.[1]

To begin with, one is reminded of Thomas Payne's, 'The cause of America is in a great measure the cause of all mankind' (*Common Sense*, 1776), or, more recently, Earl Robinson: 'All races and religions: that is America' (*Composition*, 1991). Both statements are as true at a given existential, political or historical moment, as those that support the opposite when referring to capitalism and imperialism. The truth about America seems to include both thoughts.

Tocqueville went to America 'in search of something more than America', that is, 'in search of democracy',[2] and wrote as a great travelling philosopher: 'I know no country, indeed, where the love of money has taken a stronger hold on the affection of men' (ibid.). Lincoln, in an 1837 speech, maintained that: 'These capitalists generally act harmoniously and in concert to fleece the people.' Later, Calvin Coolidge would write: 'The chief business of the American people is business' (Speech, 1925). And, in the midst of a fully-fledged consumer society, the great Henry Miller observed that 'The national vice is waste.'

To the encounters of humanism and capitalism may be added those of democracy and imperialism. Both reveal differences that underscore American disregard of Plotinus, but the force with which the contrary theses are upheld.

It is impossible to speak of the world and of America in the world without conflicting theses. Walt Whitman, great poet and lover of technology, mass production and democracy, one of the first and most aggressive defenders of the homosexual's right to pleasure, wrote: 'I shall use the word America and democracy as convertible terms.' (*Democratic Vistas* 1871) On April 3, 1914, the United States House of Representatives appropriated the text that appeared in the American Creed (1871): 'I believe in the United States of America as a Government of the people, by the people and for the people'. Taken word for word, this text is one of the most revolutionary in the world. Four years earlier, Theodore Roosevelt had already declared, 'Our country – this great republic – means nothing unless it means the triumph of a real democracy, the triumph of popular government, and, in the long run, of an economic system under which each man shall be guaranteed the opportunity to show the best that there is in him' (1910).

These and many other declarations contrasted with authoritarian, anti-labour, colonialist, imperialist measures of a most different type. The contradiction was not only a form of duplicity or hypocrisy, although it might also be that. It was America's relation to combined ideals and realities that must be understood to mean America in its variations and constants.

In a speech in Minnesota (September 1901), Theodore Roosevelt himself would prescribe, 'Speak softly and carry a big stick.' His policy of 'the big stick' was one of those that caused Latin America the most suffering. James Monroe adopted the motto 'Don't tread on me', and warned that he would consider any foreign intervention on the American continent as a threat to the peace and security of the United States. His declaration was the basis for the sadly famous 'Monroe Doctrine', that by fiat declared all of Latin America to be part of the US 'zone of influence'. Since the Louisiana Purchase and above all since the war against Mexico, the idea of 'America for the Americans' had been changing into a war cry and an imperial proclamation. The ruling class and broad sectors of the population internalized and expressed a supposed historical destiny. 'Our manifest destiny is to possess the whole of the Continent, which Providence has given us for the great experiment of liberty'.[3] The juncture of the two logical postulates, the logic of freedom and the logic of empire, embraced all forms of thought, expression and action. That great Democratic president, Franklin D. Roosevelt, said of perpetual Nicaraguan dictator Anastasio Somoza, 'He may be a son-of-a-bitch, but he's our son-of-a-bitch' (*The New Yorker* 1945).

Something common appears amidst all this opposing variety: the moral-political struggles of the United States reveal a real and ideal American who continues to reconstruct the conquest of America and its at once imaginary, propagandistic and practical ideals.

G. K. Chesterton once gave up, saying that no one was going to make a fool out of him. Assuming his role as a humourist, he wrote: 'There is nothing wrong with Americans except their ideals. The real American is all right, it is the ideal American who is all wrong.' (*The New York Times* 1931) Beyond the excellent sally lies the need to distinguish between the real and the ideal Americans.

One America (USA) Made of Many Americas

Today, in referring to the Americanization of the world, one thinks only of US influence on the world. This with respect to nations. In terms of classes, even Martí spoke of a 'working America'. He distinguished it from 'the America of the rich and powerful people who own and control that country'.

Generally, when speaking of Americanization one thinks, without clarifying, of the America of the rich and powerful and the mass culture they have imposed. 'The United States – writes Katzenstein – is unrivaled in exporting its popular culture to all corners of the globe, from fast-food chains to movies and rock music.' (Katzenstein 1996) The largest cultural industry in the world is American. It dominates communications, information, propaganda, publicity, entertainment and among the latter cinema, television and Internet.[4]

From Fordism to Toyotism, from electricity, skyscrapers, automobiles, showers, the 'watercloset' and 'do it yourself', to cell phones and personal computers, there is always something predominantly American.[5]

There are ideas that come from Japan, Germany, France, England or Italy that develop until their origins are forgotten and they seem native to America. The history of 'high tech' involves much of that, like the atomic bomb, and the 'new sciences' of cybernetics and self-regulated, adaptive and chaotic complex systems.

There is no doubt that many discoveries and innovations, that many scientists and technicians came from Europe, that they are from the 'West'; however, they became Americans and many of their creations were born in 'America' or became American. Something similar occurs with Japan. If Fordism was American-American, Toyotism became 'lean production' and acquired American nationality.

From 'assembly lines' to 'flexible production', capable of adapting to different media – among uncertainties –, US companies, their managers and their technicians have become very rich via original Japanese contributions, but at the same time they have enriched them and shared them among managers and officials of their multinational and transnational networks. Today the sciences of company organization most used throughout the world are American.

The foremost Americanization applies to the dominant culture among the masses, as well as to serial and state-of-the-art technologies. It includes two no less important lines but of which little is said except among specialists and connoisseurs: techno-scientific culture and neoconservative ideology. Without them no clear enough idea of the Americanization that comes from the dominant forces of America can be had.

American, the Science *par excellence*

The United States became the world's pole of scientific research during the Second World War. In 1945 the atomic bomb was the most overwhelming proof of US military and scientific power, as well as of the conjunction of forces devoted to that objective. Since the thirties, research had been enriched by scientists fleeing from Nazi-Fascist Europe. The exodus of those scientists increased during the war. Soon it was combined with growing support for research from the American government. The military and scientists created structurally linked organizations to achieve their purposes. These were part of a system the objective of which was to win the war by perfecting techniques for communications, intelligence, data processing, targeting, precision in the calculation and destruction of the enemy, along with preservation and improvement of their own forces.

'Knowledge by objectives' became the paradigm for triumphant military-scientific research. Its recognized home is America. Its implications encompassed all the sciences, including the social sciences. The latter were not only renewed as ideologies but also as technologies. As ideologies they were believed to be useful to the constituencies, the majorities and the underdeveloped, to the citizenry, to social welfare and to the 'development' of backward countries. As technologies they perfected their efficiency and versatility to the maximum with important advances in the techno-sciences. In both ways they contributed to strengthening the dominant systems. Before the Second World War, science had studied creation and had used its 'laws' for political and technical purposes, but since then, the objective of science was the creation of systems with functional laws and adaptive behaviour. The United States was originally the world's centre for the new paradigm. Americans and Europeans, scientists and the military, constructed research into self-regulated, self-creating, self-reproducing, homeostatic and morpho-genetic systems. Knowledge was organized around will. The practice of that knowledge was used deliberately to perfect it.

It was a two-way street. The organizations, apparatus and systems created to carry out objectives were organized according to knowledge. To know and master was linked in all disciplines, and expressed the disciplines themselves, starting at mathematics, passing through physics, chemistry and biology all the way to sociology, computers, linguistics and semiotics.

Overcoming the ruling system's contradictions was not left only to myth (as believed by many of its Marxist-Leninist adversaries and Newtonian-Baconian parties): it turned into the modelling of open and cybernetic systems.

The great movement began in the war and gained creative impetus in the fifties. Later, nearing the sixties, research was enriched by the study of complex systems. In these systems, interfaces and synapses were observed between phenomena that previously appeared to be isolated or completely opposed.

In the material, life and humane sciences, order arose out of chaos and even chaos out of order. All those discoveries had the enormous political and military implications for a science made for and by the establishment.

Soon, not only did American centres stand out in research on complex systems, but European and Japanese centres as well. In all of them niches were formed where a new science and a new culture developed, different from Newtonian-Cartesian-Baconian science and from the humanism born with Modernity (Lee 1996). In this 'new science' and 'third culture', as Brockman calls it, the United States was still in an advanced and privileged position with unquestionable world influence. The Americanization of the world includes state of the field science.

The great techno-scientific and political innovation occurred in narrow and permanent links within the ruling interests in America and the world. Both the cybernetic systems of the fifties and the complex systems of the sixties took the US and world system of ruling and appropriation as a constant. What is paradoxical is that even with all those limitations, the social sciences, seen from any perspective of class, nation, region or civilization, were enriched since Talcott Parsons and his *Structure of Social Action* (1937) and later his *The Social System* (1951), gave primacy to 'rational efficiency' in human behaviour and to control mechanisms that redirect action to achieve objectives.

Parsons grasped an entire movement that changed the world and gestated in the United States: cybersciences. Parsons' ideology, rightly criticized by C. Wright Mills, did not hinder him from making great techno-scientific discoveries of a sociology very useful to the establishment and that directly or indirectly would change the level of organization of struggles.

Later, including Europe and Japan, there was a move toward the new study of dynamic systems with proposals on chaos in order and on the chaos that comes from order which causes turbulences, disequilibriums and destabilization which it is sought to bring under control in *systems policy* and *systems war*.

Soon emergent structures also appeared, attractors and fractals, possible approaches and bifurcations that began to have an increasingly important place in research and ended in the general study of dissipative systems and their dissipative structures and a new meeting with the self-regulated systems coming from cybernetics.

Like it or not, that is how limits to certainty and uncertainty were proposed in deterministic, probabilistic, cybernetic and open and dissipative systems. Since then, no one who understands the current system of domination and appropriation as a variable and historical system can study it in depth and act on it and against it efficiently without mastering ideas from 'America', a techno-scientific 'melting-pot', high-tech and with state-of-the-field, cutting-edge research.[6]

The New Science and the New Conservatives

The 'new science' is part of a large movement of domination and appropriation of America and the world. Those who stand out in this movement are the 'neo-conservatives, the men who are changing American politics', writes Peter Steinfels (1979) in a noteworthy book on the subject.

Perhaps the Americanization of the masses makes one forget the Americanization of the elite. If the latter does not occupy as varied and extensive a region of the planet as the former, its ascendancy appears universally in techno-science and is greatly felt in the most influential areas of the United States. A large part of the governing and military elite of Latin America have studied in US schools and universities.

The degree of American neo-conservative sophistication is noteworthy. Not only have they included the most recent contributions to the techno-scientific revolution in their intellectual baggage, combining them with the great humanist Anglo-Saxon and Anglo-American culture, but they have adopted and adapted many of the forms of reasoning and expression of radical thought. Among their weapons, they have adopted concepts of the new left and even of critical Marxism, not to mention concepts of the Civil Rights movements and counter-culture.

An earlier tradition had led American conservative thinking to integrate science and scientism as an instrument of action and rationalization. Also, some time before, it had appropriated liberalism in a way similar to the way in which it expresses itself today through neo-liberalism.

Neo-conservative thought rose in the sixties as an answer to the radicalization of liberal thought, and to an emergence of movements of the new left, which spoke of a true democratic revolution. The neo-conservatives used the Marxist's own arsenal and so-called critical thought to destroy the democratic and traditional left that was also on the offensive. In this way they attacked communism, revolutionary nationalism and even social democracy. Their tactics consisted in pointing out the corruption and contradictions of those movements and in establishing bridges and identities with the Civil Rights movement and counter-culture.

The enrichment of neo-conservative discourse took advantage of the growing misery of the old left movements. By supporting the criticism of the new left, they managed to attract and co-opt a large number of leaders among academics, students, blacks, Marxists, populists and democrats. With them, the neoconservative movement added to the legitimate denunciation of the corruption of the American Welfare State, of the drop in academic standards in public universities, of the barbarization of culture. Acts of vulgarity, obscenity, drug addiction, verbal and mental violence, immorality and corruption were also targets of their attacks with arguments that appealed to the non-conservative public. To the coherence of the different discourse was added the updated handling of techno-science as an ideology, instrument and rationalization of a new policy of domination and appropriation, that began by recovering ground lost in the concessions of Rooseveltian democracy.

In fact, the neo-conservatives organized a great offensive against all mediations that had made it possible to govern on an internal and international level. They restructured mediations and repressions beginning with ideologies and languages. They centred their offensive on the end of ideologies, maintaining that their politics were based on science. They strengthened their positions with a blueprint for democracy the discourse of which included the arguments of critical thinking, the new left, and the Civil Rights movement. They manipulated these arguments and techno-scientific languages to simultaneously disqualify, on those grounds as well, and with proof, the discourses of communists, populists and social democrats. Their subterfuge consisted of hiding the social system by means of the political system. The serious flaws in the political system allowed them to distract the unwary from the immense power they had over the government, society and economy. In a few years they managed to turn the political system into the target at which all the missiles of all forces were aimed. In this way, they were determined to weaken any government in its social and social economic roles, without saying, however, that their most important objective was to strengthen it in its roles of privatizing, domination and appropriation. With noteworthy firmness, they rejected the explanation of social, cultural, political and economic problems in terms of the social system. They exonerated the ruling class and the capitalist system, and explained the evils of society as a product of immorality, corruption and inefficiency. At the same time they declared that the 'neo-liberal' politics they proposed, strictly obeyed science and was beyond ideologies.[7] But their 'truths' did not lead them to the totalitarian thought of Nazism or the communist countries. Their truths did not lead them to impose an 'official' truth. Their totalitarianism was more sophisticated. The ruling classes reserved the right to judge, over and above the heads of governments. Beneath the ruling classes and governments, they did not prohibit criticism of thought. They simply did not listen and discouraged it with systems of incentives and sanctions that structured rational choice. By not imposing an official truth in all areas, they did not impose a governmental orthodoxy associated with one single philosophical system. The varieties of neo-conservative thought together with their quasi-Marxists and quasi-fundamentalists, allowed them, in terms of religion, to accept Protestants, Jews and even Mohammedans and Catholics, as long as they in fact supported neoclassical orthodoxy and the politics of adjustment, and verbally or existentially rejected the theologies of liberation. This multiple ideological position allowed ruling elites to be above the limitations of official truth. It situated them as a power behind, or above, the government, where the chief executive had to report to them, was renewable, and not all its mistakes or corruptions were ignored all the time and in every circumstance.

The moral and technocratic axis worked in practical terms. The neo-conservative elite reserved the right to moral judgement and the evaluation of the efficiency and scientific and political capacity of their senior officials and their collaborators. They imposed values, dimensions, variables, methods and techniques to evaluate. First of all, they created and 'socialized' the theses that operate as scientific and political paradigms: 1) Not to posit the struggle in ideological terms (the end of ideologies); nor in terms of historical changes (the end of history), nor in terms of attention to the demands of the masses (against the massification of education and University, discreetly forgetting television and the 'media'), nor in terms of collective action (but of individual rational choice). 2) To uphold with expert responsibility or respecting the experts, that problems are of high complexity, and that only neoclassical and neo-liberal social and political sciences are sciences. To conclude with firmness and serenity that economic problems are technical and that neither politicians (or 'political dinosaurs'), nor peoples (or 'barbarians'), should intervene; neither governments nor bureaucracies (or 'public service') should ever accept an excess of demands that might unbalance the system. 3) To denounce as the enemy, irresponsible politicians and irresponsible people who demand 'social justice', or accuse them of being demagogues and inept, good-hearted naive people, social climbers with covert interests. To uphold, on the other hand, that the great American politicians, or friends of the Americans, are tough minded and make hard decisions. They defend America and the world from the threats of those who do not respect the established order for false reasons, either humanitarian, egalitarian or catastrophic.[8]

The virtues of neo-conservatives do not stop there. There are other virtues in language and in style, in the supreme mastery of the scientific and political double discourse and in the organization of leadership for functional changes in the system.

Language and power have always been linked. Words, terminology, normative and forbidden discourses are part of the history of great empires. In the American empire, political correctness has reached high levels in regulating concepts and reasoning. It works in processes of socialization and inhibition, of esteem and execration, rejection and taboo.

The global dominion in the techno-sciences and in neo-liberalism, just as in communications and the electronic mass media, give the United States a predominant role in the global academy of the language of technocrats and transnational elites. Their influx is not limited to the cute admirers of Walt Disney, that great master of child masses.

The 'educated American mind' cultivates a particularly rich language. Even more so, it makes the mastery of its own language a mastery of the world. The neo-conservative person is deliberately polemical and tries not to lose the offensive position. In persuasion he dominates the art of combining the 'personal tone' with 'general principles' and the latter with 'empirical, qualitative, documentary, graphic and statistical evidence'. But he is not married to exactitude nor to rigour, as the technocrats are. He handles irony, ambiguity, paradoxes and a sense of humour in a noteworthy fashion.[9] He intuits that 'Up to half of the scientific papers in the United States may be contaminated by data manipulation.' (O'Neill 1991) He knows that deep down he is compromised by the excessive lying about Vietnam, Watergate, Irangate, but shares 'the commitment to the lie' of the Nixon administration and the pride in lying about Lt.Col. Oliver North, who 'defended his decisions to lie to members of the United States Congress'. He also knows how to keep a poker face when he is reminded of Ronald Reagan's two hundred errors-lies in his first term of office.[10]

In any case, a fact even more admirable is neo-conservative double discourse, not only political, but scientific double discourse. The systematic forms it has reached in texts of the first, second and third cultures – humanist, scientific and techno-scientific – show an uncommon rigour in handling matrices with indicators of indicators. Numerous and important researches with mathematical modellings and numerical simulations, while indicating the possible behaviour of the included variables, at the same time indicate the behaviour of other even more significant variables, which are the interesting ones. The phenomenon of a double analysis deserves methodological and technical study. Double analysis leads to double talk, allusion, the indirect comment, analogy and metaphor to their greatest depths. But it removes them from rhetoric and inserts them in model simulations and numerical and non-numerical scenarios. It belongs to a double interpretation of 'knowledge by objectives' and of 'side effects', or of effects of the second, third or 'n' degrees. Many of these are declared 'undesired effects', when in fact they are the really desired and wisely hidden effects, with a maximization of 'earnings' and minimization of 'losses'.

One more point of neo-conservative thought that must be mentioned, closely linked to a new type of systemic policies, is the one Hopkins and Wallerstein synthesize when they write: 'one state – let us say the United States – credibly and continuously demonstrates its 'leadership' by shaping systematic structures and having its systemic policies almost always accepted not merely by weaker states but by strong ones as well'. The World Bank and the International Monetary Fund, among others, are shaping the systemic structures of the world today. They establish the rules and evaluate the results of their application. Thousands of officials apply them and observe their appliance. The United States and the neo-conservatives are not the only ones to come up with and implement these changes. They establish the rules. The problem is as Hobsbawm (1998) says: 'The position of the United States has no precedent. It is the only power with global interests and the only world power.' The problem worsens, because, as the same author points out: 'the US empire, different from the British one, aspires to transform the world into its own image and likeness'. In the midst of all that sophistication, Superman appears: 'I fight for justice, freedom and the American way.'

From the Cold War to the Clash of Civilizations

What is American is also defined by what is anti-American. During the long Cold War (1917–89), un-American activities served to distinguish the supporters of the United States and the 'Free World' from the 'agents of the International Communist conspiracy, serving a foreign power'.

The definition of what was anti-American was contrasted with the defence of democracy, of liberty, of independent and critical thinking. Within its scope lay the masses and the elites, the media, the laboratories and the academy. It was part of a genuine struggle between two world powers in which the military, economic, technological and scientific superiority of the United States was from time to time threatened by the USSR and by the mobilization of workers and oppressed peoples which the USSR claimed to lead.

In hegemonic and ideological terms, the Cold War was based on Manichaean postulates in which each power claimed to represent on the one hand, democratic values, and on the other, socialist values. These claims entered a critical period in the sixties in both the United States and the USSR. The systems of domination operative within each superpower and within its radius of influence revealed not only many falsehoods with respect to 'the war for the minds of men' as Eisenhower put it, but the instability of both systems.

The US's capability of response was far superior to the USSR's. The USSR's behaviour appeared to be self-destructive. In fact, it was part of a reshuffling of forces within their borders and in the world capitalist system. The United States controlled the final phase of the engagement. It unleashed a general offensive against Social democracy, against populist and nationalist-revolutionary governments and against the Communists. Reagan's Star Wars accentuated the USSR's disequilibriums. Its leadership, headed by Gorbachev, opted to put an end to the Cold War. Behind its borders, he set in motion a plan for democratization bereft of any popular, worker, employee, technician or intellectual support. The government had de-politicized those sectors by means of an official and unreal ideology. The change of course served to privatize and denationalize public and social enterprises and to launch a new stage of accumulation, domination and appropriation by the so-called military-bureaucratic 'Mafias' which became the elites among the American-style democratic politicians and businessmen or technocrats desirous of achieving the efficiency of private enterprise. The old and new leaders began to ape Americans, one day setting up a Donald Duck in Red Square, another showing off their Americanized vulgarity by eating at McDonalds; meanwhile the taste for rock and T-shirts spread among the youth.

The taste of defeat was laced with signs of a great power in ruins. In its profoundest experiences, some of its inhabitants were able to go beyond the remains of nationalism and patriotism and contemplate the failure of the true socialist blueprint. Some of them went back to the time of Rosa Luxemburg and the German revolution that did not take hold, others to the break with Trotsky.

Others to the time of the Soviets that never governed. Still others, pointed to a Socialist and anti-capitalist democracy, the features of which are difficult to define.

In any event, the Cold War not only ended in the victory of the United States over the USSR, it also put an end to domination of the minds of many Americans who ceased to believe in entrepreneurial and official propaganda about 'American democracy' and 'the American way of life'. Denunciation of US crimes in Vietnam and in the world, the struggle for peace and against military intervention deeply affected American mass consciousness, including that of its youth. Something similar took place in movements against racism and discrimination against the segments of the population of African, Latino, Asian, Islamic origin, and the native Americans and their lands and civil rights. It changed a whole generation of Anglo-Americans as well as the new generations of African, Latino, Asian and Islamic Americans.

The change did not only appear among radical groups, nor was it solely limited to counter-cultural movements. A new America began to be born with a new blueprint for its own country and for other countries and peoples and for the world. Never before had the United States boasted a more universal and young community of all races, cultures, ideologies and religions as developed then. Among co-options and capitulations, the struggle against the style of intellectual, moral, personal oppression of 'the American way of life', consolidated a new movement that was simultaneously civic, intellectual and academic, and involved African-Americans, the young, the old, women, men and homosexuals who adopted as their own, as in no other imperialist nation in the world, the struggle against American imperialism. The movement is hardly perceptible in comparison to its magnitude. It may be part of a new historical stage of struggles for democracy and for socialism as well.

The Cold War was not only an 'entente' between the superpowers to manipulate its own peoples, as Wallerstein and Chomsky propose. It was also a war. The end of the Cold War, and the evident triumph of capitalism in the entire world, gave rise to a savage offensive against countries of colonial origin, against ethnic groups, against the poor, campesinos and urban dwellers. It diminished the achievements of industrial workers and middle sectors. It reduced or nullifed social and national rights, many of which were already vitiated by bureaucracies and communist, nationalist and labour Mafias.

Globalization and neo-liberalism constituted the ideological warp and woof of the new war. In combination or alternatively, this war uses multiple ideological weapons such as the struggle for democracy and human rights, the struggle against terrorism and the drug trade, and the war for modernization and for civilization. In the political, military, technological and ideological war, the dominant forces of the United States and a majority of its allies and subordinates, today express and apply the essential theses of American neo-conservative thinking, even unwittingly.

The United States is the most aggressive superpower in applying and disseminating the new structural systems of domination and appropriation. Globalization American style includes the principal policies adopted and promoted by the great powers and obedient governments. American neo-liberalism not only tries to impose and succeeds in imposing American ways of governing, but American neo-liberal and neo-conservative concepts and discourses have become words-acts, discourse-structures of the superpower's hegemonized globalization.

The similarity between globalization and Americanization is so great that Americanization is often taken for globalization. The phenomenon requires a study of dissemination, socialization and universalization of institutions like: 1) those characteristic of political leaders and the American style of governing; 2) control of the labour movement, the combination of union clientelism and gangsterism, and the de-ideologizing of workers; 3) racism and regulating racial integration; 4) macroeconomic manipulation of integrated and non-integrated, documented or undocumented, legal and illegal immigrant workers; 5) flexible hiring of workers and professionals for short periods; 6) organized crime and its links with banking, money laundering, and government and 'cooperation with the District Attorney and other authorities'; 7) application of unjustified sanctions by false accusations of dumping or health violations or quality regulations, and the imposition of protectionist policies against those who previously were the objects of imposed opening of protected markets; 8) deregulation of the economy and establishment of zones of influence for entrepreneurial, financial, military and political complexes; 9) governmental subsidizing of large complexes and their members; 10) rules of negotiation, concession and commitment among elite groups; 11) application of 'shock tactics' that increase domination of weak countries and appropriation of their enterprises and wealth; 12) scenarios of destabilization that make it possible to overthrow disobedient governments.

Many globalizing and neo-liberal structures and measures have been pioneered by the United States. When adopted by other powers and governments, they make sure to apply the American seal to them. Many claim to adopt them on their own initiative, presenting as sovereign acts decisions that were forced on them by the International Monetary fund, World Bank, US Treasury Department or Department of State. There is thus an Americanization that is proud and another that is shameful, tending often to be hidden with the agreement and complicity of the American government itself. Simulated altercations between viceroys and the empire, so that the former will appear to govern sovereign states respected by both, are not exceptional.

In any event, the United States again prepares to lead a blueprint for Western Civilization. It does so aggressively by means of modernity, the market, and democracy, or supposedly in defensive ways, now withstanding new 'forces of evil' consisting of drug traffickers, terrorists, fundamentalists, insurgents, whom it punishes in its role of the world's policeman or as a State terrorist.

The United States situates itself in a new categorization of the world that some have called: 'The West and the rest.' This new proposal does not suggest that Western civilization become universal and that all civilizations join in a single one. It suggests, following Huntington, a policy of non-intervention in other civilizations and of Western and American hegemony in the struggle of civilizations. With the ease characteristic of this theoretician of the Trilateral and of 'limited democracy', he states in his most recent and touted book, that 'an international order based on civilizations is the surest safeguard against World war' (Huntington 1997).

Huntington's theses should not be underrated. Like Fukuyama's they represent the establishment that dominates the United States. Their most important feature is that they explicitly posit two types of struggle and cover up a third. The first, is the struggle between the West, with a nucleus consisting of the United States, Europe and NATO, augmented by Canada and Central Europe and including a Westernized Latin America with allies like Russia and India, all against China, Japan and most of Islam. The second is the struggle of all civilizations united among themselves and with the 'transnational corporations producing economic goods' against barbarians consisting of the 'transnational criminal mafias, drug cartels and terrorist gangs violently assaulting civilization'. The third, systematically hidden by Huntington, is the struggle for domination and appropriation of surplus and wealth, that affects the interests of peoples and workers, a struggle indeed, that until now, has not had the benefit of either world organizations of peoples in revolt, nor of proletarian internationalisms, like those that sought to organize revolutionaries, socialists and communists in the central and peripheral nations, nor answers that make it possible to glimpse an end to neo-liberalism in the short run, much less an end to capitalism.

In any event, all these struggles rather than positing the problem of Americanization, posit the relation between Americanization and the expansion of forces that in the United States and in the world impose processes of neo-liberal and capitalist domination and appropriation.

Huntington greedily proposes a distribution of the world among all the civilizations, as long as they identify with the transnational corporations that produce goods. Civilizations and transnational corporations will be hegemonized by the West with the United States in the lead. The US for its part, will commit itself to being respectful of the spiritual and material, political and intellectual values of the other civilizations.

Modernization, like Americanization or Westernization, of the world is unnecessary and may be counter-productive. A treaty of respect between civilizations and transnational corporations would make it possible to build a global capitalism composed of many capitalisms, a world order made of many states with their elites in power and their beliefs. The blueprint does not abjure the transformation of the world by the United States. As a nation, the United States is already the melting pot of civilizations. The New World Order may be in the image of the United States, the melting pot of several civilizations hegemonized by the US and identified with the transnational corporations.

Building an American-style World

The new theoretical and ideological framework for the construction of the world, American style, does not recognize phenomena of world domination and appropriation. In the double social science, these phenomena appear mediated by variables that determine the effectiveness of enterprises and governability of nations. Domination and appropriation of the world is expressed and clearly specified in political-military, private enterprise and/or secret reports. These reports are accepted and published as a function of national security, a fundamental concept that, since Hobbes, sanctifies the exact theory and rigorous method of the sovereign's double standard. The construction of an American-style world implies acts of violent domination and predatory and parasitic appropriation that go unrecognized as a general characteristic of the system. Only critics of the system write about them, among them such noteworthy researchers as Paul Baran, Paul Sweezy, Harry Magdoff, even C. Wright Mills, Noam Chomsky and Immanuel Wallerstein, to mention only a few of the most distinguished.

The problem of recognizing phenomena of domination and appropriation of the world by the United States is posited today with great energy. Of the multiple political-military phenomena that characterize it, the most solid is the policy of changing 'structural systems'. That policy is led by the United States and practised by all the great powers and great enterprises, *motu proprio* or at the behest of United States.

It is expressed as neo-liberalism or globalization and carried forward according to the rules established by the World Bank, International Monetary Fund and the Department of the Treasury for the deregulation of the economy and the thinning down of the State. It is cemented by means of external indebtedness following the fiscal crises of governments, and makes those of the world's periphery considerably more dependent by making the governments of the great powers, the US included, dependent on their respective creditors.

The new structural systems are particularly functional in terms of the processes of globalization and dependency. Once they are imposed by adjustment agreements, they provide the American empire with the opportunity to demand the application of measures that make them even more functional and dependent. The application of those measures is complemented by a techno-scientific offensive in academic circles in which American paradigms become hegemonic. This is complemented by support in the mass media and cultural, political, general and focused advertising. They are institutionalized in military and civil education and in the promotion of modernizing and governable democratic blueprints with local variants of the American democratic style. The blueprint is consolidated by means of power networks that include governments, transnational and associated American and domestic enterprises, many of which are subcontracted or co-opted though with varying degrees of autonomy. The logic of American-style democracy and the American-style market, is complemented by the logic of security, though the latter be not explicit at all times and only came to the fore in critical moments. It operates in both covert and open repressive and predatory forms.

The process of Americanization by means of systems of structures is advanced in Latin America but includes many of the world's countries where US influence is dominant. In central Europe and Russia, it operates increasingly with the agreement of local elites and mafias. As to the great powers of Western Europe, Daniel Singer (1998) has shown that the Maastricht monetary union is the crowning victory of the free market and a blueprint from which 'to build Europe à l'Americaine'. 'It is designed,' he writes, 'to reduce public expenditure on health, old age pensions and other welfare services while encouraging private insurance.' According to Singer, in a subsequent stage the construction of Europe, American style, will impose 'the flexibilization of labor' and attack the rights of workers. The dominant European forces are already preparing for the banquet.

The change is far from obedient solely to a phenomenon of Americanization. The similarity of Europe to America is due to its own blueprints of European dominant classes. It is they who impose the globalizing and neo-liberal models upon their own countries. The difference is in the response. Unlike its US counterpart, the European working class is capable of formidable response, as it proved recently in Paris and a little later, in Italy. The difficulty lies in that part of the workers and popular movements, not only oppose conservative governments with progressive and revolutionary ideologies but they also oppose social democratic governments with fascist ideologies that drive them to extremes with neo-liberal policies. The American style of controlling workers and the citizenry falls short. In Europe it is again a matter of ideologies: fascists, nazis, communists and progressive and revolutionary movements are reborn in the process of researching and systemizing 'the alternative'.

To posit all these problems in terms of Americanization is to misrepresent the reality of the United States, Europe and the world. Aside from ways of acting, struggling and living, there are problems of domination and appropriation within each country that do not allow attribution of the conduct of its social actors exclusively to external factors. In terms of international problems and world conflicts, the simple explanation by differences of civilization, culture or ethnicity is yet another form of deception. In the case of the United States, the common notion of what is American not only hides the existence of an Islamic and Asian population, besides other African, Latin and European communities that make that country a multinational collective, but also hides the divisions of actors of appropriation and domination from their beneficiaries and their victims. In the case of Islam, purely civilizing and religious reasoning hides differences between the caliphs, sheikhs and ayatollahs who are supporters of Islamic fundamentalism and the popular, progressive and democratic forces furiously persecuted by them.

Anti-Americanism may hide in the world's periphery, and even in some central countries – imperialism is a system of domination and appropriation, of depredation and parasitism combined with local partners. The problem is not resolved with an anti-Americanism that is no more than superficial and aggressive xenophobia. What is required is a distinction between the greatness and meanness of America in the history of world domination and appropriation. It leads to contingencies based on probable and systematic tendencies.

'The social forms that have driven US interventionism in the past, will continue to spawn interventions in the aftermath of the Cold War . . .' (Rossman Shalon 1993). Some forms and rationalizations will change, but the Americanization that follows upon military intervention will continue. Priority will be given to the search for collective intervention, but the unilateral variety will not be discarded. Arguments will continue to be the same, at least in part, while partly adjusting to circumstances: the defence of human rights, the struggle against terrorism and the drug mafias, the defence of US interests and their corporations.

Shared appropriation of privatized public enterprises and properties, and the increase of transferences in favour of the United States by service of external debt, by the deterioration of the terms of trade, by remittance of growing profits earned on the basis of highly flexibilized labour, will be features of the Americanization of the world with its partners and local contractors. The impact of policy on the planet as a whole and on the immense majority of the world's population, will increasingly pose the problem of alternatives. Some of them are already coming from America.

The American Tradition and 'Meaningful Democracy'

The Americanized world may explode like the special effects in a Hollywood thriller; it may also find a happy end, an end to horrors.

The current world is headed for environmental suicide or a global animal farm with the physical elimination of all who are superfluous. That is the choice facing the dominant forces. There is only one road to finding an alternative: meaningful democracy, that is, democracy that grants priority to problems of the majorities who govern (Noam Chomsky). Democracy for all, radical and plural that proposes to overcome the practical problems of a less unjust and freer world.

The alternative to the world-system of domination and appropriation must be set forth with socialist blueprints of liberation and democracy. In some places on earth, some blueprints will predominate over others in diverse ways. It is probable that communists will reappear among the socialists; among those in favour of liberation, the populists; among the democrats, the liberals. But on a world level it would be the democrats that tend to prevail among the most diverse progressive and radical forces, particularly those with democratic governments of the people, with ideological and religious pluralism and with less unjust policies of appropriation. The fact that all the regimes that to this day tried to resolve the problems of freedom and justice have failed due to the absence of democracy for all will make this the central and universal objective of each nation, region and community.

The importance of the blueprint for meaningful and plural democracy for all will be that much greater to the extent that it tends to posit the conceptual and structural redefinition of that term, not only in hegemonic countries like the United States, Europe and Japan, but in most of the ex-USSR and its ex-allies in India, Indonesia, Latin America and Africa, in Islam and in China. The blueprint for meaningful democracy will have to become the goal of increasingly more significant movements of citizens, workers and peoples identified with the most diverse religions, ideologies and civilizations.

America's contribution is most important to the universal blueprint. Any future movement in favour of a meaningful democracy will have the US Declaration of Independence, its Constitution, the First Amendment and the Reconstruction Amendments as part of its cultural background, its mobilizing referent, its mythic motor. It may have recourse to traditions of struggle in Biblical texts, politically updated by whites and blacks in their civic acts and by Niebuhr in his theological work. The libertarian traditions of civil disobedience of the great Henry David Thoreau will be part of an active critique of electoral systems that remain games with a slightly moral tinge. Civil disobedience as suppression of the payment of taxes to the government and other forms of non-violent protests will be features of conduct inherited from and created by multitudes of Americans who fought against wars and unjust orders. A mass of direct actions, sit-ins and mobilizations have already influenced the entire world, they even reached Gandhi's India and the universal thought of Fanon. The most recent contributions of Martin Luther King, Malcolm X and the New Left of 68, have already contributed to forging an alternative democratic blueprint to organize multitudes with practical, technical and political meaning in small and large networks.

Noam Chomsky's voice is only the tip of the American iceberg that will make new history and live it not only in thought and organization, but in music and literature, in the arts and sciences, in the imagination and the power of civil and political society, with its citizens, its workers, and its peoples. Other forms of radical American thought, like critical thinking and Marxism, cultivated more and more in that country and which Americans express with particular force, will be inserted into the construction of meaningful democracy in many other countries and civilizations.

The American style of being radical that characterizes millions of Americans, will contribute to forging the collective and individual personality of those who by experience know how to confront a regime that combines liberal and charitable humanitarianism with the accusative glare of a culpable judge, and with corruption, co-option and repression. The power of the American personality is to be found in the rebellions of its people.

The internal and international blueprint will also confront the need for firm decision that systematically appears in the behaviour of elites and the American military entrepreneurial complex. That decision is clearly visible in numerous events that have occurred in the countries of the world's semi-periphery and periphery. As Chomsky says, any true democracy 'grants priority to its own population instead of to the investors of the United States'. But the simple possibility that a democracy might exist and become uncontrollable releases in the US military entrepreneurial complex a process of open and covert diplomatic economic, financial and media intervention, all for the purpose of pressure, intimidation, destabilization and support of the counter-tendencies generated by any meaningful democracy.[11] The phenomenon enjoys a long history. It occurred before the Cold War and continues to this day. It appears systematically every time a meaningful democracy appears. It is a kind of law that operates every time a democratic government wins that grants priority to population problems, opportunities for real civil participa-tion in the organs of government and social, cultural, economic and political decision-making. Intervention occurred in Greece in 1947 to overthrow the government; in Italy in 1948 to prevent the Communists from winning; in Japan in 1947 with the collaboration of Japanese fascism, in the suppression of democratic forces; in Korea in 1945 against a popular government overthrown with the support of an invasion and with 100,000 fatalities; in Guatemala in 1954 with the 'Glorious victory' that sank that little indigenous country into a half a century of terror; in Iran in 1953; in Laos in 1960; in the Dominican Republic in 1963 and 1965; in Brazil in 1964; in Grenada in 1983; in Chile in 1973; in Nicaragua from 1978 to 1992; against Cuba from 1959 to 1998. All these interventions were explained as part of the struggle against international Communism, but it took place before and after the Cold War with diverse kinds of rationalizations that obliged one to think of a more general reason that appears in all periods and circumstances. That explanation is given by Chomsky. It is based on a non-negotiable logic: 'What the United States wants is "stabil-ity"', writes Chomsky, 'meaning security for the upper classes and large foreign investment.' That logic is substantiated over and over again in the declarations of the US government and its spokesmen. There is no doubt about it. It con-stitutes the policy of Americanization as imperialism and a struggle against meaningful democracy. The peoples of the world will have to confront it and overcome it.

To this end, they must build from their locales, nations, regions; from their assemblies and networks of systems of democratic structures that dominate the world and make 'government of the people, by the people' and with the people; the motto the United States House of Representatives made its own, solely as a declaration, as long ago as 3 April 1918.

If in the past the government could be of the people and for the people, 'But could not in any operational way be by the people', as Hobsbawm says, 'that difficulty has been overcome today as a practical possibility and already belongs to universal necessity.[12] 'For better of for worse in the twentieth century, the common people entered history as actors in their own collective right',[13] says Hobsbawm.

The Americanization of the world will be part of the world's peoples' entry into history and will have to contribute to breaking the inflexible logic of neo-conservatives and neo-liberals and their predecessors and successors.

If in the past the struggle for democracy was mediated by the United States and the struggle for Socialism was mediated by the USSR, today the struggle for meaningful democracy will mediate in the struggle for Socialism, which, with liberty, will make it meaningful. Both will be the task of citizens, peoples and workers, with Americans in their ranks.

Notes

1   Pocho: derogatory, for Mexicans who mimic Americans.
2   See *Democracy in America*, vol. 1, p. 24.
3   John Sullivan, 1845.
4   For an excellent synthesis see A. Mattelart, op. cit.
5   See Manuel Castells, *The Information Age: Economy, Society and Culture, vol. 1: The Rise of the Network Society*. Cambridge, MA: Blackwell, 1996.
6   See Richard Lee, 'Structures of Knowledge', in Terence H. Hopkins and Immanuel Wallerstein (eds), *The Age of Transition: Trajectory of the World System, 1945–2025*, London: Zed, 1996.
7   See Jürgen Habermas, *The New Conservatism*, Cambridge, MA: MIT Press, 1992, p. 25.
8   See Peter Steinfels, *The Neo-conservatives*, New York: Simon and Schuster, 1979.
9   See Peter Steinfels, *The Neo-conservatives*, New York: Simon and Schuster, 1979, pp. 21 and 27.
10 See J. A. Barnes, *A Pack of Lies. Towards a Sociology of Lying*. Cambridge: Cambridge University Press, 1994, pp. 56, 26, 27, 23, 32.
11 See Noam Chomsky, *What Uncle Sam Really Wants*, Tucson, Ariz.: Odonian Press, 1992.
12 See Eric Hobsbawm, *The Age of Extremes: A History of the World 1914–1991*, New York: Vintage, 1996, p. 579.
13 See Eric Hobsbawm, *The Age of Extremes: A History of the World 1914–1991*, New York: Vintage, 1996, p. 582.

References

Hobsbawm, Eric. 'Las hegemonias de Gran Bretana y Estados Unidos, y el Tercer Mundo', in *Analisis Politico*, Bogota, Universidad Nacional de Colombia (Jan./Apr.) 1998, pp. 7–8.

Hopkins, Terence H. and Wallerstein, Immanuel. 'The World System: Is There a Crisis?', in Hopkins and Wallerstein (eds), *The Age of Transition: Trajectory of the World System, 1945–2025*, London: Zed, 1996.

Huntingdon, Samuel P. *The Clash of Civilizations*, New York: Simon and Schuster, 1997, p. 321.

Katzenstein, Peter J. (ed.), *The Culture of National Security*, New York: Columbia University Press, 1996.

Lee, Richard. 'Structures of Knowledge', in Terence H. Hopkins and Immanuel Wallerstein (eds), *The Age of Transition: Trajectory of the World System, 1945–2025*, London: Zed, 1996, 197–200.

Mattelart, Armand, *La mondialisation de la communication*, Paris: Presses Universitaires de France, 1996, p. 48.

O'Neill, Graeme. 'Truth hurts US scientific facts', *Age*, June, 1991.

Rossman Shalon, Stephen. *Imperial Alibis: Rationalizing US Intervention after the Cold War*, Boston: South End Press, 1993, pp. 194–7.

Singer, Daniel, *The Nation*, New York, 1998.
Steinfels, Peter, *The Neo-conservatives*, New York: Simon and Schuster 1979.
The New York Times, 1, 2, 1931.
The New Yorker, 27 May 1945.
Whitman, Walt, *Democratic Vistas*, 1871.
Valéry, Paul, *Cahiers II*, Paris: La Pleiade, 1974, p. 1171.

Part V

Conclusion: American Centuries?

Chapter 22

Multiple Themes, One America; One Theme, Multiple Americas

David Slater and Peter J. Taylor

One feature of this book which readers may have noticed is that the titles of each section end with a question-mark. The different components which have contributed to making the twentieth century the 'American Century' have been highly contested in practice, and in the chapters we have commissioned. For the latter we have given our authors a free hand to express particular positions with respect to what has made the USA so dominant. The result is a great variety in both themes and opinions on those themes. However, the one concept which we have not problematized is the 'American Century' itself. No question-mark here, the USA has seen off 'imperial century', 'German century', 'people's century' and 'socialist century' to be left with the only credible claim to the hundred years.

The idea of an American Century implies that the bundle of practices described in this book add up to more than the sum of its parts. This is what we explore in this conclusion. We have drawn out what we believe to be the critical issues brought up by our authors – we will not comment upon particular positions – and have ordered them into four arguments. First, there is the question of the seductiveness of the USA which we consider as 'projected America'. Of course, there are very many other 'Americas' which are considered in the second section. Third, there is the matter of power: this book is about a variety of powers which we need to explicate. Finally, there is the future: will the twenty-first century be a second American century?

Projected America

This book has been largely about what we might call 'projected America'. It is written from outside the source of powers it reports on; it is constituted as stories written by the recipients of those powers. Our writers need never have actually visited the USA, America has come to them. But what is this 'America' they have experienced? All projections have their biases, what are the distortions in this case?

Strangely, this is quite a difficult question to answer. On the one hand there are the immensely positive images of the USA encapsulated by such phrases as the American Dream, but there have also been 'American nightmares' in the twentieth century. And this is not just a matter of selectivity in the projection of America, the negative side has not been hidden, particularly the racism. An old joke relating to US servicemen stationed in East Anglia after the war illustrates the visibility of this other side of America. A local resident, asked about his opinion of his new American soldier-neighbours answered that he liked them very much – they were friendly, generous and particularly nice to children. He added, however, that he did not like the 'white bastards' they had brought with them (Brogan, 1959, 14). Of course, you did not have to directly observe the racism, it was a major theme in American literature and film where the 'South' was the archetypal menacing 'other'. The nightmares came to a head in the 1960s with the assassinations of President Kennedy and Martin Luther King, and with the urban race riots and the war against Vietnam.

The key point is that the American Dream as a world dream survived these and other negatives. The seductiveness of things American remained as the USA continued to be both a real and virtual magnet for peoples across the world – the Chinese students building their own 'statue of liberty' in Tiananmen Square before the massacre being a classic case in point. This despite the USA tradition of supporting dictators in all three continents of the 'Third World'. The Chinese students may have been foolhardy but they were not ignorant: they would have been conversant with US support for unpleasant neighbouring dictators such as Marcos in the Philippines and Suharto in Indonesia. But this does not seem to have mattered. The USA as 'America' is the 'Teflon world power', bad things do not stick to it.

How can this be? The balance between the 'good' and the 'bad' with the victory of the former can be usefully illustrated by the immediate conversion of Kennedy's assassination from American tragedy to world tragedy. On the one hand we have a short term of office which was a near disaster in foreign policy terms: the invasion of Cuba in the Bay of Pigs fiasco, the Cuban missile crisis which nearly destroyed the world, and the starting of what was to become the USA's most disastrous war, the conflict in Vietnam. Although nobody wants to see a leader assassinated in office, the depth of shock and sympathy was out of all proportion to the death. This is shown by the oft-quoted statement for this generation that they all remember precisely where they were and what they were doing when they first heard the terrible news. Clearly Kennedy represented much more than an assassinated leader whatever view we take of his foreign policy. As a bright young man with a beautiful wife and young children, Kennedy epitomized the good side of America. Mixing ordinariness with glamour in a social mix long since pioneered in Hollywood, Kennedy's 'clan', dubbed Camelot, offered a fresh image of what was usually the business of lacklustre old men, 'high politics'.

Projected America has taken onto itself an array of properties which the vast majority of the world's population value and yearn for. Behind the front row of the politician's 'progress', 'democracy', and 'freedom', there are basic attributes of 'modernity' which are understood in terms of everyday life: 'opportunity', 'achievement', 'openness', 'newness', and above all 'affluence'. These 'good things' have been indelibly linked to America in the twentieth century, this is the projected America which can do no wrong even when it goes wrong. The 'Teflon' can even operate on those at the sharp end of the coercive side of US power: when a central American peasant farmer scrawls in brackets, under the statement "Yankee Go Home", "And Take Me With You", he is illustrating the great ambiguity at the heart of American power projections. It is the existence of projected America which justifies our acceptance of the idea of 'American Century'.

Opening up America

Throughout this collection the term 'America' has been used in a variety of contexts. Predominantly, America is taken to signify the United States of America, and whilst this usage has its own history and legitimacy, it is important to be aware of the difference that emerges when one America is displaced by multiple Americas. Historically, the Jeffersonian notion of America, as the United States, having a 'hemisphere to itself', can be contrasted to José Martí's suggestion that 'our America', that is, Latin or Hispanic America, needed to be distinguished from a 'Greater America' which included North America. Moreover within both the North and the South of the Americas, the presence of indigenous peoples with different foundations and cultural trajectories give rise to another entry into what can be meant by 'America', just as the proliferation of multicultural identities born by succeeding waves of immigration opens up America to a plurality of meanings. There are thus multiple ways of naming America, and maintaining such a diversity can help us avoid the limits of interpretative closure promoted by projected America.

Within the United States, the development of a multicultural society and the dissemination of difference assumes a variety of forms. For example, Spanish-speaking Americans are as heterogeneous as the United States itself; the breakdown is roughly 63 per cent Mexicans, 12 per cent Puerto Ricans, 8 per cent Cubans, 12 per cent Central Americans and 'diverse' Latin Americans and 5 per cent from the Dominican Republic. Each group brings with it a different history and different affinities. Spanish-speaking America is already the world's fifth-largest Hispanic nation, and within ten years only Mexico will have more Spanish speakers. Recently described as the 'keenest recruits to the American dream', the buying power of Latinos has risen 65 per cent since 1990 to $348 billion in 1998, or more than the GNP of Mexico. In some cases too, the Latino immigrant can become the personification of the American dream, as exemplified by Roberto Goizueta, a chemical engineer who left Cuba with a hundred shares in Coca-Cola, and rose to be one of the company's most powerful chairmen, increasing Coke's stockmarket value from $4 billion to $150 billion in his lifetime with the company.

It is part of America's folklore, inscribed on the Statue of Liberty – 'give me your tired, your poor, your huddled masses yearning to breathe free' – part of the nation's sense of itself, that it will take in the poor and oppressed of other lands. And yet historically there have been many bouts of anti-immigration fervour; until 1875 anyone from anywhere could freely settle in the United States, and then in 1882 the Chinese were excluded, in 1907 the Japanese were restricted and in 1924 quotas by national origin were set and the Border Patrol established. In 1965 the quotas of origin were abandoned in favour of family reunification or occupational skills, and in 1990 the ban on Communists was dropped, to be followed six years later by the Illegal Immigration Act, which to some extent responded to an anti-immigrant mood captured in polls – Americans are tired of those 'motley newcomers', Colombian maids, Vietnamese waiters, Ethiopian taxi drivers who scrape a living in the world's richest country. Yet throughout these years the immigration figures have responded not to legislative deterrent but to the seductiveness of the American Dream as described above: when America prospers immigration rises, and of course the projection of America is a crucial part of this scenario.

The idea of American exceptionalism is an important link between the American Dream within and without the USA. However, in practice it can also have very different implications. In 1946, the head of Paramount told the *New York Times* that 'in the film industry we recognize the need to inform people in foreign lands about the things that have made America a great country' (quoted in Cockburn 1995, p. 2). And this sense of greatness, the destiny that has been discussed in a number of our chapters, feeds the idea of American 'exceptionalism'. For Seymour Martin Lipset (1996), it is possible to connect American exceptionalism to a series of everyday examples; Americans are much more likely to believe in both God and the devil, to be opposed to excessive interference by the state, to be proud of their country, to be socially egalitarian, to believe that hard work gives them a chance to change their circumstances, and to commit crime. There is also a strongly-developed moralism that leads to a tendency to treat wars as crusades rather than struggles for power, and for Lipset, Americans, like the English before them, have an extraordinary gift for locating their own self-interest, then convincing themselves that it coincided with God's – the difference being that the Americans do it better. Equally, rugged individualism and a deep sense of moral responsibility can also lead into resistance to the state, as occurred during the US–Mexican War when some Americans saw it as their duty to join the Mexican army and fight for a just cause, or in more recent times when soldiers took a stand against US army atrocities in the Vietnam War. This is an 'Americanism' against the USA which is currently expressed in the militia movement and their suspicions of all things governmental. In the twentieth century, however, the exceptionalism of America is also very much about its power capacities.

Constellations of Power

Looking at the different kinds of capacity, and their projections, know-how, the arsenal of democracy and the dream factory, it is evident that the ability to effect, change, transform and influence is differentially present in the world. In today's Russia, for instance, whilst Snickers chocolate bars, Coca-Cola soft drinks and Marlboro cigarettes are the staples of provincial markets, the number of American films shown in Russian cinemas fell from a peak of 215 in 1994 to 67 in 1995 and 85 in 1996. On the other hand one American product commands universal acclaim: the dollar. Distrustful of the rouble, Russians hold an estimated $30 billion in cash, much of it, so it is suggested, under their beds (*The Economist*, 22 March 1997, p. 56). The 'Americanization' of Russia can, on occasion, however, provoke a little hostility. In 1997, in a sizeable exercise in the former Soviet states of Central Asia, Americans flew directly from their bases in the United States and parachuted into Kazakhstan – 'this shows we can go anywhere' said an American general, but when it came to planned American-Russian manoeuvres near Vladivostock in the summer of 1998, the exercises were shifted away from the city after residents protested that they were an affront to sovereignty and national pride.

The military capacity of the United States, post-Cold War, not only gives it a pre-eminence unequalled in previous periods, but with the dawning of another military revolution based on the application of information technology to weapons, the United States could go way ahead of its enemies and allies alike. These changes involve the gathering of huge amounts of data, their processing so that relevant information is displayed on a screen, and then destroying targets at a much greater distance and with much greater accuracy than was previously possible. Such changes favour attack rather than defence, and identifiable, easy-to-hit targets – whether military bases, tanks, concentration of troops, arms factories or infrastructure, military or civilian – are increasingly vulnerable to weapons such as cruise missiles steered by satellite beams, and this has been sharply demonstrated in recent attacks on Iraq, Afghanistan and the Sudan. Not only is the United States well ahead in the development and deployment of the new technologies, but overall defence spending continues to dwarf that of other major powers, and figures for 1997 show that military spending still represents more than 85 per cent of the average annual military expenditure of $304 billion (at 1997 values) for the Cold War (1948–91) period (Achcar, 1998, pp. 95–6) Military capacity is also reflected in the way the United States has been able to expand the presence of NATO eastwards into what were Soviet spheres of influence, in the way, in different parts of the world – Bosnia, Iraq, Haiti, the Middle East – US strategies of intervention, mediated through the United Nations, or enacted alone have constructed geopolitical situations more favourable to its global hegemony. Little can be done from now on without the leadership of the last surviving superpower: the Gulf War, the Arab–Israeli peace negotiations, the World Trade Organization, the IMF, the peace agreement in Bosnia, intervention in Kosovo. 'The buck stops here', read the famous sign on President Harry S. Truman's desk; it certainly does now.

Military capacity is backed by the world's most powerful economy, which in the 1990s has been particularly buoyant – median family income is up to nearly $40,000, although between 1979 and 1994 families in the top 20 per cent increased their share of income from 42 to 46 per cent (UNDP, 1998, p. 61), unemployment is below 5 per cent, and GDP growth is over 3.5 per cent per annum. When Bill Clinton first went abroad in 1968, trade accounted for about 10 per cent of American GDP, whereas today it accounts for over 30 per cent. Moreover, the American share of total industrial production by the 24 OECD countries was larger in 1990 than in 1970, and in industry the hourly production of German and Japanese workers is only 80 per cent of that of their American colleagues (Valladão, 1998, pp. 117, 119). This is also of course a question of quality as much as quantity. Of the three great exporting nations, the classic image of number three is the Mercedes Benz car, a marvellous piece of basically pre-1940 technology; the key image of number two is the Sony Walkman and the VCR, shrewdly marketed products of 1970s technology, and the pivotal images of number one are Microsoft, a Boeing 747 and Hollywood's latest megahit – which economy is the most seductive? America has the advantage of an enormous domestic market, a language that is becoming ubiquitous, and a genius for marketing, but more subliminally, its global image is of the nation that not only reached modernity first, inventing trends from blue jeans to rock 'n' roll, but also of the nation that continues to be on the cutting edge of contemporary cultural change. And this image often ties in with a perception of America as the heartland of optimism, social mobility and energy.

Power as capacity is always rooted in a relational context, and the idea of 'Americanization' reflects a certain kind of power relation. In the context of US–UK relations, there is very often a sense on the British side that Americanization is associated with 'invasion' – in recent newspaper reports we find headings like 'US bookstore invasion has begun', or with takeovers in the electricity sector, 'US assault continues', or in the TV industry, 'Britain is now being flooded with American films and TV programmes'. Here we really have come full circle from the beginning of the century when Furness (1902) wrote his *The American Invasion*. He was concerned for the USA 'conquering the world's markets . . . in manufactured articles' (p. 1). Today, there are those like David Puttnam (1997) who argue that Hollywood firepower has blown away the opposition, establishing a 'tyranny of cinema', a position given some support by the fact that in the European Union the United States claimed 70 per cent overall of the film market in 1996, up from 56 per cent in 1987. However, the more Hollywood becomes preoccupied by the global market, the more it produces blockbusters made to play as well in Manila as Manhattan. Such films are driven by special effects, and they go for general subjects which everybody can identify with – there is nothing specifically American about boats crashing into icebergs or asteroids that threaten to obliterate human life.

But, of course, the competition is not being completely defeated. The strength of the local also always needs to be remembered – for example, in the field of TV production, in every European country in 1997 the most popular television programme was a local production. Also in the domain of pop music, long supposed to provide the soundtrack to America's cultural hegemony, local performers are increasingly important – in Germany, for example, the world's third largest music market after the United States and Japan, local performers account for 48 per cent of yearly sales, double the percentage five years ago; in Spain

58 per cent of the total $1 billion music sales are generated by Spanish and Latin American artists, and in the French market, French rock groups account for nearly half the country's total sales. But again, to look at these figures from another perspective, the simple fact of there being successful French rock bands is a triumph of the American way of popular culture.

Starting with power capacities, but not neglecting relations of power, the USA has been a unique force – a constellation of powers – in the twentieth century. The strength of its projections of power have varied by form, time and place but as we come to the end of the century there remains no dispute about, in a favourite American phrase, 'Who's Number 1'.

The Future: Another American Century?

The question of another American Century is a large debate of special concern to Americans. If, indeed, they are not just different like every other nation, but innately exceptional in the ways proposed, it follows that now 'unbound', there will be no stopping future American successes. However, those who see history in more cyclical terms have argued that American exceptionalism is only a feature of the twentieth century, there have been other cases before, notably British exceptionalism in the nineteenth century, and therefore there can be other non-American exceptionalisms in the future. It may well be that neither of these positions will provide accurate predictions of power relations in the twenty-first century.

The constellation of power which has been the American Century derives from a structural situation in which political, economic and cultural power relations and capacities were largely produced and reproduced within the territorial 'containers' of nation-states (Taylor, 1994). In a sense, therefore, the USA was the greatest container, one which spilt over across the rest of the world in the twentieth century. But it is precisely this containerization which is under threat by new transnational processes which go under the name of globalization. This is not to suggest that the end of the state is nigh (territoriality is too valuable a political strategy to be abandoned), but rather that the powers of states relative to other institutions is likely to diminish. Even today, it has been suggested that, in the light of the powers of global financial markets, there remains only one state which still maintains its sovereignty, the USA (Martin and Schumann, 1997, p. 216).

This image of the USA as the last sovereign state in a transnational world may be over-developed for now, but it does provide a new view on the position of the USA in the twenty-first century. At first, it seems to suggest a defensive posture, the USA as bastion of territoriality. This certainly does not indicate a leading role in the future. But this is too simple. Many of the most powerful players in contemporary globalization are American institutions, and not just as arms of the US state. New York, for instance, is the leading 'global city' in the network which defines world cities (Sassen, 1991). In some ways globalization represents the culmination of Americanization as the USA has become more

and more involved in world markets. Hence, we might expect a quite complicated global future, not of political mosaics versus electronic networks, but of
mixes of spaces of flows and of territories which vary over time and place. The
USA should remain the most powerful territory in the system for some time to
come but it will not dominate in the way it has done in the recent past. In fact
no country will be able to replicate the special positions which Britain and the
USA have held in the last two centuries. The twenty-first century will be much
more convoluted, puzzling and perplexing than any century for at least half a
millennium.

References

Achcar, G. (1998) 'The Strategic Triad: The United States, Russia, and China', *New Left Review*, 228, March/April, pp. 91–126.

Brogan, D. W. (1959) 'From England', in F. M. Joseph (ed.), *As Others See Us*. Princeton, NJ: Princeton University Press.

Cockburn, A. (1995) 'Fatal Attraction', *The Guardian*, Friday 12 May, pp. 2–4.

Furness, C. (1902) *The American Invasion*. London: Simpkin, Marshall, Hamilton, Kent & Co.

Lipset, S. M. (1996) Interview entitled, 'America's Man of Stark Contrasts', *The Times Higher*. London, April 5, p. 13.

Martin, H.-P. and Schumann, H. (1997) *The Global Trap*. London: Zed.

Puttnam, D. (1997) 'Terminated, too', *The Guardian*, Friday May 9, pp. 8–9.

Taylor, P. J. (1994) 'The State as Container: Territoriality in the Modern World-system', *Progress in Human Geography*, 16, 151–62.

UNDP (United Nations Development Programme) (1998) *Human Development Report 1998*. New York and Oxford.

Valladão, A. G. A. (1998) *The Twenty-First Century will be American*. London and New York: Verso.

Index

Note: page numbers in *italics* refer to chapter notes, those in **bold** refer to figures and those in ***bold italics*** to tables.